Secondary Xylem Biology

Origins, Functions, and Applications

Secondary Xylem Biology
Origins, Functions, and Applications

Edited by

Yoon Soo Kim
Department of Wood Science and Engineering
Chonnam National University
Gwangju, South Korea

Ryo Funada
Faculty of Agriculture
Tokyo University of Agriculture and Technology
Fuchu, Tokyo, Japan

Adya P. Singh
Manufacturing and Bioproducts Group
Scion (New Zealand Forest Research Institute)
Rotorua, New Zealand

AMSTERDAM • BOSTON • HEIDELBERG • LONDON
NEW YORK • OXFORD • PARIS • SAN DIEGO
SAN FRANCISCO • SINGAPORE • SYDNEY • TOKYO
Academic Press is an Imprint of Elsevier

Academic Press is an imprint of Elsevier
125 London Wall, London EC2Y 5AS, UK
525 B Street, Suite 1800, San Diego, CA 92101-4495, USA
50 Hampshire Street, 5th Floor, Cambridge, MA 02139, USA
The Boulevard, Langford Lane, Kidlington, Oxford OX5 1GB, UK

Cover Image Credit: Transverse section of a young radiata pine stem imaged with confocal fluorescence microscopy using natural fluorescence. Areas of compression wood can be seen in brighter green color. There are two areas of traumatic tissue formed as a result of a wound response. The stem is about 5-mm diameter. Image courtesy of **Dr Lloyd Donaldson**, Microscopy & Wood Identification, Senior Scientist – Plant Cell Walls & Biomaterials, Scion, New Zealand.

Notices
Knowledge and best practice in this field are constantly changing. As new research and experience broaden our understanding, changes in research methods, professional practices, or medical treatment may become necessary.

Practitioners and researchers must always rely on their own experience and knowledge in evaluating and using any information, methods, compounds, or experiments described herein. In using such information or methods they should be mindful of their own safety and the safety of others, including parties for whom they have a professional responsibility.

To the fullest extent of the law, neither the Publisher nor the authors, contributors, or editors, assume any liability for any injury and/or damage to persons or property as a matter of products liability, negligence or otherwise, or from any use or operation of any methods, products, instructions, or ideas contained in the material herein.

British Library Cataloguing-in-Publication Data
A catalogue record for this book is available from the British Library

Library of Congress Cataloging-in-Publication Data
A catalog record for this book is available from the Library of Congress

ISBN: 978-0-12-802185-9

For information on all Academic Press publications
visit our website at http://store.elsevier.com/

Typeset by Thomson Digital

Printed and bound in the United States of America

Working together
to grow libraries in
developing countries

www.elsevier.com • www.bookaid.org

Contents

Part II
Function and Pathogen Resistance of Secondary Xylem

Part III
Economic Application of Secondary Xylem

11. Secondary Xylem for Bioconversion

Shiro Saka, Hyeun-Jong Bae

12. Wood as Cultural Heritage Material and its Deterioration by Biotic and Abiotic Agents

Yoon Soo Kim, Adya P. Singh

16. Rapid Freezing and Immunocytochemistry Provide New Information on Cell Wall Formation in Woody Plants

Keiji Takabe, Jong Sik Kim

17. Distribution of Cell Wall Components by TOF-SIMS

Dan Aoki, Kaori Saito, Yasuyuki Matsushita, Kazuhiko Fukushima

Contributors

Dan Aoki, Department of Biosphere Resources Science, Graduate School of Bioagricultural Sciences, Nagoya University, Furo-cho, Chikusa-ku, Nagoya, Japan

Hyeun-Jong Bae, Department of Bioenergy Science & Technology, Chonnam National University, Buk-gu, Gwangju, South Korea

Lorena Balducci, Département des Sciences Fondamentales, University of Quebec in Chicoutimi, 555, Boulevard de l'Université, Chicoutimi (QC), Canada

Shahanara Begum, Faculty of Agriculture, Tokyo University of Agriculture and Technology, Fuchu, Tokyo, Japan; Faculty of Agriculture, Bangladesh Agricultural University, Mymensingh, Bangladesh

Ingo Burgert, Swiss Federal Institute of Technology Zürich (ETH Zürich), Institute for Building Materials, Zürich; Swiss Federal Laboratories for Materials Science and Technology (EMPA), Applied Wood Materials, Dübendorf, Switzerland

Etienne Cabane, Swiss Federal Institute of Technology Zürich (ETH Zürich), Institute for Building Materials, Zürich; Swiss Federal Laboratories for Materials Science and Technology (EMPA), Applied Wood Materials, Dübendorf, Switzerland

Katarina Čufar, Department of Wood Science and Technology, Biotechnical Faculty, University of Ljubljana, Rozna dolina, Ljubljana, Slovenia

Geoffrey Daniel, Department of Forest Products/Wood Science, Swedish University of Agricultural Sciences, Uppsala, Sweden

Lloyd A. Donaldson, Manufacturing and Bioproducts Group, Scion (New Zealand Forest Research Institute), Rotorua, New Zealand

Dieter Eckstein, Centre of Wood Sciences, University of Hamburg, Leuschnerstr, Hamburg, Germany

Benhua Fei, International Centre for Bamboo and Rattan, Beijing, China

Jörg Fromm, Institute for Wood Biology, University of Hamburg, Hamburg, Germany

Kazuhiko Fukushima, Department of Biosphere Resources Science, Graduate School of Bioagricultural Sciences, Nagoya University, Furo-cho, Chikusa-ku, Nagoya, Japan

Ryo Funada, Faculty of Agriculture, Tokyo University of Agriculture and Technology, Fuchu, Tokyo, Japan

Zhimin Gao, International Centre for Bamboo and Rattan, Beijing, China

Jožica Gričar, Department of Forest Yield and Silviculture, Slovenian Forestry Institute, Ljubljana, Slovenia

Kyung-Hwan Han, Department of Horticulture, Michigan State University, East Lansing, MI, USA; Department of Forestry, Michigan State University, East Lansing, MI, USA

Md. Rahman Hasnat, Faculty of Agriculture, Tokyo University of Agriculture and Technology, Fuchu, Tokyo, Japan

Risto Jalkanen, Management and Production of Renewable Resources, Natural Resources Institute Finland, Rovaniemi, Finland

Daniel E. Keathley, Department of Horticulture, Michigan State University, East Lansing, MI, USA

Tobias Keplinger, Swiss Federal Institute of Technology Zürich (ETH Zürich), Institute for Building Materials, Zürich; Swiss Federal Laboratories for Materials Science and Technology (EMPA), Applied Wood Materials, Dübendorf, Switzerland

Jong Sik Kim, Department of Forest Products, Swedish University of Agricultural Sciences, Uppsala, Sweden

Won-Chan Kim, School of Applied Biosciences, College of Agriculture and Life Sciences, Kyungpook National University, Daegu, South Korea

Yoon Soo Kim, Department of Wood Science and Engineering, Chonnam National University, Gwangju, South Korea

Jae-Heung Ko, Department of Plant and New Resources, Kyung Hee University, Yongin, Korea

Gerald Koch, Thünen Institute of Wood Research, Leuschnerstr, Hamburg, Germany

Kayo Kudo, Faculty of Agriculture, Tokyo University of Agriculture and Technology, Fuchu, Tokyo; Institute of Wood Technology, Akita Prefectural University, Noshiro, Akita, Japan

Silke Lautner, Faculty of Wood Science and Technology, Eberswalde University for Sustainable Development, Eberswalde, Germany

Eryuan Liang, Key Laboratory of Alpine Ecology and Biodiversity, Institute of Tibetan Plateau Research, Chinese Academy of Sciences, Beijing; CAS Center for Excellence in Tibetan Plateau Earth Sciences, Beijing, China

Zhijia Liu, International Centre for Bamboo and Rattan, Beijing, China

Yasuyuki Matsushita, Department of Biosphere Resources Science, Graduate School of Bioagricultural Sciences, Nagoya University, Furo-cho, Chikusa-ku, Nagoya, Japan

Vivian Merk, Swiss Federal Institute of Technology Zürich (ETH Zürich), Institute for Building Materials, Zürich; Swiss Federal Laboratories for Materials Science and Technology (EMPA), Applied Wood Materials, Dübendorf, Switzerland

Eri Nabeshima, Faculty of Agriculture, Tokyo University of Agriculture and Technology, Fuchu, Tokyo; Faculty of Agriculture, Ehime University, Matsuyama, Ehime, Japan

Satoshi Nakaba, Faculty of Agriculture, Tokyo University of Agriculture and Technology, Fuchu, Tokyo, Japan

Widyanto Dwi Nugroho, Faculty of Agriculture, Tokyo University of Agriculture and Technology, Fuchu, Tokyo, Japan; Faculty of Forestry, Universitas Gadjah Mada, Yogyakarta, Indonesia

Yuichiro Oribe, Tohoku Regional Breeding Office, Forestry and Forest Products Research Institute, Takizawa, Iwate, Japan

Peter Prislan, Department of Wood Science and Technology, Biotechnical Faculty, University of Ljubljana, Rozna dolina, Ljubljana, Slovenia

Ping Ren, Key Laboratory of Alpine Ecology and Biodiversity, Institute of Tibetan Plateau Research, Chinese Academy of Sciences; University of the Chinese Academy of Sciences, Beijing, China

Sergio Rossi, Département des Sciences Fondamentales, University of Quebec in Chicoutimi, 555, Boulevard de l'Université, Chicoutimi (QC), Canada

Markus Rüggeberg, Swiss Federal Institute of Technology Zürich (ETH Zürich), Institute for Building Materials, Zürich; Swiss Federal Laboratories for Materials Science and Technology (EMPA), Applied Wood Materials, Dübendorf, Switzerland

Kaori Saito, Division of Diagnostics and Control of the Humanosphere, Research Institute for Sustainable Humanosphere, Kyoto University, Uji, Kyoto, Japan

Shiro Saka, Graduate School of Energy Science, Department of Socio-Environmental Energy Science, Kyoto University, Yoshida-honmachi, Sakyo-ku, Kyoto, Japan

Yuzou Sano, Laboratory of Woody Plant Biology, Research Faculty of Agriculture, Hokkaido University, Sapporo, Japan

Uwe Schmitt, Thünen Institute of Wood Research, Leuschnerstr, Hamburg, Germany

Jeong-Wook Seo, Centre of Wood Sciences, University of Hamburg, Leuschnerstr, Hamburg, Germany; Department of Wood and Paper, Chungbuk National University, Naesudong-ro, Seowon-gu Cheongju, Chungbuk, South Korea

Adya P. Singh, Manufacturing and Bioproducts Group, Scion (New Zealand Forest Research Institute), Rotorua, New Zealand

Tripti Singh, Manufacturing and Bioproducts Group, Scion (New Zealand Forest Research Institute), Rotorua, New Zealand

Horst Stobbe, Institute of Arboriculture, Brookkehre Hamburg, Germany

Keiji Takabe, Division of Forest and Biomaterials Sciences, Graduate School of Agriculture, Kyoto University, Kyoto, Japan

Frank W. Telewski, Department of Plant Biology, W.J. Beal Botanical Garden, Michigan State University, East Lansing, MI, USA

Jin Wang, International Centre for Bamboo and Rattan, Beijing, China

Yusuke Yamagishi, Faculty of Agriculture, Tokyo University of Agriculture and Technology, Fuchu, Tokyo; Faculty of Agriculture, Hokkaido University, Sapporo, Japan

Preface

INTRODUCTORY REMARKS

Wood (secondary xylem) is the most important sustainable and renewable material on this planet from an economic as well as an environmental perspective, serving as a raw material for the processing of a wide range of useful products. Wood is the final product of complex integrated physiological, biochemical, and molecular activities accompanying the development and differentiation of cambial derivative cells.

During the course of discussions with close international colleagues, the stimulus and need for a book arose that can bring together up-to-date information not only on processes related to wood formation but also on aspects of functions and applications, and thus can serve as an important text or source of reference for undergraduate and postgraduate students in wood biology. The information available on these aspects is scattered and fragmentary and not covered in a single volume.

This book is divided into four major parts.

The first part deals with various endogenous and exogenous effects on secondary xylem formation – information crucial for understanding xylogenesis.

U. Schmitt gives an overview of seasonal cambial activity, environmental control of related processes, and also the importance of cambial activity in the restoration of tissues after wounding.

R. Funada outlines the sequences of xylogenesis in trees, from reactivation of the cambium in the early spring by temperature to programmed cell death, leading to maturation of the secondary xylem. The dynamics of cortical microtubules closely related to the orientation and localization of newly deposited cellulose microfibrils are also covered.

E. Liang reviews the effect of moisture stress on the times and dynamics of xylem formation, with information on how drought in the spring could largely delay the onset of xylogenesis, leading to smaller numbers of xylem cells during the growing season, and likely early cessation of xylem differentiation in water-limited environments, such as in the Himalayan regions.

J. Fromm provides an overview of the major stress types affecting wood formation, with information on the negative consequences of chemical and physical environmental stresses the physiology, biochemistry, and structure. Among various abiotic stresses, nutrient deficiency, drought, temperature, soil salinity, and air pollution are mainly highlighted.

F. Telewski introduces flexure wood formed by mechanical loading on the trunk or branches of a tree. The alterations in the physical and chemical structures of the secondary xylem by loading result in the formation of flexure wood characteristics, with an increase in xylem production and cellulose microfibrillar angle and a decrease in the elastic modulus.

L. Donaldson describes the reaction wood formed as a geotropic response of trees and shrubs, which generally occurs in leaning stems and branches. Anatomical, physical, and chemical properties of reaction wood, in comparison with normal wood, are described. In addition, the wood qualities of reaction wood are briefly mentioned.

The second part of the book deals with the function and resistance of the secondary xylem.

Pits in the secondary wall in woody plants play an important role in the conduction of water in living trees and penetration of treatment liquids into timbers. Linking the structure of pits with the physiology provides insights into the regulation mechanisms of pit membrane in water flow against progressing cavitation. Y. Sano reviews recent studies on the structure of bordered pit membranes and its relevance to resistance against cavitations. The relationship between micromorphological characteristics and conduit cavitation resistance is clarified for pit membranes in conifers and angiosperms.

G. Daniel outlines the main morphological changes produced in wood cell walls following colonization and decay by brown-, white-, and soft-rot fungi, resulting from biomineralization of wood's main structural components. Modes of wood degradation by wood decay fungi are described, with examples from light and electron microscopic studies. The enzymatic and nonenzymatic systems used by wood decay fungi are also briefly reviewed.

Compared to wood-decaying fungi, bacteria can tolerate more extreme conditions, such as highly toxic preservatives and extremely low levels of oxygen. A.P. Singh reviews the micromorphological changes in wood cell walls attacked by wood-degrading bacteria, with an emphasis on the ultrastructural aspects of the micromorphological patterns produced.

The third part of the book deals with the economic utilization of woody plants. Many attempts are being made to maximize the added-value of lignocellulosics, such as genetical design of woody plants, bioconversion of woody material for renewable energy, and bioinspired functionalization of wood. In addition, wood as cultural heritages and bamboo as a substitute for woody biomass are treated.

Molecular biology has been employed for the modification and improvement of secondary xylem, particularly targeting cell wall characteristics. K.-H. Han outlines the genetic regulation of the biosynthesis of secondary cell walls, with a focus on genes encoding secondary wall-associated cellulose synthases, enzymes involved in lignin and hemicelluloses synthesis, and transcriptional regulators of secondary wall biosynthesis. Woody biomass can be converted into biofuels and useful biochemicals to replace fossil resources, using environmentally benign processes.

S. Saka briefly covers chemical pretreatments of lignocellulosic biomass and details enzymatic bioconversion. The obstacles to enzymatic bioconversion of woody biomass are also pointed out.

Wood is a natural biomaterial with intrinsic evolutionary optimization of its formation and structure, and the knowledge has served in developing high-performance engineering applications. I. Burgert describes the recent developments and advances in generating bioinspired wood products.

An intimate human link, representing human life and values, is embedded in wood from time immemorial. Viewing wood as cultural heritages, Y.S. Kim describes the influences of biotic and abiotic agents on the anatomical, physical, and chemical characteristics of wooden cultural heritages.

Unlike woody plants, bamboo does not produce a cambium, which is responsible for the production of secondary xylem. However, bamboo shares many similarities with woody plants, while marked differences occur in the cell wall ultrastructure. B. Fei outlines the anatomical, biological, and chemical characteristics of bamboo and the recent progress made in bamboo molecular biology, in relation to extending the potential of bamboo as an important biomass resource and as a substitute for wood biomass.

The fourth part deals with advanced techniques for investigating secondary xylem biology and wood ultrastructure. G. Daniel covers the diverse microscopy techniques, giving examples and pointing out limitations, with particular emphasis on sample preparation for studying secondary xylem biology. K. Takabe describes the novel rapid freezing and freeze substitution method that has provided new information on cell wall formation in woody plants. In combination with immunocytochemistry, detailed information on the localization of enzymes involved in the biosynthesis of cell wall components has been provided. K. Fukushima focuses on the application of time-of-flight secondary ion mass spectrometry in studying the main polymer components of woody plant cells as well as inorganics and low-molecular weight extractives, which are detectable with submicron lateral resolution by this technique.

The main aim of this book has been to provide a comprehensive coverage of areas relevant for understanding wood biology in a single volume, which can be useful as a text in undergraduate and postgraduate courses. However, satisfactorily fulfilling this aim is not without difficulties, mainly because of differing writing styles and levels of treatment of the topics covered. Nevertheless, we hope that the contents and presentations in this book stimulate further exploration of knowledge on wood biology.

We are much indebted to the authors who shared with us their valuable time and enthusiasm in writing their chapters. The book could not have been completed without their passion and kindness. It was Dr H.-J. Bae of Chonnam National University who provided the initial stimulus and continued pushing for the preparation of this volume. Prof R. Funada pointed out how this kind of book in English is urgently needed for university-level wood biology courses.

We are deeply grateful to the publishers for their constant encouragement and support and to Mary Elisabeth for the editorial work. Our grateful thanks also go to the anonymous reviewers for their valuable comments and constructive criticisms. We are so fortunate to have had enthusiastic support and backing from our families and are very grateful for their patience and tolerance throughout the preparation of this book.

Part I

Development of Secondary Xylem

Chapter 1

The Vascular Cambium of Trees and its Involvement in Defining Xylem Anatomy

Uwe Schmitt*, Gerald Koch*, Dieter Eckstein**, Jeong-Wook Seo**,†,
Peter Prislan‡, Jožica Gričar§, Katarina Čufar‡, Horst Stobbe¶,
Risto Jalkanen††

*Thünen Institute of Wood Research, Leuschnerstr, Hamburg, Germany; **Centre of Wood Sciences,
University of Hamburg, Leuschnerstr, Hamburg, Germany; †Department of Wood and Paper,
Chungbuk National University, Naesudong-ro, Seowon-gu Cheongju, Chungbuk, South Korea;
‡Department of Wood Science and Technology, Biotechnical Faculty, University of Ljubljana, Rozna
dolina, Ljubljana, Slovenia; §Department of Forest Yield and Silviculture, Slovenian Forestry Institute,
Ljubljana, Slovenia; ¶Institute of Arboriculture, Brookkehre Hamburg, Germany; ††Management and
Production of Renewable Resources, Natural Resources Institute Finland, Rovaniemi, Finland

Chapter Outline

INTRODUCTION AND OUTLINE

The growth of perennial plants from tall upright trees up to small prostrate dwarf shrubs is a complex of interlinked processes resulting in a three-dimensional body whereby meristematic tissues play a crucial role. Among the various fractions of meristematic tissues, we focus on the vascular cambium as a coherent lateral sheet of only a few cell layers in thickness between the secondary phloem and the secondary xylem, spreading from the roots, through the stem, up to the tips of the branches. During its active period, the vascular cambium delivers phloem cells to the outside and xylem cells to the inside through cell divisions.

Secondary Xylem Biology. http://dx.doi.org/10.1016/B978-0-12-802185-9.00001-2

The phellogen or cork cambium, another lateral meristem, is not considered here. Larson (1994) defines the vascular cambium as follows:

> *The cambium performs its meristematic task of producing daughter cells that differentiate to specialized tissue systems. Its derivatives vary either in form, or function, or rate of production at different positions on the tree, with age of the tree and with season of the year.*

The cambium contains two cell types, that is, fusiform cambial cells and ray initials. The elongated fusiform cambial cells are responsible for the production of all axially oriented cell types, such as tracheids and axial parenchyma in gymnosperms as well as fibers, vessels, and axial parenchyma in angiosperms. Bailey (1920, 1923) has already determined the lengths of fusiform initials varying between 0.17 mm in *Robinia* and 8.7 mm in *Sequoia*. The nearly iso-diametric ray initials deliver all cells composing the rays.

Several detailed descriptions of the cambium, mainly its structure/function relationships, have already been published (e.g., Iqbal, 1990; Evert, 2006). Among all of them, Larson's textbook *The Vascular Cambium* (1994) provided a comprehensive state-of-the-art survey at that time. More information was added on the cellular aspects of wood formation by Fromm (2013).

The cambium of trees is a very powerful tissue producing an enormous amount of biomass. An estimate by FAO (2012) for the global above-ground woody biomass is 434 billion m^3. Also in Germany, the production of woody biomass, being around 11 m^3 per ha and year, is impressive (Polley et al., 2009).

The cambium is not only responsible for the quantitative side of wood formation, varying over time, but it also determines taxa-specific anatomical features, stable over time. In the complex process of cambial activity, external and internal influences are interacting (Fig. 1.1). In the following, three examples will underline the relevance of the cambium for a successful performance of trees around the world: first, the seasonal activity and its control by environmental influences, second, the cambium's involvement in the restoration of tissues after injuries, and third, the cambium-dependent shaping of taxa-specific wood anatomical characteristics.

SEASONAL VARIATION OF CAMBIAL ACTIVITY

The cambium of trees outside the belt of tropical rain forests generally undergoes a seasonal activity cycle with a dormant and an active period around each year. Such a cycle also becomes well visible on the fine-structural level with distinct cytoplasmic changes (Fig. 1.2).

The seasonal variation of cambial activity is illustrated considering Scots pine (*Pinus sylvestris*) in northern Finland as an example. The study trees have been growing 80 km ("tree line" = site 1) and 300 km ("Arctic Circle" = site 2) south from the northern tree line (Seo et al., 2013). North Finland stands for the circumpolar boreal forest belt and the sites 1 and 2 represent two climatically different environments.

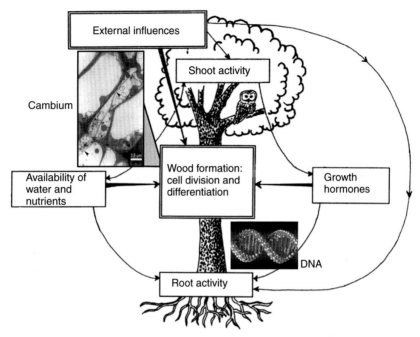

FIGURE 1.1 **Interaction between external influences and shoot and root activity as well as wood formation.** From the internal factors only the genetic make-up and the growth hormones are explicitly mentioned; inset in the left corner, dividing cambium cell; arrow heads point to new cell walls. *(Simplified after a model by Denne and Dodd (1981).)*

The intra-annually accumulated output of the cambial activity of Scots pine in five consecutive study years (2000–2004) follows an S-shaped function (Fig. 1.3). The rate of growth at both sites is highest in the second half of June and the first half of July. During these 4 weeks, 2/3 of the total annual growth is formed. This applies even for the unusually cool and moist summer of 1996 at the same two sites (Schmitt et al., 2004).

Depending on the actual weather conditions during the 5-year study period, the trees at the "Arctic Circle" start radial growth between the end of May and mid-June; earlywood passes into latewood during the first half of July and amounts to about 75% of the total annual tree-ring width at both sites. Radial growth ends between end of July and mid-August. At the "tree line" site, radial growth starts significantly later and ends slightly earlier (Fig. 1.4). Thus, the cambium of Scots pine is active for around 9 weeks at the "Arctic Circle" site and 7 weeks at the "tree line" site.

If we put these observations in the context of an entire growth phenological cycle throughout a year, we could conclude the following sequence of events. The winter buds break in the first half and height growth starts in the second half of May (Salminen and Jalkanen, 2007). Growth in thickness follows around end of May/early June when the heat sum, in terms of degree days, has reached

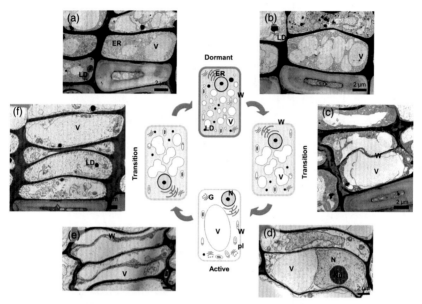

FIGURE 1.2 Schematic diagram of cytoplasmic changes in cambial cells of *F. sylvatica* during a seasonal cycle. (a and b) Dormant cells in winter; numerous vacuoles, endoplasmic reticulum (ER) mostly smooth, Golgi apparatus (G) with few secretory vesicles, numerous lipid droplets (LD). (c) Transition to activity in late winter to early spring, showing elongation and fusion of vacuoles following the resumption of cyclosis, rough endoplasmic reticulum, active Golgi apparatus; nucleus (N) with nucleolus (Nu), plasmodesma (pl). (d and e) Active cells in spring or early summer, with a large vacuole (V). (f) Transition to rest in autumn; fragmentation of the vacuole (V) and thickening of the cell wall (W). *(From Prislan et al. (2013a).)*

12.5% of the long-term, site-specific sum of degree days (Seo et al., 2008). Bud break and onset of growth in height and girth differ between years and latitude, proving a flexible and immediate response to the annually changing temperature. By this capability, the trees take advantage of an above-average warm spring to improve their site dominance (Bailey and Harrington, 2006) but also to avoid late frost damage (Hannerz, 1999). Growth in height and girth culminate clearly before the warmest period of the year, which is in the second half of July. According to Rossi et al. (2008), maximum growth appears to converge toward the summer solstice so that trees can safely complete cell-wall formation before an untimely frost may happen in the early autumn. Height growth finishes by the end of June/early July, when the heat sum has accumulated to approx. 41% (Salminen and Jalkanen, 2007). Soon after, the earlywood passes into latewood. Growth in thickness ceases by the end of July/mid-August with a heat sum of approx. 80% of the long-term heat sum.

Xylem and phloem formation, as well as cambium and leaf phenology and their relation to weather factors, were also studied in beech (*Fagus sylvatica*) trees growing at two sites in Slovenia at different altitudes (400 and 1200 m.a.s.l.) and during three consecutive years from 2008 to 2010. Leaf unfolding, onset of

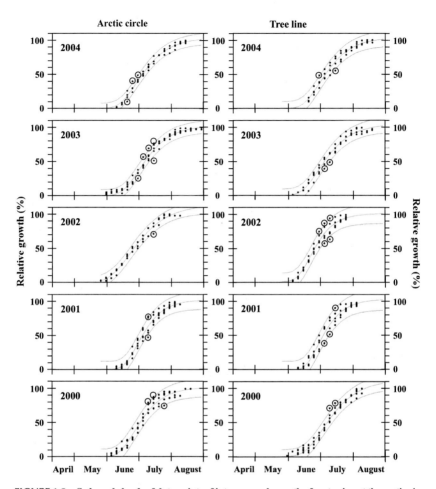

FIGURE 1.3   S-shaped clouds of data points of intra-annual growth of scots pine at the arctic circle and near tree line during five consecutive growing seasons and superimposed Gompertz functions with their upper and lower 95% confidence limits. Data points outside these limits (circled) were defined as outliers, so that they were eliminated from further analysis. *(From Seo et al. (2011).)*

cambial cell production, and increased number of active phloem cells occurred mid-April at low elevation and in the first week of May at higher elevation. Maximum rate of xylem cell production occurred from 20 May until 9 June at low elevation and about 2 weeks later at high elevation. Maximum rate of phloem cell production occurred more than one month earlier at both sites. Cessation of xylem and phloem cell production was observed around 19 August at low-elevation site and around 10 days earlier at high-elevation site. Differentiation of the last-formed xylem cells was concluded by mid-September at both plots. The year-to-year variability of these phenological phases was not statistically significant but the differences between the sites were. Temperature and "degree days" before the occurrence of most of the observed phases differed

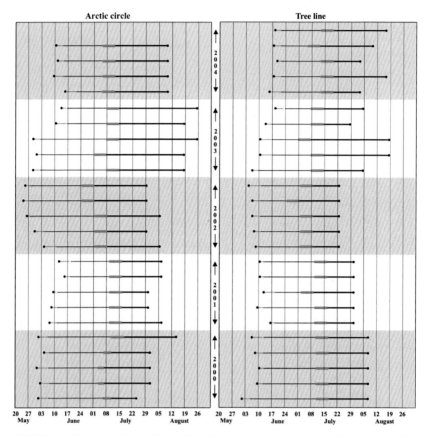

FIGURE 1.4   Duration of wood formation of scots pine at the arctic circle and at tree line during five vegetation periods, 2000–2004. Dotted line, estimated range for the onset of wood formation; thin line, earlywood formation; grey thick line, transition from early- to latewood; black line, latewood formation. *(From Seo et al. (2011).)*

significantly between the sites, thus demonstrating that the differences in xylem and phloem formation between sites can be attributed to a high intraspecific plasticity of beech (Prislan et al., 2013b).

MECHANICAL INJURY OF THE CAMBIUM AND ITS RESTORATION

All plants are prone to mechanical injuries lifelong, caused by logging, traffic accidents, pruning, insects, pathogens, or abiotic factors, leading to cell death along the superficial wound edges, but at the same time wound tissue is formed in deeper layers. Such reactions and the involvement of the cambial zone are presented on the macro- and microscopic level.

Independent on depth and size of wounds, the strategy of trees aims at a prompt protection of the inner tissue to avoid water loss and the penetration by microorganisms as well as to stop air embolism in the water-conducting cells. To achieve this goal, passive resistance and active responses impede the spread of wound-associated effects (Dujesiefken and Liese, 2015). Passive resistance refers to boundaries that are already in existence at the time of wounding, such as cell walls, rays, and heartwood portions. However, active responses are always related to living parenchyma cells in bark as well as in wood tissue. They become stimulated from a nearby wound and start with the production of substances of mostly phenolic character, which in hardwoods are released into the lumens of neighboring fibers and vessels to block them. In addition, parenchyma cells are able to synthesize suberin being deposited as an inner layer of their own walls to further strengthen the barrier against water loss. This mechanism is well known for bark (e.g., Biggs, 1985; Trockenbrodt and Liese, 1991; Trockenbrodt, 1994; Oven et al., 1999) and wood (e.g., Biggs, 1987; Duchesne et al., 1992; Schmitt and Liese, 1993). All these active processes finally lead to the formation of a so-called boundary layer. This is a narrow but highly effective, mostly discolored zone at a certain distance around a wound separating dead outer tissue and living inner tissue thus protecting the living tissue against wound-associated impacts.

Whenever wounds are set rather close to the cambium or even reach inner woody tissue, the cambial zone with its meristematic cells becomes affected, too. Cambium cells have thin walls and are therefore rather sensitive to any mechanical injuries and to drying. A narrow zone of cambial tissue at and close to a wound rapidly degenerates, whereby the extent of degeneration depends on the depth and size of damage, season, tree species, current tree vitality, and tree age (e.g., Manion 1991; Fink, 1999; Roloff, 2004). For example, dry and hot summer periods as well as cold winters are less favorable to keep the amount of degenerated tissue as low as possible, whereas wet and cool summers mostly lead to only a few cell layers dying because of rapid and intensive wound reactions.

The cambium is not involved in the formation of a boundary layer around a wound. It rather takes over responsibility for the closure of wound surfaces through the formation of callus tissue. In principle, two strategies are possible for closing a wound: the first leads to the formation of a lateral callus and the second to the formation of a surface callus (Stobbe et al., 2002). In most cases a lateral callus develops to gradually close the wound. During the vegetation period, within a few days after wounding large isodiametric cells with thin, nonoriented new walls appear at the wound edges along the transition zone between the xylem and phloem. These cells develop mainly from redifferentiated young phloem cells, but undifferentiated xylem cells and cambium cells are also part of this process. Such a tissue already forms a microscopically well-visible early callus. Thereafter, oriented cell divisions occur toward the wound surface leading to an expanding callus (Fig. 1.5).

Within this young callus, a regeneration of the cambium can be observed, which reforms as a band crossing the young callus tissue as a tangential extension. In some wounds, however, the reformation of the cambial zone occur in phloem areas by dedifferentiated, flattened parenchyma cells finally fusing to a continuous band of cambium cells. The new cambium in all cases then restarts regular xylem and phloem formation (Frankenstein et al., 2005) (Fig. 1.6).

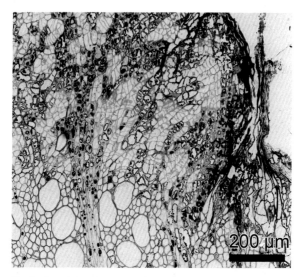

FIGURE 1.5   *F. sylvatica*. Light micrograph of an early stage of a lateral callus with parenchymatous cells showing cell divisions oriented toward the wound surface. *(From Grünwald et al. (2002).)*

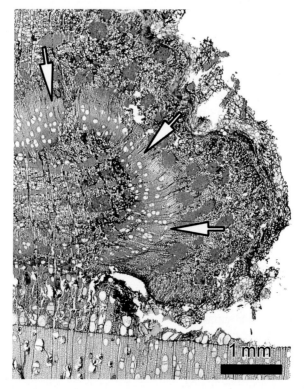

FIGURE 1.6   *Populus* spp. Light micrograph of a growing lateral callus and a regenerated cambium (arrows) producing callus xylem and phloem. Micrograph: C. Frankenstein.

As compared with undisturbed cambium, the wound cambium is character-ized by a higher cell division activity leading to a distinctly faster growth of the callus. Year by year, the callus covers more of the original wound surface. Even when covering the entire wound surface, such a lateral callus leaves a narrow, axially oriented central cleft.

The second, less often found strategy leads to the formation of a so-called surface callus. This phenomenon was variously described in the past and termed "reproduction of new bark and wood tissue (Hartig, 1844), surface or superfi-cial callus" (Fenner, 1949) or simply surface callus (Dujesiefken et al., 2001). Formation of such a callus can only be initiated when living phloem and/or cambium and/or differentiating xylem remain on the wound surface, whereby the differentiating xylem often seems to play a major role as compared with cambium and inner phloem. On the cellular level, the most important precondition for the formation of a surface callus is that undifferentiated cells in the developing xylem cells without secondary wall, cambium cells as well as developing phloem cells remain on the wound surface. If wounds are set deeper into already differentiated xylem, no surface callus can be formed. Un-differentiated xylem cells as well as the remaining cambium cells are capable of carrying out cell divisions, which are initiated through wounding to build up an early callus tissue on parts or along the entire wound surface consisting of large, vacuolated, and thin-walled parenchymatous cells (Fig. 1.7).

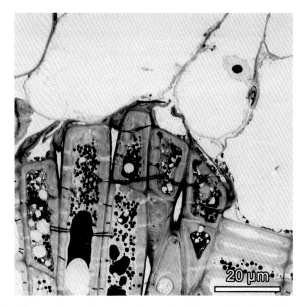

FIGURE 1.7 *Tilia americana.* Electron micrograph of the transition from differentiated xylem cells (ray cells and fibers) present at the time of wounding (lower part) and formation of the wound-associated large, thin-walled parenchymatous cells composing the early callus tissue.

Within this young callus tissue, a wound periderm develops first in the outer portion and then a wound cambium reforms in the inner portion. Both new tissues finally lead to the formation of new xylem and new phloem as well as to a protecting outer bark. During the vegetation period, a surface callus becomes well visible already several weeks after wounding (Fig. 1.8), whereby during the dormant season this process can take several months. Surface callus formation can be promoted by covering the entire wound with a black plastic wrap (Fig. 1.9). This treatment keeps living cells alive by protecting them against drying and UV radiation. Unlike wound closure by a lateral callus, a surface callus does not show a central cleft (Figs 1.10 and 1.11).

FIGURE 1.8   Surface callus of an oak tree (*Quercus robur*) 9 weeks after wounding. Large amount of the wound surface covered by callus tissue.

FIGURE 1.9   Black plastic wrap on the stem of *Tilia* covering the wound to promote surface callus formation.

FIGURE 1.10   Stem surface of a beech tree (*F. sylvatica*) showing a wound overgrown by a lateral callus from both sides after several years; the callus surface still shows a central cleft.

FIGURE 1.11   Stem of a beech tree (*F. sylvatica*) showing a wound closed by a surface callus after several years; note that the callus surface appears as a homogenous tissue without a central cleft.

MICROSCOPIC XYLEM FEATURES DEFINED BY THE CAMBIUM

Wood identification is an important field of activity, for example, for the use of timbers well suited for various applications, in the frame of the European Timber Regulation, which came into force to mainly prevent illegal logging (requiring documents containing the exact botanical name of the imported wood) or in the frame of the International Convention on the Trade of Endangered Species (Koch et al., 2011).

Some wood anatomical details are already laid down during the division of the cambial cells and the following deposition of young, undifferentiated xylem mother cells. One of these features is the storied arrangement of axially oriented xylem cell types and the storied ray pattern. Also, the size of the rays is already determined through the ray initials in the cambium.

Here, we list a selection of those details that are a major basis for wood identification, such as storied structures as well as size of rays and specially formed cell types, for example, tile cells and sheath cells.

Storied structures are an important differentiating and identifying feature, particularly with regard to the enormous variation in the anatomy of tropical wood species. Storied structures are generally defined as "cells arranged in tiers" or "horizontal series of cell arrangements" of rays, axial parenchyma, fibers, and vessel elements. The formation of storied structures is strongly initiated by the arrangement of storied cambial fusiform initials, which are aligned horizontally with laterally neighboring cells of similar length (Carlquist, 1988). During cambium ontogeny, the storied pattern develops gradually and its expression depends upon the age of the tree and on the diameter of the primary vascular ring at the time of cambium initiation (Larson, 1994). This process has been studied intensely by many authors who additionally suggest that storied pattern development is initiated by longitudinal anticlinal divisions of fusiform initials and the absence of intrusive growth of their daughter cells (Bailey, 1923; Butterfield, 1972; Larson, 1994). According to this view, the storey of fusiform initials should be regarded as the formation of cells developed from one ancestral procambium (Myskow and Zagórska-Marek, 2004). Further investigations also refer to the excessive intrusive growth of fusiform initials and their pseudotransversal shortening divisions as a dynamic process of storied structure formation (Zagórska-Marek, 1984). Based on this knowledge, it can be stated that the formation of a storied pattern is not only a rigid process of longitudinal and anticlinal cell divisions but rather can be used as distinct anatomical structures, which are very helpful for routine wood identification.

Macroscopically Visible Storied Structures

Especially the storied arrangement of rays or "tiers of rays" is already visible at low magnifications or even with the unaided eye or a hand lens; it appears as fine horizontal striations or "ripple marks" on the tangential surface (Figs 1.12 and 1.13).

Overall, the presence of macroscopically visible storied ray structures is described for five of the currently defined 57 plant orders (according to Bresinsky et al. (2008)), among them Zygophyllales, Fabales, Malvales, Lamiales, and Sapindales (Table 1.1).

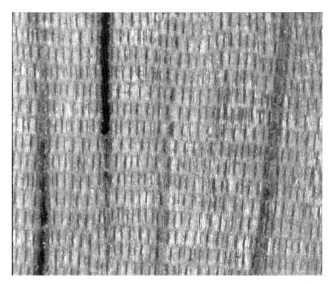

FIGURE 1.12   Tangential surface (magnifying lens 12×) of *Daniellia ogea* with distinct arrangement of regularly storied rays (tiers of rays).

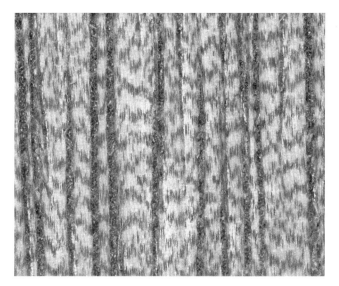

FIGURE 1.13   Tangential surface (magnifying lens 12×) of *Dacryodes* with irregularly storied rays (wavy arrangement of rays).

Microscopically Visible Storied Structures

On the microscopic level, storied structures can be clearly identified for the various individual cell types or tissues: fibers and vessel elements, rays, and axial parenchyma. In general, storied structures should be determined on tangential

TABLE 1.1 Taxonomic Classification of Plant Orders/Families and Individual Wood Species with the Presence of Macroscopically Visible Storied Ray Structures

Order	Family	Species (representative)
Zygophyllales	Zygophyllaceae	*Guaiacum officinale/Guaiacum sanctum*
Fabales	Fabaceae, Caesalpinioideae	*Afzelia bipindensis/Afzelia bella*
		Apuleia leiocarpa
		Dicorynia guianensis
		Microberlinia bisculata/Microberlinia brazzavillensis
	Fabaceae, Faboideae	*Dalbergia nigra*
		Dipteryx odorata
		Millettia laurentii
Malvales	Dipterocarpaceae	*Dryobalanops aromatica*
		Shorea laevis
		Shorea spp., subg. *Shorea*
	Malvaceae	*Heritiera simplicifolia*
		Triplochiton scleroxylon
Sapindales	Meliaceae	*Entandrophragma cylindricum*
		Entandrophragma utile
		Swietenia macrophylla
Lamiales	Bignoniaceae	*Tabebuia ipe*

According to Höwler (2011).

sections. This important diagnostic feature can be recorded singly or in combination as in some woods all elements are storied (e.g., *Millettia laurentii*, *Dalbergia maritima* (Figs 1.14 and 1.15)) while in others only some elements are storied.

According to the "IAWA list of microscopic features for hardwood identification" (Wheeler et al., 1989), the following definitions of storied structures are given and distinguished for microscopic wood identification:

- Presence of storied structures;
- All rays storied;
- Low rays storied, high rays nonstoried;
- Axial parenchyma storied and/or vessel elements storied;
- Fibers storied;
- Rays and/or axial elements irregularly storied;
- Number of ray tiers per axial millimeter.

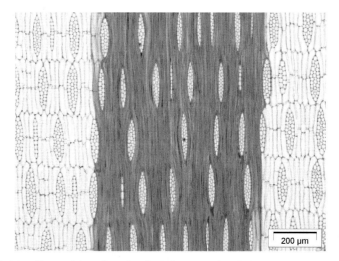

FIGURE 1.14   **Tangential section of *Millettia laurentii* with distinct storied wood structures (axial and ray parenchyma) as a result of a storied cambium.**

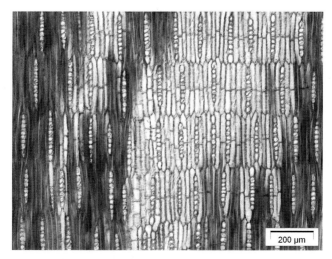

FIGURE 1.15   **Tangential section of *Dalbergia maritima* with distinct storied wood structures (axial and ray parenchyma) as a result of a storied cambium.**

By using these characters, it should be considered that the occurrence of storied structures is variable within species and samples. For instance, in some samples of *Swietenia* (Meliaceae) rays are definitely storied, in others irregularly storied, and still in others rays are not storied.

Width and Size of Rays

In addition to the distinctive storied structures, the type and dimensions of rays are also strongly defined by cambial cell divisions.

On the macroscopic level, size and frequency of rays vary considerably and may also be a very useful diagnostic feature, particularly the very wide and high rays (Fig. 1.16). Ray width is determined in transverse or tangential sections and can roughly be subdivided into two macroscopically distinguishable size classes:

- Rays narrow – Generally not detectable with the unaided eye because of their small size. All softwoods, characterized by very narrow (nearly all uniseriate) rays belong to this category. Narrow rays, though not necessarily uniseriate, are also found in most hardwoods, for example, American mahogany (*Swietenia macrophylla*, Meliceae), sapele (*Entandrophragma cylindricum*, Meliaceae), hevea (*Hevea brasiliensis*, Euphorbiaceae).
- Rays broad (also in combination with narrow rays) – Easily visible with the unaided eye, about 0.5 mm (or more) in width, for example, red oak and white oak (*Quercus*, Fagaceae) as well as beech (*F. sylvatica*, Fagaceae).

Aggregate rays also count as wide rays. They constitute rays composed of a number of individual rays so closely associated with one another that they appear macroscopically as a single broad ray (Fig. 1.17). The individual rays are separated by axial elements, mostly fibers, for example, in alder (*Alnus* spp., Betulaceae) and hornbeam (*Carpinus betulus*, Betulaceae).

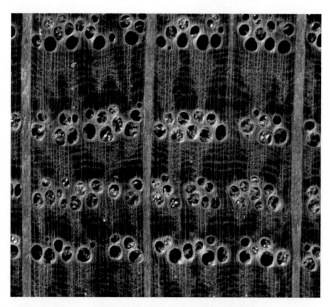

FIGURE 1.16 **Transverse surface (magnifying lens 12×) of *Quercus robur* with distinct broad rays easily visible with the unaided eye, about 0.5 mm (or more) in width.**

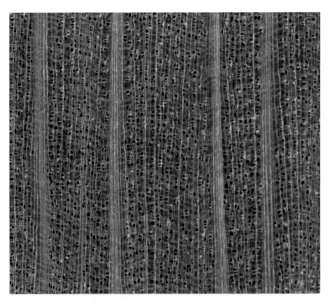

FIGURE 1.17   **Transverse surface (magnifying lens 12×) of *Carpinus betulus* with aggregate rays which appear macroscopically as single large rays.**

On the microscopic level, the cellular composition of the rays as well as their width and size represent additional important characters for wood identification. Especially the differentiation between procumbent ray cells and upright and/or square ray cells is of high diagnostic value. Generally, upright and square cells, if present in combination with procumbent cells, are located in the marginal rows, that is, those rows at the top and bottom of a ray and procumbent cells are located in the center of a ray. Furthermore, the number of upright and/or square ray cells can be clearly counted. Using ray width, the differentiation between "rays exclusively uniseriate" and "larger rays" (ray width 1–3 cells, larger rays commonly 4- to 10-seriate or larger rays commonly > 10-seriate (Figs 1.18 and 1.19)) is easily possible (according to the IAWA list (Wheeler et al., 1989)).

Special Cell Types

Two special characters of very high diagnostic value are the presence of tile and sheath cells, which are also directly defined by cambial cell divisions.

Tile cells (Fig. 1.20) represent a special type of apparently empty upright (rarely square) ray cells occurring in intermediate horizontal series usually interspersed among procumbent cells, for example, in *Durio, Pterospermum* (Sterculiaceae), and in some species of *Grewia* (Malvaceae).

Sheath cells (Fig. 1.21) are defined as ray cells located along the sides of broad rays (more than three-seriate) as viewed on tangential sections; they are larger (generally taller than broad) than the central cells, for example, in *Ceiba*

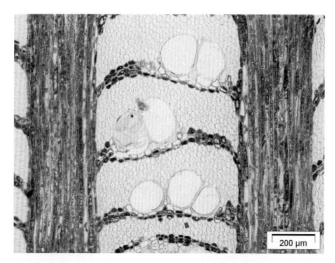

FIGURE 1.18 Transverse section of *Roupala montana* with large rays commonly >10-seriate.

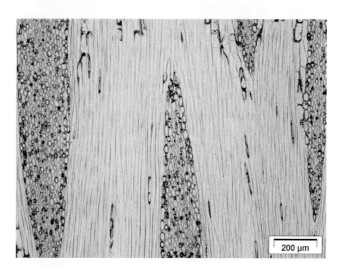

FIGURE 1.19 Tangential section of *Roupala montana* with rays of two distinct sizes (small/uniseriate and large rays).

pentandra (Bombacaeae), *Dipterocarpus lowii* (Dipterocarpaceae), and *Eribroma oblonga* (Sterculiaceae).

The recognition of both characters "occurrence of tile or sheath cells" (associated with wood rays) strongly reduce the number of remaining species (species that fulfil these characters) by using the established computerized databases "Commercial Timbers" (delta intkey system) or "Insidewood." For example, if you enter the occurrence of tile cells in the database "Commercial Timbers," only six taxa remain out of the 367 listed and described commercial

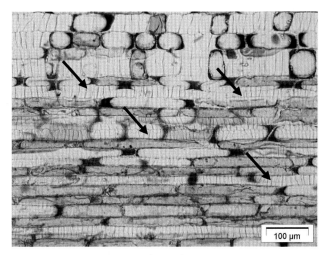

FIGURE 1.20   Radial section of *Durio* with tile cells – upright (rarely square) ray cells occurring in intermediate horizontal series.

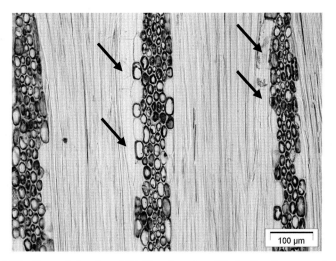

FIGURE 1.21   Tangential section of *Heritiera javanica* with sheath cells – parenchyma cells that are located along the sides of broad rays.

timbers confirming the high diagnostic value of this character (Richter and Dallwitz, 2000).

The selected and described characters are of very high importance for microscopic wood identification, but only represent a selection of wood anatomical characters, which are directly defined by the vascular cambium. Further important characters are, for example, fiber length and structure of axial parenchyma.

CONCLUSIONS

The afore-mentioned topics should demonstrate the diverse potential of the vascular cambium. As to its involvement in xylem formation, we traced its intra-annual dynamics of activity along a latitudinal and an altitudinal transect whereby climate turned out to determine the onset, progress and end of wood formation.

Cambial cells, together with undifferentiated xylem and phloem cells, are also involved in the regeneration processes after injuries by forming an early callus tissue exclusively consisting of parenchyma cells of which some are converted to cambial cells delivering phloem cells to the outside and xylem cells to the inside.

Besides these processes, which are largely controlled by external influences, the cambium contains internal information to express specific anatomical characteristics stable over time, such as storied rays and/or storied axial cells, which are key features for differentiating wood taxa from one another.

Thus, the vascular cambium is not only responsible for the quantitative part but also for the qualitative part of wood formation.

REFERENCES

Bailey, I.W., 1920. The cambium and its derivative tissues. II. Size variations of cambial initials in gymnosperms and angiosperms. Am. J. Bot. 7, 355–367.

Bailey, I.W., 1923. The cambium and its derivatives. IV. The increase in girth of the cambium. Am. J. Bot. 10, 499–509.

Bailey, J.D., Harrington, C.A., 2006. Temperature regulation of bud-burst phenology within and among years in a young Douglas-fir (*Pseudotsuga menziesii*) plantation in western Washington, USA. Tree Physiol. 26, 421–430.

Biggs, A.R., 1985. Suberized boundary zones and the chronology of wound response in tree bark. Phytopathology 75, 1191–1195.

Biggs, A.R., 1987. Occurrence and location of suberin in wound reaction zones in xylem of 17 tree species. Phytopathology 77, 718–725.

Bresinsky, A., Körner, C., Kadereit, J.W., Neuhaus, G., Sonnewald, U., 2008. Strasburger – Lehrbuch der Botanik. Begründet von E. Strasburger, thirty-sixth ed. Spektrum Akademischer Verlag/ Heidelberg.

Butterfield, B.G., 1972. Developmental changes in the vascular cambium of *Aeschynomene hispida* Willd. New Zealand J. Bot. 10, 373–386.

Carlquist, S., 1988. Comparative Wood Anatomy. Systematic, Ecological, and Evolutionary Aspects of Dicotyledon Wood. Springer/Berlin–Heidelberg, New York.

Denne, M.P., Dodd, R.S., 1981. The environmental control of xylem differentiation. In: Barnett, J.R., Biggs, A.R. (Eds.), Xylem Cell Development. Castle House, Turnbridge Wells/UK, pp. 236–255.

Duchesne, L.C., Hubbes, M., Jeng, R.S., 1992. Biochemistry and molecular biology of defense reactions in the xylem of angiosperm trees. In: Blanchette, R.A., Biggs, A.R. (Eds.), Defense Mechanisms of Woody Plants Against Fungi. Springer Series on Wood Science. Springer/ Berlin Heidelberg, New York.

Dujesiefken, D., Liese, W., 2015. The CODIT Principle. Implications for Best Practices. Intern. Society of Arboriculture, Champaign, Illinois/USA.

Dujesiefken, D., Stobbe, H., Kowol, T., 2001. Der Flächenkallus – eine Wundreaktion von Bäumen nach Rücke- und Anfahrschäden. Forstw. Cbl 120, 80–89.

Evert, R.F., 2006. Esau's Plant Anatomy: Meristems, Cells, and Tissues of the Plant Body: Their Structure, Function, and Development, third ed. John Wiley and Sons.

FAO, 2012. FAOSTAT-Forestry Database [online]. Food and Agriculture Organization of the United Nations (FAO) http://faostat.fao.org/.

Fenner, C., 1949. Shade method of direct cambium-to-bark development. National Shade Tree Conf. Proc. 18, 131–138.

Fink, S., 1999. Pathological and Regenerative Plant Anatomy. Handbuch der Pflanzenanatomie, 14/6, Gebrüder Bornträger, Berlin Stuttgart/Germany.

Frankenstein, C., Schmitt, U., Waitkus, C., Eckstein, D., 2005. Wound callus formation – a microscopic study on poplar (Populus tremula L. × Populus tremuloides Michx.). J. Appl. Bot. Food Qual. 79, 44–51.

Fromm, J. (Ed.), 2013. Cellular Aspects of Wood Formation. Plant Cell Monographs, vol. 20, Springer/Heidelberg, New York.

Grünwald, C., Stobbe, H., Schmitt, U., 2002. Entwicklungsstufen der seitlichen Wundüberwallung von Laubgehölzen. Forstw. Cbl. 121, 50–58.

Hannerz, M., 1999. Evaluation of temperature models for predicting bud burst in Norway spruce. Can. J. For. Res. 29, 9–19.

Hartig, T., 1844. Bericht über gelungene Versuche zur Reproduktion neuer Holz- und Rindenschichten aus dem Holzkörper der Bäume. Verhandlungen des Vereins zur Beförderung des Gartenbaus in den königlich preußischen Staaten 17: 329–331.

Höwler, K., 2011. Zeigt Stockwerkbau der Holzelemente eine phylogenetische Tendenz? BSc Thesis, Department of Biology and Wood Science, University of Hamburg.

Iqbal, M (Ed.), 1990. The Vascular Cambium. Research Studies Press, Taunton/England.

Koch, G., Richter, H.G., Schmitt, U., 2011. The data base CITESwoodID – Computer-aided identification and description of CITES-protected trade timbers. IAWA J. 32, 213–220.

Larson, P.R., 1994. In: Timell, T. (Ed.), The Vascular Cambium: Development and Structure. Springer Series in Wood Science. Springer, Berlin/Germany.

Manion, P.D., 1991. Tree Disease Concepts. Prentice-Hall, Englewood Cliffs/USA.

Myskow, E., Zagórska-Marek, B., 2004. Ontogenetic development of storied ray pattern in cambium of Hippophae rhamnoides L. Acta Soc. Bot. Pol. 73, 93–101.

Oven, P., Torelli, N., Shortle, W.C., Zupančič, M., 1999. The formation of ligno-suberized layer and necrophylactic periderm in beech bark (Fagus sylvatica L.). Flora 194, 137–144.

Polley, H., Hennig, P., Schwitzgebel, F., 2009. Holzvorrat, Holzzuwachs, Holznutzung in Deutschland. AFZ-Der Wald 64, 1076–1078.

Prislan, P., Čufar, K, Koch, G., Schmitt, U., Gričar, J., 2013a. Review of cellular and subcellular changes in the cambium. IAWA J. 34, 391–407.

Prislan, P., Gričar, J., de Luis, M., Smith, K.T., Čufar, K., 2013b. Phenological variation in xylem and phloem formation in Fagus sylvatica from two contrasting sites. Agr. Forest Meteorol. 180, 142–151.

Richter, H.G., Dallwitz, M.J., 2000 (onwards). Commercial timbers: descriptions, illustrations, identification, and information retrieval. In English, French, German, Portuguese, and Spanish, www.delta-intkey.com/wood/.

Roloff, A., 2004. Bäume–Phänomene der Anpassung und Optimierung. Ecomed, Landsberg/Germany.

Rossi, S., Deslauriers, A., Gričar, J., Seo, J.-W., Rathgeber, C.B.K., Anfodillo, T., Morin, H., Levanič, T., Oven, P., Jalkanen, R., 2008. Critical temperatures for xylogenesis in conifers of cold climates. Glob. Ecol. Biogeogr. 17, 696–707.

Salminen, H., Jalkanen, R., 2007. Intra-annual height increment of *Pinus sylvestris* at high latitudes in Finland. Tree Physiol. 27, 1347–1353.

Schmitt, U., Liese, W., 1993. Response of xylem parenchyma by suberization in some hardwoods after mechanical injury. Trees 8, 23–30.

Schmitt, U., Jalkanen, R., Eckstein, D., 2004. Cambium dynamics of *Pinus sylvestris* and *Betula* spp in the northern boreal forest in Finland. Silva Fenn. 38, 167–178.

Seo, J.-W., Eckstein, D., Jalkanen, R., Rickebusch, S., Schmitt, U., 2008. Estimating the onset of cambial activity in Scots pine in northern Finland by means of the heat-sum approach. Tree Physiol. 28, 105–112.

Seo, J.-W., Eckstein, D., Jalkanen, R., Schmitt, U., 2011. Climatic control of intra- and inter-annual wood-formation dynamics of Scots pine in northern Finland. Environ. Exp. Bot. 72, 422–431.

Seo, J.-W., Eckstein, D., Olbrich, A., Jalkanen, R., Salminen, H., Schmitt, U., Fromm, J., 2013. Climate control of wood formation: illustrated for Scots pine at its northern distribution limit. Cellular aspects of wood formation. In: Fromm, J. (Ed.), Plant Cell Monographs 20. Springer, Berlin Heidelberg/Germany, pp. 159–185.

Stobbe, H., Schmitt, U., Eckstein, D., Dujesiefken, D., 2002. Developmental stages and fine structure of surface callus formed after debarking of living lime trees (*Tilia* sp.). Ann. Bot. 89, 773–782.

Trockenbrodt, M., 1994. Light and electron microscopic investigations on wound reactions in the bark of *Salix caprea* L. and *Tilia tomentosa* Moench. Flora 189, 131–140.

Trockenbrodt, M., Liese, W., 1991. Untersuchungen zur Wundreaktion in der Rinde von *Populus tremula* L. und *Platanus* × *acerifolia*. Angew. Bot. 65, 279–287.

Wheeler, E., Baas, P., Gasson, P.E. (Eds.), 1989. IAWA list of microscopic features for hardwood identification. IAWA Bull.n.s. 10, 219–332.

Zagórska-Marek, B., 1984. Pseudotransverse divisions and intrusive elongation of fusiform initials in storeyed cambium of *Tilia*. Can. J. Bot. 62, 20–27.

Chapter 2

Xylogenesis in Trees: From Cambial Cell Division to Cell Death

Ryo Funada*, Yusuke Yamagishi*,**, Shahanara Begum*,†, Kayo Kudo*,††,
Eri Nabeshima*,‡, Widyanto Dwi Nugroho*,‡‡, Md. Rahman Hasnat*,
Yuichiro Oribe§, Satoshi Nakaba*

*Faculty of Agriculture, Tokyo University of Agriculture and Technology, Fuchu, Tokyo, Japan;
**Faculty of Agriculture, Hokkaido University, Sapporo, Japan; †Faculty of Agriculture,
Bangladesh Agricultural University, Mymensingh, Bangladesh; ††Institute of Wood Technology,
Akita Prefectural University, Noshiro, Akita, Japan; ‡Faculty of Agriculture, Ehime University,
Matsuyama, Ehime, Japan; ‡‡Faculty of Forestry, Universitas Gadjah Mada, Yogyakarta,
Indonesia; §Tohoku Regional Breeding Office, Forestry and Forest Products Research Institute,
Takizawa, Iwate, Japan

Chapter Outline

INTRODUCTION

Wood, which is a renewable resource, has been used for thousands of years as a raw material such as timber, furniture, pulp and paper, chemicals, and fuels. In addition, since wood is a major carbon sink, it is expected to play an important role in removing the excess atmospheric CO_2 that is generated by the burning of fossil fuels. Moreover, wood has recently been used as a resource for bioethanol. Therefore, there is still great demand for wood as a biomaterial and source of bioenergy.

Wood is produced by the vascular cambium (cambium) of the stems of trees (Catesson, 1994; Larson, 1994; Funada, 2000, 2008). The cambium is defined as the actively dividing layer of cells that lies between, and gives rise to the secondary xylem and phloem. The periclinal division of cambial cells leads to an increase in stem diameter of trees. The division of cambial cells produces the secondary phloem on the outside and the secondary xylem on the inside.

Secondary Xylem Biology. http://dx.doi.org/10.1016/B978-0-12-802185-9.00002-4

The amount of secondary xylem cells produced is usually much higher than the amount of secondary phloem cells. Thus, matured secondary xylem cells are usually used as wood. The quantity of wood depends on the amounts of secondary xylem cells produced by division of cambial cells.

The cambial activity of trees exhibits seasonal cycles of activity and dormancy, which are known as annual periodicity in temperate and cool zones. This periodicity plays an important role in the formation of wood and reflects the environmental adaptivity of trees, for example, their tolerance to cold in winter in cool and temperate zones. Therefore, details of the cell biological and physiological aspects of the regulation of cambial activity in trees are of considerable interest.

Differences in wood quality, in particular mechanical properties of wood, are largely due to differences in wood structure. Wood structure is determined by the process of xylem differentiation (xylogenesis), such as cell enlargement and cell wall thickening of cambial derivatives. However, its precise process is not yet fully understood. Therefore, in order to create new woods with desirable quantities and qualities by biotechnological techniques, more detailed cellular and molecular information is needed on wood formation.

CHANGES FROM CAMBIAL DORMANCY TO ACTIVITY

Cambial activity ceases in the autumn or winter seasons and enters the dormant stage. Cambial dormancy consists of two stages, namely, rest and quiescence (Catesson, 1994; Larson, 1994) (Fig. 2.1a). The resting stage is maintained by conditions within the tree and it is followed by the quiescent stage, which is controlled by environmental conditions. The resting stage of dormancy is a physiological state wherein the cambium cannot divide, even under favorable growth conditions such as adequate temperature, water supply, light, and plant hormones. During the first 2–4 weeks of dormancy, the cambium is unable to produce new cells even when the plant hormone auxin (indole-3-acetic acid; IAA) is supplied under favorable environmental conditions (Sundberg et al., 1987).

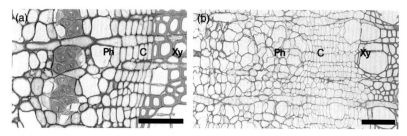

FIGURE 2.1   **Light micrographs of transverse sections showing dormant (a) and active cambium and differentiating secondary xylem (b) of the main stem in *P. sieboldii* × *P. grandidentata*. Ph, secondary phloem; C, cambium; Xy, secondary xylem.** Scale bars = 50 μm (a) and 100 μm (b).

After exposure to natural or artificial chilling, the cambium gradually regains the ability to produce new cells in response to IAA under appropriate environmental conditions. When the cambial cells become fully responsive to IAA in this way, the cambium is deemed to be in the quiescent stage of dormancy, which is imposed by adverse external factors. During the quiescent stage of cambial dormancy, the cambium is able to divide when exposed to appropriate environmental conditions. The transition from rest to quiescence involves structural, histochemical, and functional changes in cambial cells (Lachaud et al., 1999; Rensing and Samuels, 2004; Samuels et al., 2006).

Cambial activity generally resumes in the early spring with a change from the quiescent dormant state to the active state (cambial reactivation). Cambial activity in trees is generally regulated by internal factors such as plant hormones (Aloni, 1991; Sundberg et al., 2000). However, with respect to the plant hormone auxin, no increase in the level of IAA was detected in the cambial region of conifers at the onset of cambial reactivation, suggesting the absence of a clear relationship between the timing of cambial reactivation and endogenous levels of IAA (Sundberg et al., 1991; Funada et al., 2001a, 2002). Therefore, other factors appear to be necessary for induction of cambial reactivation.

Numerous studies have demonstrated that the timing of cambial reactivation is controlled by temperature (Begum et al., 2013). Localized heating of stems over a range of temperatures from 22°C to 30°C for 5 or 6 days in cold winter induces localized reactivation of the cambium in evergreen conifers, such as *Pinus contorta* (Savidge and Wareing, 1981), *Picea sitchensis* (Barnett and Miller, 1994), *Cryptomeria japonica* (Oribe and Kubo, 1997; Begum et al., 2010b, 2012b), *Abies sachalinensis* (Oribe et al., 2001, 2003), *Picea abies* (Gričar et al., 2006), and *Abies firma* (Begum et al., 2012b). These observations suggested that an increase in temperature of stems might be a limiting factor for cambial reactivation during the quiescent dormancy of evergreen conifers.

Localized heating of stems during dormancy also induces the division of cambial cells in a deciduous diffuse-porous hardwood hybrid poplar *Populus sieboldii* × *P. grandidentata* (Fig. 2.1b; Begum et al., 2007). Localized heating for 4 weeks induces cambial reactivation that occurs earlier than natural cambial reactivation. Moreover, 2 months of localized heating results in xylem differentiation in heated poplar stems (Begum et al., 2007). In addition, localized heating for 6 weeks induces earlier cambial reactivation than natural cambial reactivation in a deciduous ring-porous hardwood *Quercus serrata* (Kudo et al., 2014). By contrast, 2 weeks of localized heating of stems of a deciduous conifer *Larix leptolepis* fails to induce cambial reactivation (Oribe and Kubo, 1997). Longer localized heating of the stems of deciduous trees, as compared to those of evergreen conifers, might be required for conversion of cambium from a quiescent to an active state. Therefore, Begum et al. (2013) proposed that the state of dormancy in deciduous trees is deeper than in evergreen conifers.

Under natural conditions, temperatures in late winter and early spring are important external stimulus of the induction of cambial reactivation and xylem differentiation in hybrid poplar and *C. japonica* in temperate and cool zones. Earlier cambial reactivation and differentiation of secondary xylem cells are observed when it is warmer than normal in late winter and early spring (Begum et al., 2008). For example, in Fuchu, Tokyo, Japan, early spring in 2007 was warmer than in 2005 and 2008 and thus, cambial reactivation occurred earlier in 2007 than in 2005 and 2008 in hybrid poplar and *C. japonica* trees (Begum et al., 2008, 2010a). Once cambial reactivation has occurred, xylem differentiation starts within 3 or 4 weeks under natural conditions. The timing of xylem differentiation appears to depend on the date of onset of cambial reactivation, which might be controlled by temperature.

Cambial reactivation in stems of hybrid poplar and *C. japonica* occurs when the maximum daily temperature exceeds 15°C for 8–10 days and 10°C or 11°C for 25–27 days, respectively. Therefore, Begum et al. (2008) proposed that the timing of cambial reactivation could be predicted from the accumulation of maximum daily temperature in degrees above a threshold value. The sum of the number of degrees centigrade in excess of a threshold value for the daily maximum temperature from January 1 to the initiation of cambial cell division was calculated (Begum et al., 2008, 2010a). This value was defined as the cambial reactivation index (CRI), as follows: $CRI = \sum (T_{md}-T_t)$, where T_{md} is the daily maximum temperature in excess of a given threshold temperature and T_t is the given threshold temperature.

In the case of hybrid poplar, CRI was 93°C in 2005 and 96°C in 2007. In the case of *C. japonica*, when 10°C was used as the threshold temperature, the CRIs of 94°C and 97°C for 2007 and 2008 were closer together than other values of CRI. Therefore, CRI_{md} based on the threshold maximum temperature might help us to predict the timing of cambial reactivation only from analyses of meteorological data. Similarly, the temporal integration of daily temperatures above a threshold value of 5°C, expressed in terms of degree-days, allows the prediction of the effect of temperature on the timing of onset of cambial activity in *Pinus sylvestris* (Seo et al., 2008).

On the other hand, Rossi et al. (2007) observed that, in three conifers (*Larix decidua*, *Pinus cembra*, and *Picea abies*), cambial activity and xylem differentiation occurred above a certain threshold value of mean daily temperature from 5.6°C to 8.5°C. In addition, Rossi et al. (2008) observed that, in *Abies balsamea*, *L. decidua*, *Pinus cembra*, *P. sylvestris*, *Pinus leucodermis*, *Pinus uncinata*, and *P. abies*, the critical average temperature for initiation of active cambial cell division ranged between 8°C and 9°C. Deslauriers et al. (2008) reported that, in the conifer *P. leucodermis*, the calculated threshold minimum, mean, and maximum daily temperatures for wood formation were approximately 5.5, 8.2, and 11.5°C, respectively. Thus, threshold temperatures appear to differ among species.

It has been reported that bud burst and the development of new leaves are related to cambial reactivation and xylem differentiation (Aloni, 1991). However, cambial reactivation in localized heated portions of the stems of hybrid poplar was not associated with bud burst, indicating that a close relationship does not always exist between the timing of bud burst and cambial reactivation (Begum et al., 2007). In addition, earlier cambial reactivation and differentiation of first vessel elements are induced by localized heating in the absence of buds in *Q. serrata* (Kudo et al., 2014). Bud growth is not essential for cambial reactivation and the differentiation of first vessel elements. Aloni (1991) also observed that vessel elements were evident in stems of the deciduous ring-porous hardwood *Melia azedarach* that had been disbudded approximately 1 month prior to bud break. Thus, the growth of the current year's buds that is closely related to increases in total amounts of endogenous IAA in cambium does not act as the trigger for cambial reactivation and the differentiation of first vessel elements. The endogenous IAA in dormant quiescent cambium might be adequate levels for cambial reactivation and the first xylem differentiation (Sundberg et al., 1991; Funada et al., 2001a, 2002).

By contrast, xylem differentiation in heated portions of stems of hybrid poplar starts after bud flushing, suggesting that some factors from developing buds and expanding new leaves might be required for the xylem differentiation (Begum et al., 2007). In *M. azedarach*, disbudding approximately 1 month prior to bud break results in differentiation of a few isolated and very narrow vessels (Aloni, 1991). Only a few narrow vessel elements are observed in the heated plus disbudded seedlings of *Q. serrata*, while many large earlywood vessel elements are found in heated seedlings without disbudding (Fig. 2.2) (Kudo et al., 2014). Therefore, buds or bud growth, which might provide a continuous supply of IAA to cambium, is required for the continuous formation of large vessel elements.

FORMATION OF CELL WALL

As soon as cambial cells lose the ability to divide, they start to differentiate into secondary phloem or xylem cells. The stages in the development of secondary xylem cells can be categorized as follows: cambial cell division, cell expansion or elongation, cell wall thickening, cell wall sculpturing (formation of modified structure), lignification, and cell death (cell autolysis) (Funada, 2000, 2008). Fusiform cambial cells differentiate into longitudinal tracheids, vessel elements, wood fibers, and axial parenchyma cells, while ray cambial cells differentiate into ray parenchyma cells and, in some conifers, such as *Pinus* and *Larix*, ray tracheids. Cells derived from fusiform cambial cells increase in length and in diameter as they approach their final shape during differentiation (Kitin et al., 1999, 2001). For example, vessel elements and earlywood tracheids increase only slightly in length but they increase considerably in radial diameter.

Since tracheids or ray parenchyma cells derived from fusiform cambial cell or ray cambial cells are aligned in a radial direction, successive aspects of xylogenesis can be observed in a radial file within a single specimen. Thus, cambial

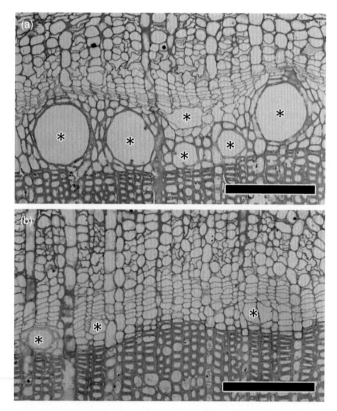

FIGURE 2.2 Light micrographs showing transverse views of the cambium and differentiating xylem of *Quercus serrata*. In the heated seedling (a), there were many wide vessel elements (asterisks). In the heated portion of heated and disbudded seedling (b), only a few narrow vessel elements were observed (asterisks). Scale bars = 200 μm.

derivatives are a suitable system to follow the process of differentiation of secondary xylem cells *in situ*.

The pressure of the protoplast against the cell wall (turgor pressure) within cells originates from the vacuole. It provides the driving force for the enlargement of cells in plants. The increase in the volume of the vacuole is derived from a gradient in the water potential between the cytoplasm and vacuole and the apoplast. When the turgor pressure in the cell exceeds the yield point of the cell wall, the cell can expand or elongate. As the cell expands or elongates, the cell wall becomes stiffer and, consequently, its yield point increases. Finally, cell expansion or elongation ceases.

The very thin and plastic cell wall that is characteristic for the stage of cell enlargement is called the primary wall. Cellulose that is highly crystalline and has very high tensile strength is the major component of the cell wall. Thus, cellulose microfibrils form a framework in the cell wall. The primary wall consists of loose aggregates of cellulose microfibrils (Abe and Funada, 2005). This structure allows expansion of the xylem cells derived from the cambium.

The orientation of cellulose microfibrils of the radial walls in differentiating tracheids changes during cell expansion (Abe et al., 1995b). The cellulose microfibrils on the innermost surface of the primary wall are not well-ordered. Most of cellulose microfibrils in the tracheids at the early stage of cell expansion are predominantly oriented longitudinally. Longitudinally oriented cellulose microfibrils might act to restrain the longitudinal elongation due to turgor pressure. Therefore, longitudinally oriented cellulose microfibrils in the primary wall of the fusiform cambial cells serve first to facilitate lateral expansion. As the cell expands, the predominant orientation of cellulose microfibrils on the innermost surface changes from longitudinal to transverse. At the final stage of cell expansion, cellulose microfibrils are oriented transversely to the cell axis. These observations suggest that it is not necessary to adopt the multinet growth hypothesis to explain the difference in orientation of cellulose microfibrils between the outer and inner parts of the primary wall in tracheids.

When cell expansion in differentiating tracheids is almost complete, well-ordered cellulose microfibrils are deposited on the inner surface of the primary wall, establishing the deposition of secondary wall (Abe et al., 1997; Abe and Funada, 2005). Once the formation of the secondary wall has begun, no further radial expansion of tracheids occurs. The secondary xylem cells of woody plants, such as tracheids, wood fibers, and vessel elements, have cell walls with a highly organized structure. Continuous deposition of the secondary wall increases the thickness of the cell wall. The thickness of the cell wall varies depending on cell function, cambial age, and the season at which the cell is formed, such as earlywood or latewood (Fig. 2.3a). In general, cells that function to support the tree, such as tracheids and wood fibers, form thick secondary walls. Thus, the ultrastructure of tracheids and wood fibers is of great importance to define the mechanical properties of wood. The cell wall supports

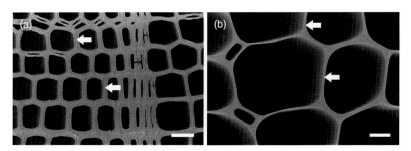

FIGURE 2.3 Scanning electron micrographs of transverse section showing earlywood–latewood tracheids of *Chamaecyparis obtusa* (a) and wood fibers of *Ochroma lagopus* (b). Arrows indicate cell walls. Scale bars = 20 μm (a) and 10 μm (b). *(Courtesy of Dr Y. Sano.)*

the heavy weight of the tree itself and functions in the transport of water from roots to leaves, which can sometimes reach more than 100 m in height. In addition, the cell wall prevents microbial and insect attack, thereby protecting the tree during its very long life that, in some cases, can exceed several thousand years. In addition, the thickness of the cell wall of wood fibers varies depending on species (Fig. 2.3b).

During formation of the secondary wall in tracheids or wood fibers, the cellulose microfibrils change their orientation progressively from a flat helix (S_1 layer) to a steep Z-helix (S_2 layer) in a clockwise rotation when viewed from the lumen side of cells. They are oriented at about 5–30° with respect to the cell axis. No cellulose microfibrils with an S-helix are observed during formation of the S_2 layer. This shift in the angles of cellulose microfibrils is considered to generate a semihelicoidal structure (Prodhan et al., 1995; Abe and Funada, 2005).

The cellulose microfibrils of the S_2 layer are closely aligned with a high degree of parallelism. When the rotational change in the orientation of cellulose microfibrils is arrested, a thick cell wall is formed as a result of the continuous deposition of cellulose microfibrils. The thickness of the secondary wall is important in terms of the properties of wood because it is closely related to the specific gravity of wood. The duration of the arrest in the orientation of cellulose microfibrils determines the thickness of the S_2 layer and, thus, the thickness of the secondary wall.

At the final stage of the formation of the secondary wall, the orientation of newly deposited cellulose microfibrils changes from a steep Z-helix to a flat helix with counterclockwise rotation when viewed from the lumen side of cells. This corresponds to a directional switch in the orientation of the cellulose microfibrils from clockwise to counterclockwise, when viewed from the lumen side, during formation of the secondary wall. The deposition of cellulose microfibrils in a flat helix results in the S_3 layer. The cellulose microfibrils in the S_3 layer are deposited in bundles. This texture differs from that of the S_2 layer where the cellulose microfibrils have a high degree of parallelism. The shift in angles of cellulose microfibrils is more abrupt during the transition from the S_3 to the S_3 layer than that from the S_1 to the S_2 layer (Abe and Funada, 2005). The rate of change in the orientation of cellulose microfibrils determines the structure of the cell wall layer.

A schematic model of the orientation of newly deposited cellulose microfibrils in a tracheid is shown in Fig. 2.4. The direction of orientation of cellulose microfibrils changes progressively with changing speed of rotation during the formation of the secondary wall (Funada, 2008).

Cellulose is synthesized by enzyme complexes (terminal complexes) in the plasma membrane (Kimura et al., 1999). Observations in a wide variety of plant cells have revealed that cortical microtubules, one of cytoskeletons, play an important role in the orientation of newly deposited cellulose microfibrils (Giddings and Staehelin, 1991; Nick, 2000; Baskin, 2001; Funada, 2000, 2002, 2008;

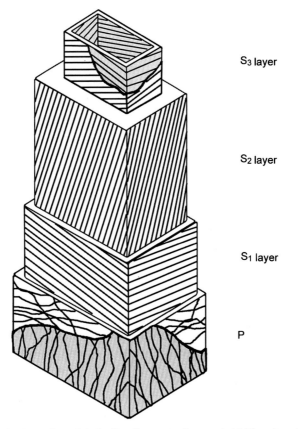

FIGURE 2.4   A schematic model of cell wall structure in a tracheid. The orientation of newly deposited cellulose microfibrils in primary wall (P) and secondary wall (S_1, S_2, and S_3 layers).

Funada et al., 2000). It has been postulated that cortical microtubules that are closely associated with the plasma membrane, guide the movement of terminal complexes because coalignment of cortical microtubules and newly deposited cellulose microfibrils has been often observed in the cells of lower and higher plants (Giddings and Staehelin, 1991). In addition, microtubule-depolymerizing agents, such as colchicine, disrupt the orientation of cellulose microfibrils. Moreover, direct visualization of cellulose synthase (CesA) in living cells of transgenic plants of *Arabidopsis thaliana* revealed that the CesA complexes moved in plasma membranes (Paredez et al., 2006). In addition, the movement of CesA complexes in linear tracks was coincident with cortical microtubules. These observations support strongly the hypothesis that cortical microtubules control the movement of cellulose synthase complexes in plasma membrane.

   During the formation of the secondary wall after the cessation of cell expansion, the abundant cortical microtubules are aligned in well-ordered arrays.

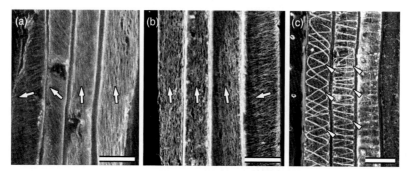

FIGURE 2.5 **Immunofluorescence images obtained by confocal laser scanning microscopy showing the orientation (a and b) and localization (c) of cortical microtubules, viewed from the lumen side of cells.** Successive changes in the orientation of cortical microtubules (arrows) from a flat S-helix to a steep Z-helix (a) and from a steep Z-helix to a flat S-helix (b) during formation of the secondary wall in differentiating tracheids of *A. sachalinensis*. Bands of helically oriented cortical microtubules (arrow heads) are visible at the final stage of formation of the secondary wall in differentiating tracheids of *T. cuspidata* (c). Scale bars = 50 μm.

Successive changes in the orientation of cortical microtubules are observed in differentiating tracheids or wood fibers during the formation of secondary walls (Fig. 2.5a and b) (Abe et al., 1994, 1995a; Prodhan et al., 1995; Furusawa et al., 1998; Chaffey et al., 1997a, 1999, 2002; Begum et al., 2012a). The orientation of cortical microtubules changes by clockwise rotation from a flat S-helix to a steep Z-helix when viewed from the lumen side. This shift in the direction of cortical microtubules is completed within three or four tracheids or wood fibers in a radial file. Then, the cortical microtubules are oriented in a steep Z-helix at almost the same angle over the next 10–15 tracheids or wood fibers of the radial file. After further differentiation, the orientation of cortical microtubules returns from the steep Z-helix to a flat S-helix in tracheids or wood fibers. This shift is completed within one or two tracheids or wood fibers in a radial file.

These observations provide strong evidence that the orientation of cortical microtubules changes progressively in a similar manner to the changes in the orientation of newly deposited cellulose microfibrils during the formation of the secondary wall. Thus, there is a very close relationship between cortical microtubules and newly deposited cellulose microfibrils. The cortical microtubules control the ordered orientation of cellulose microfibrils in the semihelicoidal cell walls of tracheids or wood fibers.

During the formation of the secondary wall in tracheids or wood fibers, the orientation of cortical microtubules changes abruptly from a steep Z-helix to a flat S-helix in contrast to the gradual change from a flat S-helix to a steep Z-helix. As shown in Fig. 2.4, the shift in angles of newly deposited cellulose microfibrils is more abrupt during the transition from a Z-helix to a flat S-helix (from the S_2 to the S_3 layer) than the transition from a flat S-helix to a steep

Z-helix (from the S_1 to the S_2 layer). The velocity of reorientation of microtubule might be closely related to the reorientation of newly deposited cellulose microfibrils. Therefore, the rotational motion of cortical microtubules reflects the thickness of intermediate layers and the structure of the secondary wall.

FORMATION OF MODIFIED STRUCTURE

Heterogeneous thickenings of the secondary wall are frequently observed in plant cells. Modifications in structure are normal features of the cell wall in the secondary xylem cells. These modifications, such as helical thickenings, pits, and perforations, are formed by localized deposition of components of the cell wall, in particular cellulose microfibrils. Their size, structure, and number are characteristic anatomical features for individual species and thus, they are frequently used for wood identification.

The secondary xylem cells develop localized ridges of parallel bundles of cellulose microfibrils on the innermost surface of the secondary wall. The cellulose microfibrils are oriented helically with respect to the cell axis. Such a thickening of the cell wall is known as helical thickening or spiral thickening. Helical thickenings are observed in tracheids or wood fibers of only a few species but they are relatively common in vessel elements.

At the final stage of formation of the secondary wall, bands of obliquely oriented cortical microtubules appear in tracheids of conifers such as *Taxus cuspidata*, in which helical thickenings are generally formed (Fig. 2.5c) (Uehara and Hogetsu, 1993; Furusawa et al., 1998). At first, these bands of cortical microtubules are approximately 3–4 μm wide. Then the bands become narrow and rope-like. Slightly disordered cortical microtubules are observed between bands. These rope-like bands of cortical microtubules are oriented helically beneath the cell wall around tracheids. As the tracheids differentiate, the cortical microtubules eventually disappear.

When rope-like bands of cortical microtubules are observed, localized ridges of cell wall materials, in particular cellulose microfibrils, are clearly visible under a transmitted light microscope (Uehara and Hogetsu, 1993; Furusawa et al., 1998). The rope-like bands are superimposed on the localized ridges of cell wall. This spatial congruence suggests that these bands of cortical microtubules might be involved in the formation of the helical thickenings. Bands of cortical microtubules are also observed in differentiating vessel elements of hardwoods such as *Aesculus hippocastanum*, in which helical thickenings are generally formed (Chaffey et al., 1999). Thus, cortical microtubules control the localized deposition of cellulose microfibrils during the formation of helical thickenings.

The application of a microtubule-depolymerizing agent, colchicine, to stems disrupts some cortical microtubules in differentiating tracheids of *T. cuspidata*. Certain tracheids lack cortical microtubules at the final stage of formation of the secondary wall (Funada et al., 2001b). Such tracheids have no helical thickenings.

These results confirm a role for localized cortical microtubules in the formation of the helical thickenings.

A similar localization of cortical microtubules was found during the formation of pits and perforation plates in differentiating secondary xylem cells (Abe et al., 1995b; Funada et al., 1997, 2001b; Chaffey et al., 1997b, 1999, 2002). Therefore, localized cortical microtubules control the localized deposition of cellulose microfibrils in the cell wall, thereby the formation of a modified structure.

As mentioned earlier, there are considerable evidences that the dynamics of cortical microtubules are closely related to the orientation and localization of newly deposited cellulose microfibrils in the differentiating secondary xylem cells. Thus, the manipulation of cortical microtubules would allow control of the structure of the cell wall, thereby improvement of wood quality.

CELL DEATH

With the onset of secondary wall deposition, the lignification begins at the intercellular layer, progressing to the primary wall and eventually to the secondary wall. When lignification has been completed, cell death occurs immediately in tracheary elements, such as tracheids and vessel elements (Funada, 2000, 2008). The loss of tonoplast might initiate collapse of vacuoles and cell death in vessel elements and wood fibers in *Populus trichocarpa* (Arend and Fromm, 2003). Secondary xylem cells that contribute to the mechanical support of the tree or to the conduction of water pass through the successive developmental stages. This process of cell death might be expected to resemble the programmed cell death that occurs in differentiating tracheary elements derived from single cells isolated from the mesophyll of *Zinnia elegans* (Fukuda, 1996, 2004).

By contrast, ray parenchyma cells derived from ray cambial cells remain alive for several years or more without immediate autolysis of cell organelles (Nakaba et al., 2006). Even after maturation, that is, secondary wall formation and lignification, they retain their organelles and remain viable. As a result, ray parenchyma cells play an important role in the storage and radial transport of materials, mediating radial continuity in the stem. In addition, these cells contribute to the formation of heartwood. The process of differentiation, and in particular, the death of long-lived ray parenchyma cells, differs from those of short-lived tracheids or vessel elements (Nakaba et al., 2006). In ray parenchyma cells, no successive death occurs even within the same radial cell lines of a ray in conifers, such as *A. sachalinensis* and *Pinus densiflora* (Fig. 2.6) (Nakaba et al., 2006, 2008, 2013). These observations suggest that the programmed death of ray parenchyma cells might not be controlled in a time-dependent manner. In addition, the timing of cell death is different among ray parenchyma cells. Cell death occurs earlier in ray parenchyma cells that are located in upper and lower cell lines of a ray in species that form no ray tracheids in the conifer *A. sachalinensis* (Fig. 2.6a). In species that form ray

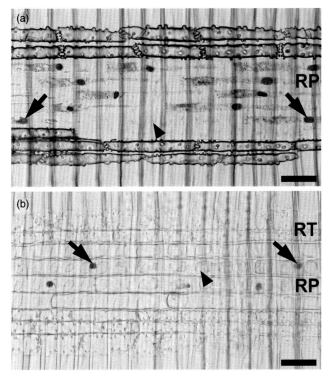

FIGURE 2.6 **Light micrographs of radial sections, stained with acetocarmine, showing nuclei in ray parenchyma cells in 10th annual ring from the cambium of** *A. sachalinensis* **(a) and 32nd annual ring from the cambium of** *P. densiflora* **(b).** Arrows indicate nuclei. Arrowheads indicate ray parenchyma cells that had lacked nuclei. The left side of each micrograph corresponds to the outer side of the tree. RP, ray parenchyma cell; RT, ray tracheid. Scale bars = 50 mm.

tracheids, such as *P. densiflora* and *Pinus rigida*, early death is observed in ray parenchyma cells that are in contact with ray tracheids (Fig. 2.6b). These observations indicate that the position within a ray and neighboring short-lived ray tracheids might affect the timing of cell death in long-lived ray parenchyma cells in conifers, therefore the function of ray parenchyma cells, such as the storage and transport of materials and the synthesis of heartwood substances.

In addition, the timing of cell death is different among types of ray parenchyma cells, such as contact cells and isolation cells (Murakami et al., 1999) in hybrid poplar (Nakaba et al., 2012). The disappearance of nuclei from ray parenchyma cells does not occur successively from the pith side. Cell death occurs earliest in contact cells, which are located within the upper or lower lines of a ray and connected to adjacent vessel elements through pits. Cell death occurs latest in isolation cells, which are located within the other cell lines of a ray and have no direct connection with vessel elements through pits. These observations indicate that the position within a ray and neighboring short-lived ray vessel element might affect the timing of cell death in long-lived ray parenchyma cells in hardwoods. Therefore, short-lived tracheary elements, such as vessel elements and ray tracheids, might be responsible for the early death of neighboring long-lived ray parenchyma cells that have direct connections with them through pits.

FUTURE PROSPECTIVES

For direct investigations of the mechanisms of differentiation of xylem cells, systems for induction of the differentiation of tracheary elements *in vitro* are very useful. The differentiation *in vitro* of isolated mesophyll cells of Z. *elegans* into tracheary elements (Fukuda and Komamine, 1980), which is an excellent model, has provided extensive information about xylem differentiation at the cellular and molecular levels (Fukuda, 1996, 2004; Mellerowicz and Sundberg, 2008; Novo-Uzal et al., 2013). Similarly, a system of differentiation of tracheary elements *in vitro*, using suspension-cultured cells of A. *thaliana*, has been exploited for analyses of xylem differentiation such as cytoskeletal dynamics and gene expression during the formation of cell walls (Oda et al., 2005; Pesquet et al., 2010; Oda and Fukuda, 2012). However, in such differentiation systems *in vitro*, the major products are tracheary elements that resemble simple types of primary xylem cells with annular, spiral, scalariform, and reticulate types of thickening of secondary walls. Therefore, we need to establish a new system to induce tracheary elements with highly organized cell wall structure that resemble the secondary xylem.

Cultured cells of a conifer, *Pinus radiata*, yielded highly developed types of tracheary elements with reticulate or pitted thickening of the secondary wall (Möller et al., 2003, 2006). In addition, Yamagishi et al. (2012, 2015) observed tracheary elements with different types of secondary wall in calli of two conifers, *Torreya nucifera* and *C. japonica* (Fig. 2.7a–c). Some differentiated cells resembled tracheary elements of primary xylem with spiral or reticulate thickening of cell walls (Fig. 2.7a and b). Other cells resembled tracheary elements of secondary xylem with thick cell walls and bordered pits. In calli of *T. nucifera*, some tracheary elements formed a highly developed structure with bordered pits and widespread secondary wall thickening, showing secondary xylem-like structure (Fig. 2.7c). Furthermore, in a differentiation system using calli of hybrid poplar (*P. sieboldii* × *P. grandidentata*), which has a model tree, Yamagishi et al. (2013) observed some tracheary elements with broad areas of cell walls and bordered pits (Fig. 2.7d). Therefore, it might be possible to induce the formation of well-developed secondary xylem-like tracheary elements from cultured cells of many woody plants. Such induction systems provide a new model for studies of the cellular and molecular mechanism of secondary xylem cells, as a consequence, of the detailed process of wood formation *in vitro*.

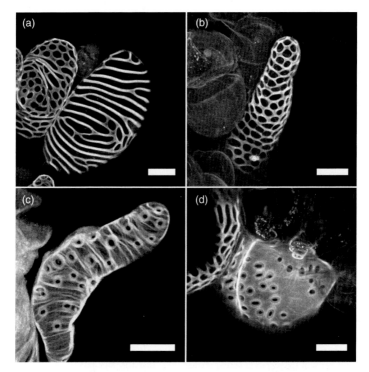

FIGURE 2.7   Confocal laser scanning micrographs showing tracheary elements in calli of *C. japonica* (a and b), *T. nucifera* (c), and *P. sieboldii* × *P. grandidentata* (d). Tracheary elements with a mixture of helical and reticulate thickening (a) and reticulate thickening (b) of secondary walls. A tracheary element with spiral thickening of inner surfaces of secondary wall and bordered pits (c). A tracheary element with broad regions of secondary wall thickening and bordered pits (d). Scale bars = 25 μm.

ACKNOWLEDGMENT

This work was supported, in part, by Grants-in-Aid for Scientific Research from the Ministry of Education, Science, Sports and Culture of Japan (nos. 19580183, 20120009, 21380107, 22.00104, 23380105, 24380090, 24.2976, 15K07508, and 15H04527).

REFERENCES

Abe, H., Funada, R., 2005. The orientation of cellulose microfibrils in the cell walls of tracheids in conifers: a model based on observations by field emission-scanning electron microscopy. IAWA J. 26, 161–174.

Abe, H., Ohtani, J., Fukazawa, K., 1994. A scanning electron microscopic study of changes in microtubule distributions during secondary wall formation in tracheids. IAWA J. 15, 185–189.

Abe, H., Funada, R., Imaizumi, H., Ohtani, J., Fukazawa, K., 1995a. Dynamic changes in the arrangement of cortical microtubules in conifer tracheids during differentiation. Planta 197, 418–421.

Abe, H., Funada, R., Ohtani, J., Fukazawa, K., 1995b. Changes in the arrangement of microtubules and microfibrils in differentiating conifer tracheids during the expansion of cells. Ann. Bot. 75, 305–310.

Abe, H., Funada, R., Ohtani, J., Fukazawa, K., 1997. Changes in the arrangement of cellulose microfibrils associated with the cessation of cell expansion in tracheids. Trees 11, 328–332.

Aloni, R., 1991. Wood formation in deciduous hardwood trees. In: Raghavendra, A.S. (Ed.), Physiology of Trees. John Wiley and Sons, New York, pp. 175–197.

Arend, M., Fromm, J., 2003. Ultrastructural changes in cambial cell derivatives during xylem differentiation in poplar. Plant Biol. 5, 255–264.

Barnett, J.R., Miller, H., 1994. The effect of applied heat on graft union formation in dormant *Picea sitchensis* (Bong.) Carr. J. Exp. Bot. 45, 135–143.

Baskin, T.I., 2001. On the alignment of cellulose microfibrils by cortical microtubules: a review and a model. Protoplasma 215, 150–171.

Begum, S., Nakaba, S., Oribe, Y., Kubo, T., Funada, R., 2007. Induction of cambial reactivation by localized heating in a deciduous hardwood hybrid poplar (*Populus sieboldii* × *P. grandidentata*). Ann. Bot. 100, 439–447.

Begum, S., Nakaba, S., Bayramzadeh, V., Oribe, Y., Kubo, T., Funada, R., 2008. Temperature responses of cambial reactivation and xylem differentiation in hybrid poplar (*Populus sieboldii* × *P. grandidentata*) under natural conditions. Tree Physiol. 28, 1813–1819.

Begum, S., Nakaba, S., Oribe, Y., Kubo, T., Funada, R., 2010a. Cambial sensitivity to rising temperatures by natural condition and artificial heating from late winter to early spring in the evergreen conifer *Cryptomeria japonica*. Trees 24, 43–52.

Begum, S., Nakaba, S., Oribe, Y., Kubo, T., Funada, R., 2010b. Changes in the localization and levels of starch and lipids in cambium and phloem during cambial reactivation by artificial heating of main stems of *Cryptomeria japonica* trees. Ann. Bot. 106, 885–895.

Begum, S., Shibagaki, M., Furusawa, O., Nakaba, S., Yamagishi, Y., Yoshimoto, J., Jin, H.O., Sano, Y., Funada, R., 2012a. Cold stability of microtubules in wood-forming tissues of conifers during seasons of active and dormant cambium. Planta 235, 165–179.

Begum, S., Nakaba, S., Yamagishi, Y., Yamane, K., Islam Md, A., Oribe, Y., Ko, J.H., Jin, H.O., Funada, R., 2012b. A rapid decrease in temperature induces latewood formation in artificially reactivated cambium of conifer stems. Ann. Bot. 110, 875–885.

Begum, S., Nakaba, S., Yamagishi, Y., Oribe, Y., Funada, R., 2013. Regulation of cambial activity in relation to environmental conditions: understanding the role of temperature in wood formation of trees. Physiol. Plant 147, 46–54.

Catesson, A.M., 1994. Cambial ultrastructure and biochemistry: changes in relation to vascular tissue differentiation and the seasonal cycle. Int. J. Plant Sci. 155, 251–261.

Chaffey, N., Barlow, P., Barnett, J., 1997a. Cortical microtubules rearrange during differentiation of vascular cambial derivatives, microfilaments do not. Trees 11, 333–341.

Chaffey, N.J., Barnett, J.R., Barlow, P.W., 1997b. Cortical microtubule involvement in bordered pit formation in secondary xylem vessel elements of *Aesculus hippocastanum* L. (Hippocastanaceae): a correlative study using electron microscopy and indirect immunofluorescence microscopy. Protoplasma 197, 64–75.

Chaffey, N., Barnett, J., Barlow, P., 1999. A cytoskeletal basis for wood formation in angiosperm trees: the involvement of cortical microtubules. Planta 208, 19–30.

Chaffey, N., Barlow, P., Sundberg, B., 2002. Understanding the role of the cytoskeleton in wood formation in angiosperm trees: hybrid aspen (*Populus tremula* × *P. tremuloides*) as the model species. Tree Physiol. 22, 239–249.

Deslauriers, A., Rossi, S., Anfodillo, T., Saracino, A., 2008. Cambial phenology, wood formation and temperature thresholds in two contrasting years at high altitude in southern Italy. Tree Physiol. 28, 863–871.

Fukuda, H., 1996. Xylogenesis: initiation, progression, and cell death. Annu. Rev. Plant Physiol. Plant Mol. Biol. 47, 299–325.

Fukuda, H., 2004. Signals that control plant vascular cell differentiation. Nat. Rev. Mol. Cell Biol. 5, 379–391.

Fukuda, H., Komamine, A., 1980. Establishment of an experimental system for the study of tracheary element differentiation from single cells isolated from the mesophyll of *Zinnia elegans*. Plant Physiol. 65, 57–60.

Funada, R., 2000. Control of wood structure. In: Nick, P. (Ed.), Plant Microtubules: Potential for Biotechnology. Springer, Berlin, pp. 51–81.

Funada, R., 2002. Immunolocalisation and visualisation of the cytoskeleton in gymnosperms using confocal laser scanning microscopy. In: Chaffey, N. (Ed.), Wood Formation in Trees: Cell and Molecular Biology Techniques. Taylor and Francis Pub, London, pp. 143–157.

Funada, R., 2008. Microtubules and the control of wood formation. In: Nick, P. (Ed.), Plant Microtubules: Development and Flexibility. Springer, Berlin, pp. 83–119.

Funada, R., Abe, H., Furusawa, O., Imaizumi, H., Fukazawa, K., Ohtani, J., 1997. The orientation and localization of cortical microtubules in differentiating conifer tracheids during cell expansion. Plant Cell Physiol. 38, 210–212.

Funada, R., Furusawa, O., Shibagaki, M., Miura, H., Miura, T., Abe, H., Ohtani, J., 2000. The role of cytoskeleton in secondary xylem differentiation in conifers. In: Savidge, R.A., Barnett, J.R., Napier, R. (Eds.), Cell and Molecular Biology of Wood Formation. BIOS Scientific Publishers, Oxford, pp. 255–264.

Funada, R., Kubo, T., Tabuchi, M., Sugiyama, T., Fushitani, M., 2001a. Seasonal variations in endogenous indole-3-acetic acid and abscisic acid in the cambial region of *Pinus densiflora Sieb. et Zucc.* stems in relation to earlywood-latewood transition and cessation of tracheid production. Holzforschung 55, 128–134.

Funada, R., Miura, H., Shibagaki, M., Furusawa, O., Miura, T., Fukatsu, E., Kitin, P., 2001b. Involvement of localized cortical microtubules in the formation of a modified structure of wood. J. Plant Res. 114, 491–497.

Funada, R., Kubo, T., Sugiyama, T., Fushitani, M., 2002. Changes in levels of endogenous plant hormones in cambial regions of stems of *Larix kaempferi* at the onset of cambial activity in springtime. J. Wood Sci. 48, 75–80.

Furusawa, O., Funada, R., Murakami, Y., Ohtani, J., 1998. Arrangement of cortical microtubules in compression wood tracheids of *Taxus cuspidata* visualized by confocal laser microscopy. J. Wood Sci. 44, 230–233.

Giddings, Jr., T.H., Staehelin, L.A., 1991. Microtubule-mediated control of microfibril deposition: a re-examination of the hypothesis. In: Lloyd, C.W. (Ed.), The Cytoskeletal Basis of Plant Growth and Form. Academic Press, London, pp. 85–99.

Gričar, J., Zupančič, M., Čufar, K., Koch, G., Schmitt, U., Oven, P., 2006. Effect of local heating and cooling on cambial activity and cell differentiation in the stem of Norway spruce (*Picea abies*). Ann. Bot. 97, 943–951.

Kimura, S., Laosinchai, W., Itoh, T., Cui, X., Linder, C.R., Brown, Jr., R.M., 1999. Immunogold labeling of rosette terminal cellulose-synthesizing complexes in the vascular plant *Vigna angularis*. Plant Cell 11, 2075–2085.

Kitin, P., Funada, R., Sano, Y., Beeckman, H., Ohtani, J., 1999. Variations in the lengths of fusiform cambial cells and vessel elements in *Kalopanax pictus*. Ann. Bot. 84, 621–632.

Kitin, P., Sano, Y., Funada, R., 2001. Analysis of cambium and differentiating vessel elements in *Kalopanax pictus* using resin cast replicas. IAWA J. 22, 15–28.

Kudo, K., Nabeshima, E., Begum, S., Yamagishi, Y., Nakaba, S., Oribe, Y., Yasue, K., Funada, R., 2014. The effects of localized heating and disbudding on cambial reactivation and formation of earlywood vessels in seedlings of the deciduous ring-porous hardwood *Quercus serrata*. Ann. Bot. 113, 1021–1027.

Lachaud, S., Catesson, A.M., Bonnemain, J.L., 1999. Structure and functions of the vascular cambium. C. R. Acad. Sci. 322, 633–724.

Larson, P.R., 1994. The Vascular Cambium: Development and Structure. Springer, Heidelberg, pp. 1–725.

Mellerowicz, E.J., Sundberg, B., 2008. Wood cell walls: biosynthesis, developmental dynamics and their implications for wood properties. Curr. Opin. Plant Biol. 11, 293–300.

Möller, R., McDonald, A.G., Walter, C., Harris, P.J., 2003. Cell differentiation, secondary cell-wall formation and transformation of callus tissue of *Pinus radiata* D. Don. Planta 217, 736–747.

Möller, R., Ball, R., Henderson, A., Modzel, G., Find, J., 2006. Effect of light and activated charcoal on tracheary element differentiation in callus cultures of *Pinus radiata* D. Don. Plant Cell Tissue Organ Cult. 85, 161–171.

Murakami, Y., Funada, R., Sano, Y., Ohtani, J., 1999. The differentiation of contact cells and isolation cells in the xylem ray parenchyma of *Populus maximowiczii*. Ann. Bot. 84, 429–435.

Nakaba, S., Sano, Y., Funada, R., 2006. The positional distribution of cell death of ray parenchyma in a conifer, *Abies sachalinensis*. Plant Cell Rep. 25, 1143–1148.

Nakaba, S., Kubo, T., Funada, R., 2008. Differences in patterns of cell death between ray parenchyma cells and ray tracheids in the conifers *Pinus densiflora and Pinus rigida*. Trees 22, 623–630.

Nakaba, S., Begum, S., Yamagishi, Y., Jin, H.O., Kubo, T., Funada, R., 2012. Differences in the timing of cell death, differentiation and function among three different types of ray parenchyma cells in the hardwood *Populus sieboldii* × *P. grandidentata*. Trees 26, 743–750.

Nakaba, S., Sano, Y., Kubo, T., Funada, R., 2013. Disappearance of microtubules, nuclei and starch during cell death of ray parenchyma in *Abies sachalinensis*. IAWA J. 34, 135–146.

Nick, P., 2000. Control of plant height. In: Nick, P. (Ed.), Plant Micro Tubules: Potential for Biotechnology. Springer, Berlin, pp. 1–23.

Novo-Uzal, E., Fernández-Pérez, F., Herrero, J., Gutiérrez, J., Gómez-Ros, L.V., Bernal, M.Á., Díaz, J., Cuello, J., Pomar, F., Pedreño, M.A., 2013. From *Zinnia* to *Arabidopsis*: approaching the involvement of peroxidases in lignification. J. Exp. Bot. 64, 3499–3518.

Oda, Y., Fukuda, H., 2012. Initiation of cell wall pattern by a Rho- and microtubule-driven symmetry breaking. Science 337, 1333–1336.

Oda, Y., Mimura, T., Hasezawa, S., 2005. Regulation of secondary cell wall development by cortical microtubules during tracheary element differentiation in *Arabidopsis* cell suspensions. Plant Physiol. 137, 1027–1036.

Oribe, Y., Kubo, T., 1997. Effect of heat on cambial reactivation during winter dormancy in evergreen and deciduous conifers. Tree Physiol. 17, 81–87.

Oribe, Y., Funada, R., Shibagaki, M., Kubo, T., 2001. Cambial reactivation in locally heated stems of the evergreen conifer *Abies sachalinensis* (Schmidt) Masters. Planta 212, 684–691.

Oribe, Y., Funada, R., Kubo, T., 2003. Relationships between cambial activity, cell differentiation and the localization of starch in storage tissues around the cambium in locally heated stems of *Abies sachalinensis* (Schmidt) Masters. Trees 17, 185–192.

Paredez, A.R., Somerville, C.R., Ehrhardt, D.W., 2006. Visualization of cellulose synthase demonstrates functional association with microtubules. Science 312, 1491–1495.

Pesquet, E., Korolev, A.V., Calder, G., Lloyd, C.W., 2010. The microtubule-associated protein AtMAP70-5 regulates secondary wall patterning in *Arabidopsis* wood cells. Curr. Biol. 20, 744–749.

Prodhan, A.K.M.A., Funada, R., Ohtani, J., Abe, H., Fukazawa, K., 1995. Orientation of microfibrils and microtubules in developing tension-wood fibres of Japanese ash (*Fraxinus mandshurica* var. *japonica*). Planta 196, 577–585.

Rensing, K.H., Samuels, A.L., 2004. Cellular changes associated with rest and quiescence in winter-dormant vascular cambium of *Pinus contorta*. Trees 18, 373–380.

Rossi, S., Deslauriers, A., Anfodillo, T., Carraro, V., 2007. Evidence of threshold temperatures for xylogenesis in conifers at high altitudes. Oecologia 152, 1–12.

Rossi, S., Deslauriers, A., Gričar, J., Seo, J.W., Rathgeber, C.B.K., Anfodillo, T., Morin, H., Levanic, T., Oven, P., Jalkanen, R., 2008. Critical temperatures for xylogenesis in conifers of cold climates. Global Ecol. Biogeogr. 17, 696–707.

Samuels, A.L., Kaneda, M., Rensing, K.H., 2006. The cell biology of wood formation: from cambial divisions to mature secondary xylem. Can. J. Bot. 84, 631–639.

Savidge, R.A., Wareing, P.F., 1981. Plant growth regulators and the differentiation of vascular elements. In: Barnett, J.R. (Ed.), Xylem Cell Development. Castle House, London, pp. 192–235.

Seo, J.W., Eckstein, D., Jalkanen, R., Rickebusch, S., Schmitt, U., 2008. Estimating the onset of cambial activity in Scots pine in northern Finland by means of the heat-sum approach. Tree Physiol. 28, 105–112.

Sundberg, B., Little, C.H.A., Riding, R.T., Sandberg, G., 1987. Levels of endogenous indole-3-acetic acid in the vascular cambium region of *Abies balsamea* trees during the activity–rest–quiescence transition. Physiol. Plant 71, 163–170.

Sundberg, B., Little, C.H.A., Cui, K., Sandberg, G., 1991. Level of endogenous indole-3-acetic acid in the stem of *Pinus sylvestris* in relation to the seasonal variation of cambial activity. Plant Cell Environ. 14, 241–246.

Sundberg, B., Uggla, C., Tuominen, H., 2000. Cambial growth and auxin gradients. In: Savidge, R.A., Barnett, J.R., Napier, R. (Eds.), Cell and Molecular Biology of Wood Formation. BIOS Scientific Publishers, Oxford, pp. 169–188.

Uehara, K., Hogetsu, T., 1993. Arrangement of cortical microtubules during formation of bordered pit in the tracheids of *Taxus*. Protoplasma 172, 145–153.

Yamagishi, Y., Sato, T., Uchiyama, H., Yoshimoto, J., Nakagawa, R., Nakaba, S., Kubo, T., Funada, R., 2012. Tracheary elements that resemble secondary xylem in calli derived from the conifers, *Torreya nucifera* and *Cryptomeria japonica*. J. Wood Sci. 58, 557–562.

Yamagishi, Y., Yoshimoto, J., Uchiyama, H., Nabeshima, E., Nakaba, S., Watanabe, U., Funada, R., 2013. In vitro induction of secondary xylem-like tracheary elements in calli of hybrid poplar (*Populus sieboldii* × *P. grandidentata*). Planta 237, 1179–1185.

Yamagishi, Y., Uchiyama, H., Sato, T., Kitamura, K., Yoshimoto, J., Watanabe, U., Nakaba, S., Funada, R., 2015. *In vitro* induction of the formation of tracheary elements from suspension-cultured cells of the conifer *Cryptomeria japonica*. Trees 29, 1283–1289.

Chapter 3

Xylogenesis and Moisture Stress

Eryuan Liang*,**, Lorena Balducci†, Ping Ren*,††, Sergio Rossi†

*Key Laboratory of Alpine Ecology and Biodiversity, Institute of Tibetan Plateau Research, Chinese Academy of Sciences, Beijing, China; **CAS Center for Excellence in Tibetan Plateau Earth Sciences, Beijing, China; †Département des Sciences Fondamentales, University of Quebec in Chicoutimi, 555, Boulevard de l'Université, Chicoutimi (QC), Canada; ††University of the Chinese Academy of Sciences, Beijing, China

Chapter Outline

INTRODUCTION (PLANTS AND WATER)

The development of plants can be summarized as a sum of the growth and differentiation of their cells, which also represents the cyclical process involving all changes during the whole lifespan: germination, growth, maturation, reproduction, and senescence (Raven et al., 2000). Long-term plant survival is assured by the constant availability of a number of growth factors – one of the most important is water.

Water plays several important roles in plants. In particular, at the cell level, water is a solvent for ions and organic molecules, a structuring agent in building proteins and nucleic acids, and a substrate for all enzymatic reactions. More importantly, cell growth is based on the turgor pressure, which, in turn, is related to the mechanical and physiological processes of water involving transport of solutes and thermal functions through evapotranspiration (Abe and Nakai, 1999; Taiz and Zeiger, 2006). Plants are constituted by a complex modular organization. The integration of exchanges between these modules is inevitably based on the transport of nutrients and metabolites dissolved in water through the interconnected vascular system. Thus, xylem anatomical features, such as lumen diameter, are the result of a trade-off between conductivity efficiency and safety (Tyree and Zimmermann, 2002; Sperry, 2003).

Secondary Xylem Biology. http://dx.doi.org/10.1016/B978-0-12-802185-9.00003-6

45

Water availability in the soil changes for local environmental variations and according to daily and seasonal cycles (Herzog et al., 1995). Plants consume a huge amount of water, which is absorbed from the soil and, in large part (90–98%), is lost with evapotranspiration (Taiz and Zeiger, 2006; Raven et al., 2000). Foliar water losses through stomata are an inevitable consequence of a key physiological process, the absorption of carbon dioxide.

In plants, water can be stored in different compartments: (1) in the phloem (2) in cell walls, and (3) in the living cells of cambium, bark, and parenchyma rays (Zimmermann, 1983). However, water may not be directly available for transpiration from all these compartments. Water stored in the living cells of bark and phloem can contribute more effectively to the daily flow of transpiration because of the elasticity of the tissues and the close connection with the xylem (Steppe et al., 2006). Thus, the water released from the bark can act as a buffer to prevent the xylem water potential from becoming very negative during periods with high transpiration rates. The internal reserve of water changes in time, decreasing during the day and being replenished during the night, when transpiration slows down, thereby forming a day–night cycle (Zweifel et al., 2006; Turcotte et al., 2009).

The current climate warming is expected to produce dramatic consequences on the frequency of summer drought, even in ecosystems where rain is abundant and well distributed throughout the growing season (Dai, 2013). Under warmer conditions, evapotranspiration of trees increases, producing greater tension forces in the water column within the xylem conduits and potential reductions of the cohesion between the water molecules. This can entail recurrent phenomena of embolism and cavitation and the loss of hydraulic conductivity, with the consequent reduction of the conducting area across the stem (McDowell, 2011; Ryan, 2011). Declines in water availability during the growing season could result in a decrease in water supply and an increase in the vulnerability of the water transport system, which is likely to substantially affect the growth in terms of xylem quality (xylem traits) and quantity (xylem amount), and, consequently, wood production (Balducci et al., 2013, 2014). Thus, it is a key issue to better identify the role of water in the process of xylem formation of trees.

In semiarid and arid regions, which occupy ~30% of the world's land surface area (Lal, 2004), plants have already experienced a progressive aridification over recent decades under global warming, thus moisture is beginning to be a critical driver of plant growth (Allen et al., 2010; Crimmins et al., 2011). Moisture stress has been shown to affect many phenological events, such as leafing and flowering in drought-sensitive areas (Peñuelas et al., 2002; Bernal et al., 2011). However, to date, few studies have linked the growth resumption, namely, the onset of xylogenesis, to moisture availability (Ren et al., 2015). In this chapter, we describe the xylogenetic process, and in particular that occurring in semiarid and arid areas, emphasizing the role of moisture in controlling the timings of xylogenesis. We also discuss future research directions for a robust prediction of xylogenesis in drought-prone areas.

THE XYLOGENETIC PROCESS

Xylogenesis is a complex process of division and differentiation of the cambium, represented by the phases of cell enlargement and cell wall thickening and lignification (Barnerr, 1981; Iqbal, 1990; Larson, 1994; Savidge, 1996). During development, the cambial derivatives alter morphologically and physiologically, finally differentiating into the specific elements of the woody tissues (Iqbal, 1990; Larson, 1994). This metabolic process is usually annual, mostly starting in spring and finishing in autumn, and plays a vital role in determination of the amount and quality of wood (Larson, 1994). The timing of xylogenesis is critical in defining the structure and functioning of forest ecosystems (Chaffey, 2002; Rossi et al., 2012). Therefore, all factors controlling or involving xylogenesis are of considerable interest (Savidge, 1996).

Apart from genetics, phylogenetics and phytohormone control of cambial activity (Savidge, 1996), abiotic factors, such as length of chilling period, photoperiod, and temperature, are critical drivers of xylogenesis (Körner and Basler, 2010). Climatic factors driving phenology depend on regional climate (Wolkovich et al., 2014). In cold-climate regions, where precipitation is abundant throughout the year and moisture is rarely a limiting factor, temperature is a major driver of xylogenesis, especially at the beginning of the growing season (Rossi et al., 2008; Li et al., 2013). An overwhelming number of studies have demonstrated that localized heating of stems during the quiescent stage induces reactivation of cambial cell division in evergreen conifers (Oribe et al., 2001; Gricar et al., 2006; Begum et al., 2013), showing additional evidence that temperature is a key factor for xylogenesis. At alpine timberline, Rossi et al. (2007) showed the existence of temperature thresholds for xylogenesis. Furthermore, the temperatures for the onset and ending of xylogenesis in conifers converged to a daily minimum threshold of 4–5°C (Rossi et al., 2008), and thus the onset of xylogenesis is expected to advance under global warming if and where water is not a limiting factor (Rossi et al., 2011). However, the close relationship between temperature and water availability requires deepening the effect of moisture on the temporal dynamics of xylogenesis, mainly in regions or ecosystems where drought events occur regularly.

TIMINGS OF XYLOGENESIS AND WATER DEFICIT

Water and its hydrostatic pressure are essential for xylem cell production and expansion (Savidge, 1996). During cell division, mitosis can occur only after a cambial cell has increased in diameter. Similarly, the enlargement phase is a turgor-driven process depending on cellular water uptake and solute accumulation into cellular vacuoles (Kozlowski and Pallardy, 2002; Fonti et al., 2010). When still enclosed by primary cell wall, the xylem cells are able to expand by absorbing water (Larson, 1994). Under water stress conditions, the pressure potential of the apoplastic water surrounding the expanded cells decrease abruptly, or even become lower than the osmotic potential of the expanded cells (Kozlowski and Pallardy, 2002). As a result, newly expanded cells lose turgor even though they have formed secondary walls (Abe et al., 2003). Therefore, the production of new cells is inhibited following the decline in cambial cell activity. To some extent, cell production can recover in case of improvement of the moisture conditions, as shown by the occurrence of intra-annual density fluctuations of false rings (Rigling et al., 2001; Cherubini et al., 2003). All of these processes can be documented by monitoring the timings of xylem formation.

Drought can have a great influence on the timing of xylogenesis, which is expected to be strongly species-specific due to the genetic and physiological traits of phenology (Davis et al., 2010). For example, two dominant Mediterranean shrub species, *Erica multiflora* and *Globularia alypum*, submitted to the same drought conditions showed contrasting spring phenologies (maybe also for xylem phenology) (Llorens et al., 2004; Bernal et al., 2011). Extreme drought events during the growing season; however, may produce intra-annual density fluctuations in tree rings (Cherubini et al., 2003; Campelo et al., 2007; De Luis et al., 2011a) or distorted and collapsed cells (Arend and Fromm 2007). In addition, water deficiency occurring in the second part of the growing season can cause an early cessation of cell division (Eilmann et al., 2011) or a light ring that is characterized by a narrow latewood band of thin-walled tracheids. (Liang and Eckstein, 2006) (Fig. 3.1). Although temperature remains the main factor for cambial reactivation even in Mediterranean climates, Vieira et al. (2014) found that water stress played its more important role in summer, by triggering an earlier conclusion of wood formation in maritime pine (*Pinus pinaster*). In an inner alpine valley, Oberhuber and Gruber (2010) also observed that cell enlargement stopped 2–3 weeks earlier in a xeric than in a mesic site. In particular, for the samplings, dry conditions during spring and summer strongly limit their cambial activity (De Luis et al., 2011b).

To date, only a few xylogenesis studies have been performed in dry sites in the inner alpine valley and Mediterranean areas, not showing influences of precipitation on the onset of xylogenesis (Camarero et al., 2010; Gruber et al., 2010; Swidrak et al., 2011). It is likely that the inner alpine valleys, with mean annual precipitation >500 mm, and the Mediterranean regions characterized by moist springs are not dry enough to affect the phenology during growth resumption (Ren et al., 2015). A monitoring of xylem formation realized in controlled environment showed that a heavy water deficit during June substantially reduced cambial activity, which needed 2–4 weeks to be restored after rewatering and resulted in the formation of a narrow tree-ring (Balducci et al., 2013). However, it is necessary to select the semiarid forests, with their regular exposure to water-limited growth conditions, to definitely test all possible links between timings of growth resumption and moisture availability.

A recent study showed that drought in spring can delay the onset of xylogenesis in a semiarid area of the northeastern Tibetan Plateau, where the

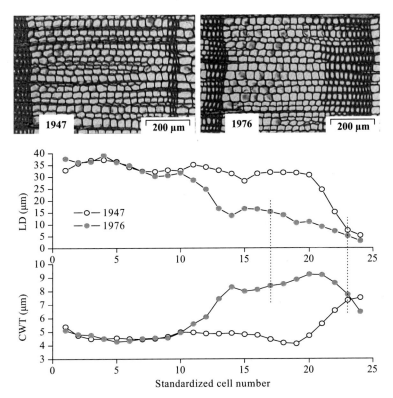

FIGURE 3.1 **Chinese pine (*Pinus tabulaeformis*) cross-sections through the light ring (LR) of 1947 and the normal ring of 1976 (top); tracheidograms of the cell-lumen diameter (LD) and the cell-wall thickness (CWT) of both tree rings (bottom).** The two dotted vertical lines show the earlywood–latewood boundary (Liang and Eckstein, 2006).

average annual precipitation of 200 mm is close to the survival limit of natural forest (Ren et al., 2015) (Fig. 3.2). By microcore collections from five mature Qilian juniper (*Juniperus przewalskii*) trees from 2009 to 2011 on the northeastern Tibetan Plateau, the authors found that drought can cause a delayed onset of xylogenesis in spring, and lead to an early cessation of xylem differentiation in summer (Ren et al., 2015) (Figs 3.3 and 3.4). In the central Himalayas, the occurrence of missing rings in timberline Himalayan birch (*Betula utilis*) indicates that drought stress controls the onset of xylogenesis. During the instrumental meteorological record, years with a high percentage of locally missing rings coincided with dry and warm premonsoon (March–May) seasons, inhibiting cambial resumption (Liang et al., 2014). A similar situation was found for alpine juniper shrub on the central Tibetan Plateaus (Liang et al., 2012). In these cases, warmer conditions during the early growing season may limit xylem growth by enhancing moisture stress. When the

FIGURE 3.2 **Landscape of nature Qilian juniper (*J. przewalskii*) forest ranging from 3800 m to 4150 m above sea level (a.s.l.) in the semiarid area of the northeastern Tibetan Plateau where Qilian juniper trees are on average approximately 500 years old, with the oldest individuals being around 1045 years old.** As shown by the meteorological station at Dulan (36°18′N, 98°06′E, 3190 m a.s.l.), 32 km from the study site, the average annual precipitation of 200 mm is close to the survival limit of natural forests. Based on weekly microcores from 2009 to 2011, Ren et al. (2015) found that precipitation in the early growing season can be a critical trigger of xylogenesis when the thermal conditions are favorable. *(Photo by Eryuan Liang.)*

drought stress is severe and lasts longer, xylogenesis cannot be initiated, and no tree-ring is produced (Liang et al., 2006, 2012, 2014). The occurrence of missing rings under extreme drought stresses provides compelling evidence of moisture-triggered xylogenesis.

XYLEM GROWTH AND MOISTURE STRESS

A line of dendrochronological studies have shown that precipitation in, or prior to, the early growing season plays an important role in determining tree-ring width in semiarid areas (Fritts, 1976; Schweingruber, 1997; Hughes et al., 2011). It has been reported that even relatively short-term maintenance of high water-deficit conditions would impact on the characteristics of tracheids by suppressing the capacity of the developing xylem to generate positive turgor, reducing wall extensibility and ultimately reducing cell expansion rates (Nonami and Boyer, 1990; Dünisch and Bauch, 1994; Abe et al., 2003; Bouriaud et al., 2005; Rossi et al., 2009). Several researches have investigated the link between cell structure and variations in precipitation or drought stress (Eckstein et al., 2004; Liang and Eckstein, 2006; Giovannelli et al., 2007; Fonti et al., 2010; Eilmann et al., 2011; Gea-Izquierdo et al., 2012; Balducci et al., 2013; Lautner, 2013). Under Mediterranean climate, xylem growth tends to show a typical bimodal pattern caused by subsequent cambial reactivations closely following the spring

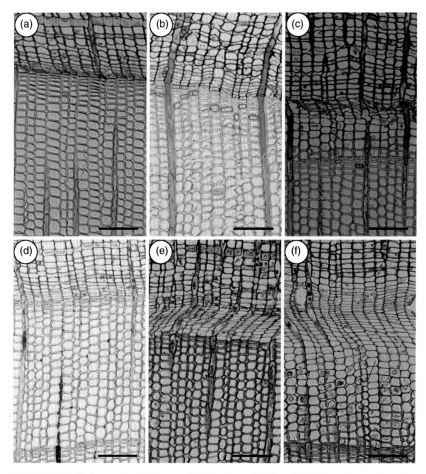

FIGURE 3.3   Xylogenesis of *J. przewalskii* between 2010 and 2011 on the northeastern Tibetan Plateau (Ren et al., 2015). Scale bar = 100 μm. (a) May 1, 2010; (b) June 10, 2010; (c) July 10, 2010; (d) April 29, 2011; (e) June 10, 2011; (f) July 13, 2011. There was no cell division in (a) and (d); Due to dry conditions in the first-half of May, no enlarging xylem cell were observed on June 10, 2010 (b), but there were four enlarging xylem cells on June 10, 2011 (e). See detailed climatic conditions in Fig. 3.4. On July 10, 2010 and July 13, 2011, the xylem cells in enlarging and cell-wall thickening phases were observed.

and autumn precipitation (Camarero et al., 2010). Water stress induces the production of tracheids with smaller lumen area in both Mediterranean and boreal climates (Rossi et al., 2009; Vieira et al., 2014). In the dry inner alpine valley, however, an irrigation experiment revealed that control, nonirrigated, trees presented tracheids with a wider lumen (a more effective water-conducting system) than control trees (Eilmann et al., 2011). In contrast to xylogenesis in cold climates (Rossi et al., 2008; Li et al., 2013), wood formation in water-limited

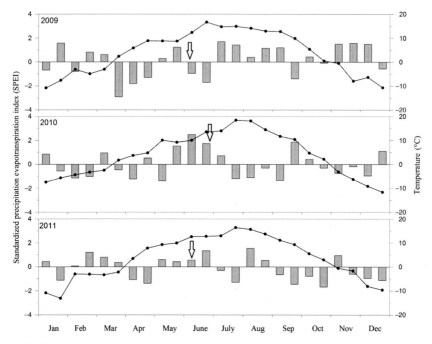

FIGURE 3.4 **The standardized precipitation evapotranspiration index (SPEI) (bars) and the mean temperature (lines) on a scale of half a month during 2009–2011 in Dulan areas of the northeastern Tibetan Plateau (Ren et al., 2015).** Arrows indicate the onset of xylogenesis of Qilian juniper.

environments is mainly determined by the rate of cell production rather than the duration of xylogenesis (Vieira et al., 2014; Ren et al., 2015). These xylogenesis studies set a physiological foundation to retrieve precipitation or moisture signals from variations in cell structure and cell size by wood anatomy and cell chronology (Baas and Schweingruber, 1987; Woodcock, 1989; Sass and Eckstein, 1995; Sass-Klaassen et al., 2007; Pumijumnong and Park, 1999; Wimmer, 2002; Garcia Gonzales and Eckstein, 2003; Vaganov et al., 2006; Fonti et al., 2010; Drew et al., 2013; Xu et al., 2013).

QUANTIFICATION OF MOISTURE AVAILABILITY

The quantification of water availability is a critical aspect to establish a reasonable connection between the timing of xylogenesis and moisture conditions. In the long-term field experiments or monitoring like irrigation and nonirrigation, or by controlling different timings of drought, an appropriate index to indicate water availability is of critical importance. Such indices include amount of precipitation, soil moisture, and water potential. However, plant water uptake depends not only on moisture, but also on evapotranspiration, which is controlled by temperature. Thus, a perspective should combine precipitation and temperature, such as the standardized precipitation evapotranspiration index (SPEI) (Vicente-Serrano et al., 2010) and air humidity (Laube et al., 2014). The SPEI, based on precipitation and temperature data, enables evaluation of water surplus and deficit on different time scales. It was successfully used to capture the effects of moisture on xylogenesis (Vieira et al., 2014; Ren et al., 2015) (Fig. 3.4). Laube et al. (2014) suggested that temperature-related air humidity might be the main trigger of the spring phenology of trees.

FURTHER RESEARCH

Adaptation of tree growth to drought stress may make the relationship between moisture stress and xylogenesis more complex. Trees subjected to deficits of water and nutrients allocate greater proportions of carbohydrate toward root growth, in order to increase the root/shoot ratio and the capacity to absorb water and minerals (Ibrahim et al., 1997; Leuschner et al., 2007). It has been suggested that gypsum crystallization water could constitute a significant water source for organisms growing on gypsum soils, particularly during dry periods (Palacio et al., 2014), and partly releasing the drought stress. Evidence has been obtained for enhanced synthesis under water stress of water-channel proteins and other proteins that may protect membranes and other important macromolecules from damage and denaturation during cell dehydration (Bewley et al., 1983; Qiu et al., 2013; Guerriero et al., 2014).

Unlike bud phenology, which can be realized directly in the field, xylogenesis requires the collection of woody samples that undergo a series of analyses in a laboratory performed by specialized or experienced people (Rossi et al., 2006). Because of the cost of such monitoring, investigations on xylogenesis generally last 1 or 2 years, and long-term chronologies of xylem phenology are basically lacking, which prevent the complete understanding of this growth process and the related environmental drivers. Moreover, climatic factors are generally intrinsically linked, making it difficult to completely separate their effects (Wolkovich et al., 2014). Thus, studies on xylem phenology of plants under natural and controlled growing conditions are necessary to separate dominant factors in driving xylogenesis from multiple climatic factors, and particularly to fill the gap between moisture availability and the onset of xylem formation. Given the complex responses of trees to drought (Zweifel et al., 2006; McDowell, 2011; Ryan, 2011), it is a major challenge to capture a moisture threshold for xylogenesis. A multidisciplinary approach including tree physiology, wood anatomy, xylogenesis, isotope, and sap flow could help to answer the key questions related to the effect of moisture on xylogenesis (Eckstein, 2004; Zweifel et al., 2006; Linares et al., 2009; Rossi et al., 2009; Battipaglia et al., 2010; Ryan, 2011; Begum et al., 2013; Cuny et al., 2013; Deslauriers et al., 2014; Ježík et al., 2015).

CONCLUSIONS

Water and soil moisture are essential factors for plants for realizing a functional and efficient system of transport – the xylem. There has been an ongoing effort to better understand how moisture stress limits the timings of xylogenesis and how moisture availability affects cell size and xylem growth. Recent works highlighted that drought in spring may delay the onset of xylogenesis and during the growing season trigger an earlier ending of wood formation. The influences of moisture stress on xylogenesis show a fingerprint in cell structure and cell size that in turn provides occasions for retrospective studies on the past climatic conditions based on developing cell chronologies (Eckstein, 2004; Liang and Eckstein, 2006; Vaganov et al., 2006) (see an example in Fig. 3.1). However, long-term monitoring of xylogenesis by a multidisciplinary approach is still necessary to capture a moisture threshold in limiting the timings of xylogenesis. Such a threshold is the key to predict the responses of xylogenesis to moisture availability.

ACKNOWLEDGMENTS

This work was supported by the National Natural Science Foundation of China (41171161). We appreciate the encouragement of Prof. Yoon Soo Kim to write this chapter.

REFERENCES

Abe, H., Nakai, T., 1999. Effect of the water status within a tree on tracheid morphogenesis in *Cryptomeria japonica* D. Don. Trees 14, 124–129.

Abe, H., Nakai, T., Utsumi, Y., Kagawa, A., 2003. Temporal water deficit and wood formation in *Cryptomeria japonica*. Tree Physiol. 23, 859–863.

Allen, C.D., Macalady, A.K., Chenchouni, H., Bachelet, D., McDowell, N., Vennetier, M., Kitzberger, T., Rigling, A., Breshears, D.D., Hogg, E.H., Gonzalez, P., Fensham, R., Zhang, Z., Castro, J., Demidova, N., Lim, J.-H., Allard, G., Running, S.W., Semerci, A., Cobb, N., 2010. A global overview of drought and heat-induced tree mortality reveals emerging climate change risks for forests. For. Ecol. Manag. 259, 660–684.

Arend, M., Fromm, J., 2007. Seasonal change in the drought response of wood cell development in poplar. Tree Physiol. 27, 985–992.

Baas, P., Schweingruber, F.H., 1987. Ecological trends in the wood anatomy of trees, shrubs and climbers from Europe. IAWA Bull. New Ser. 8, 245–274.

Balducci, L., Deslauriers, A., Giovannelli, A., Rossi, S., Rathgeber, C.B., 2013. Effects of temperature and water deficit on cambial activity and woody ring features in *Picea mariana* saplings. Tree Physiol. 33, 1006–1017.

Balducci, L., Deslauriers, A., Giovannelli, A., Beaulieu, M., Delzon, S., Rossi, S., Rathgeber, C.B.K., 2014. How do drought and warming influence survival and wood traits of *Picea mariana* saplings? J. Exp. Bot. 66, 377–389.

Barnerr, J.R., 1981. Xylem Cell Development. Castle House Publications, Tunbridge Wells.

Battipaglia, G., De Micco, V., Brand, W.A., Linke, P., Aronne, G., Saurer, M., Cherubini, P., 2010. Variations of vessel diameter and delta ^{13}C in false rings of *Arbutus unedo* L. reflect different environmental conditions. New Phytol. 188, 1099–1112.

Begum, S., Nakaba, S., Yamagishi, Y., Oribe, Y., Funada, R., 2013. Regulation of cambial activity in relation to environmental conditions: understanding the role of temperature in wood formation of trees. Physiol. Plant 147, 46–54.

Bernal, M., Estiarte, M., Peñuelas, J., 2011. Drought advances spring growth phenology of the Mediterranean shrub *Erica multiflora*. Plant Biol. 13, 252–257.

Bewley, J.D., Laesen, K.M., Papp, J.E., 1983. Water-stress-induced changes in the pattern of protein synthesis in maize seedling mesocotyls: a comparison with the effects of heat shock. J. Exp. Bot. 34, 1126–1133.

Bouriaud, O., Leban, J.-M., Bert, D., Deleuze, C., 2005. Intra-annual variations in climate influence growth and wood density of Norway spruce. Tree Physiol. 25, 651–660.

Camarero, J.J., Olano, J.M., Parras, A., 2010. Plastic bimodal xylogenesis in conifers from continental Mediterranean climates. New Phytol. 185, 471–480.

Campelo, F., Nabais, C., Freitas, H., Gutiérrez, E., 2007. Climatic significance of tree-ring width and intra-annual density fluctuations in *Pinus pinea* from a dry Mediterranean area in Portugal. Ann. For. Sci. 64, 229–238.

Chaffey, N.J., 2002. Wood Formation in Trees – Cell and Molecular Biology Techniques. Taylor and Francis, London.

Cherubini, P., Gartner, B.L., Tognetti, R., Bräker, O.U., Schoch, W., Innes, J.L., 2003. Identification, measurement and interpretation of tree rings in woody species from Mediterranean climates. Biol. Rev. Camb. Philos. Soc. 78, 119–148.

Crimmins, T.M., Crimmins, M.A., Bertelsen, C.D., 2011. Onset of summer flowering in a Sky Island is driven by monsoon moisture. New Phytol. 191, 468–479.

Cuny, H.E., Rathgeber, C.B.K., Senga Kiessé, T., Hartmann, F.P., Barbeito, I., Fournier, M., 2013. Generalized additive models reveal the intrinsic complexity of wood formation dynamics. J. Exp. Bot. 64, 1983–1994.

Dai, A., 2013. Increasing drought under global warming in observations and models. Nature Clim. Change 3, 52–58.

Davis, C.C., Willis, C.G., Primack, R.B., Miller-Rushing, A.J., 2010. The importance of phylogeny to the study of phenological response to global climate change. Philos. Trans. R. Soc. Lond. B. Biol. Sci. 365, 3201–3213.

De Luis, M., Novak, K., Raventós, J., Gričar, J., Prislan, P., Čufar, K., 2011a. Climate factors promoting intra-annual density fluctuations in Aleppo pine (*Pinus halepensis*) from semiarid sites. Dendrochronologia 29, 163–169.

De Luis, M., Novak, K., Raventós, J., Gričar, J., Prislan, P., Čufar, K., 2011b. Cambial activity, wood formation and sapling survival of *Pinus halepensis* exposed to different irrigation regimes. For. Ecol. Manag. 262, 1630–1638.

Deslauriers, A., Beaulieu, M., Balducci, L., Giovannelli, A., Gagnon, M., Rossi, S., 2014. Impact of warming and drought on carbon balance related to wood formation in black spruce. Ann. Bot. 114, 335–345.

Drew, D.M., Allen, K., Downes, G.M., Evans, R., Battaglia, M., Baker, P., 2013. Wood properties in a long-lived conifer reveal strong climate signals where ring-width series do not. Tree Physiol. 33, 37–47.

Dünisch, O., Bauch, J., 1994. Influence of soil substrate and drought on wood formation of spruce (*Picea abies* L. karst) under controlled conditions. Holzforschung 48, 447–457.

Eckstein, D., 2004. Change in past environments – secrets of the tree hydrosystem. New Phytol. 163, 1–4.

Eilmann, B., Zweifel, R., Buchmann, N., Graf Pannatier, E., Rigling, A., 2011. Drought alters timing, quantity, and quality of wood formation in Scots pine. J. Exp. Bot. 62, 2763–2771.

Fonti, P., von Arx, G., Garcia-Gonzalez, I., Eilmann, B., Sass-Klaassen, U., Gartner, H., Eckstein, D., 2010. Studying global change through investigation of the plastic responses of xylem anatomy in tree rings. New Phytol. 185, 42–53.

Fritts, H.C., 1976. Tree Rings and Climate. Academic Press, New York.

Garcia Gonzales, I., Eckstein, D., 2003. Climatic signal of earlywood vessels of oak on a maritime site. Tree Physiol. 23, 497–504.

Gea-Izquierdo, G., Fonti, P., Cherubini, P., Martin-Benito, D., Chaar, H., Canellas, I., 2012. Xylem hydraulic adjustment and growth response of *Quercus canariensis* willd. to climatic variability. Tree Physiol. 32, 401–413.

Giovannelli, A., Deslauriers, A., Fragnelli, G., Scaletti, L., Castro, G., Rossi, S., Crivellaro, A., 2007. Evaluation of drought response of two poplar clones (*Populus* × *canadensis* Mönch 'I-214' and *P. deltoides* Marsch. 'Dvina') through high resolution analysis of stem growth. J. Exp. Bot. 58, 2673–2683.

Gricar, J., Zupancic, M., Cufar, K., Koch, G., Schmitt, U., Oven, P., 2006. Effect of local heating and cooling on cambial activity and cell differentiation in the stem of Norway spruce (*Picea abies*). Ann. Bot. 97, 943–951.

Gruber, A., Strobl, S., Veit, B., Oberhuber, W., 2010. Impact of drought on the temporal dynamics of wood formation in *Pinus sylvestris*. Tree Physiol. 30, 490–501.

Guerriero, G., Sergeant, K., Hausman, J.-F., 2014. Wood biosynthesis and typologies: a molecular rhapsody. Tree Physiol. 34, 839–855.

Herzog, K.M., Häsler, R., Thum, R., 1995. Diurnal changes in the radius of a subalpine Norway spruce stem: their relation to the sap flow and their use to estimate transpiration. Trees-Struct Funct 10, 94–101.

Hughes, M.K., Swetnam, T.W., Diaz, H.F., 2011. Dendroclimatology: Progress and Prospects. Springer-Verlag, Dordrecht.

Ibrahim, L., Proe, M.F., Cameron, A.D., 1997. Main effects of nitrogen supply and drought stress upon whole-plant carbon allocation in poplar. Can. J. For. Res. 27, 1413–1419.

Iqbal, M., 1990. The Vascular Cambium. Research Studies Press, Taunton.

Ježík, M., Blaženec, M., Letts, M.G., Ditmarová, L., Sitková, Z., Střelcová, K., 2015. Assessing seasonal drought stress response in Norway spruce (*Picea abies* (L.) Karst.) by monitoring stem circumference and sap flow. Ecohydrology 8, 378–386.

Körner, C., Basler, D., 2010. Phenology under global warming. Science 327, 1461–1462.

Kozlowski, T.T., Pallardy, S.G., 2002. Acclimation and adaptive responses of woody plants to environmental stresses. Bot. Rev. 68, 270–334.

Lal, R., 2004. Carbon sequestration in dryland ecosystems. Envirom. Manag. 33, 528–544.

Larson, P.R., 1994. The Vascular Cambium: Development and Structure, Springer Series in Wood Science. Springer-Verlag, Berlin.

Laube, J., Sparks, T.H., Estrella, N., Menzel, A., 2014. Does humidity trigger tree phenology? Proposal for an air humidity based framework for bud development in spring. New Phytol. 202, 350–355.

Lautner, S., 2013. Wood formation under drought stress and salinity. In: Fromm, J. (Ed.), Cellular Aspects of Wood Formation. Springer-Verlag, Berlin, pp. 187–202.

Leuschner, C., Moser, G., Bertsch, C., Röderstein, M., Hertel, D., 2007. Large altitudinal increase in tree root/shoot ratio in tropical mountain forests of Ecuador. Basic Appl. Ecol. 8, 219–230.

Li, X., Liang, E., Gricar, J., Prislan, P., Rossi, S., Cufar, K., 2013. Age-dependence of xylogenesis and its climatic sensitivity in Smith fir on the south-eastern Tibetan Plateau. Tree Physiol. 33, 48–56.

Liang, E., Eckstein, D., 2006. Light rings in Chinese pine (*Pinus tabulaeformis*) in semiarid areas of north China and their palaeo-climatological potential. New Phytol. 171, 783–791.

Liang, E., Liu, X., Yuan, Y., Qin, N., Fang, X., Huang, L., Zhu, H., Wang, L., Shao, X., 2006. The 1920s drought recorded by tree rings and historical documents in the semi-arid and arid areas of northern China. Clim. Change 79, 403–432.

Liang, E., Lu, X., Ren, P., Li, X., Zhu, L., Eckstein, D., 2012. Annual increments of juniper dwarf shrubs above the tree line on the central Tibetan Plateau: a useful climatic proxy. Ann. Bot. 109, 721–728.

Liang, E., Dawadi, B., Pederson, N., Eckstein, D., 2014. Is the growth of birch at the upper timberline in the Himalayas limited by moisture or by temperature? Ecology 95, 2453–2465.

Linares, J.C., Camarero, J.J., Carreira, J.A., 2009. Plastic responses of *Abies pinsapo* xylogenesis to drought and competition. Tree Physiol. 29, 1525–1536.

Llorens, L., Peñuelas, J., Estiarte, M., Bruna, P., 2004. Contrasting growth changes in two dominant species of a Mediterranean shrubland submitted to experimental drought and warming. Ann. Bot. 94, 843–853.

McDowell, N.G., 2011. Mechanisms linking drought, hydraulics, carbon metabolism, and vegetation mortality. Plant Physiol. 155, 1051–1059.

Nonami, H., Boyer, J.S., 1990. Wall extensibility and cell hydraulic conductivity decrease in enlarging stem tissues at low water potentials. Plant Physiol. 93, 1610–1619.

Oberhuber, W, Gruber, A., 2010. Climatic influences on intra-annual stem radial increment of *Pinus sylvestris* (L.) exposed to drought. Trees 24, 887–898.

Oribe, Y., Funada, R., Shibagaki, M., Kubo, T., 2001. Cambial reactivation in locally heated stems of the evergreen conifer *Abies sachalinensis* (Schmidt) masters. Planta 212, 684–691.

Palacio, S., Azorín, J., Montserrat-Martí, G., Ferrio, J.P., 2014. The crystallization water of gypsum rocks is a relevant water source for plants. Nature Commun. 5:article number 4660, doi:10.1038/ncomms5660.

Peñuelas, J., Filella, I., Comas, P., 2002. Changed plant and animal life cycles from 1952 to 2000 in the Mediterranean region. Glob. Change Biol. 8, 531–544.

Pumijumnong, N., Park, W.K., 1999. Vessel chronologies from teak in northern Thailand and their climatic signal. IAWA J. 20, 285–294.

Qiu, Z., Wan, L., Chen, T., Wan, Y., He, X., Lu, S., Wang, Y., Lin, J., 2013. The regulation of cambial activity in Chinese fir (*Cunninghamia lanceolata*) involves extensive transcriptome remodeling. New Phytol. 199, 708–719.

Raven, P.H., Evert, R.F., Eichhorn, S.E., Bouharmont, J., 2000. Biologie Végétale. De Boeck Université, Paris.

Ren, P., Rossi, S., Gricar, J., Liang, E., Cufar, K., 2015. Is precipitation a trigger of the onset of xylogenesis in *Juniperus przewalskii* on the northeastern Tibetan Plateau? Ann. Bot. 115 (4), 629–639.

Rigling, A., Waldner, P.O., Forster, T., Brasker, O.U., Pouttu, A., 2001. Ecological interpretation of tree-ring width and intra-annual density fluctuations in *Pinus sylvestris* on dry sites in the central Alps and Siberia. Can. J. For. Res. 31, 18–31.

Rossi, S., Menardi, R., Anfodillo, T., 2006. Trephor: a new tool for sampling microcores from tree stems. IAWA J. 27, 89–97.

Rossi, S., Deslauriers, A., Anfodillo, T., Carraro, V., 2007. Evidence of threshold temperatures for xylogenesis in conifers at high altitudes. Oecologia 152, 1–12.

Rossi, S., Deslauriers, A., Griçar, J., Seo, J.-W., Rathgeber, C.B.K., Anfodillo, T., Morin, H., Levanic, T., Oven, P., Jalkanen, R., 2008. Critical temperatures for xylogenesis in conifers of cold climates. Glob. Ecol. Biogeog. 17, 696–707.

Rossi, S., Simard, S., Rathgeber, C., Deslauriers, A., De Zan, C., 2009. Effects of a 20-day-long dry period on cambial and apical meristem growth in *Abies balsamea* seedlings. Trees 23, 85–93.

Rossi, S., Morin, H., Deslauriers, A., Plourde, P.-Y., 2011. Predicting xylem phenology in black spruce under climate warming. Glob. Change Biol. 17, 614–625.

Rossi, S., Morin, H., Deslauriers, A., 2012. Causes and correlations in cambium phenology: towards an integrated framework of xylogenesis. J. Exp. Bot. 63, 2117–2126.

Ryan, M.G., 2011. Tree responses to drought. Tree Physiol. 31, 237–239.

Sass, U., Eckstein, D., 1995. The variability of vessel size of beech (*Fagus sylvatica* L.) and its ecophysiological interpretation. Trees 9, 247–252.

Sass-Klaassen, U., Chowdhury, Q., Sterck, F.J., Zweifel, R., 2007. Effects of water availability on the growth and tree morphology of *Quercus pubescens* Willd. and *Pinus sylvestris* L. in the Valais, Switzerland. TRACE – Tree Rings in Archaeology, Climatology and Ecology, vol. 5, Proceedings of Dendrosymposium, April, 20–22, 2006, Tervuren, Belgium.

Savidge, R.A., 1996. Xylogenesis, genetic and environmental regulation – a review. IAWA J. 17, 269–310.

Schweingruber, F.H., 1997. Tree Rings and Environment: Dendroecology. Paul Haupt Publishers, Berne.

Sperry, J.S., 2003. Evolution of water transport and xylem structure. Int. J. Plant Sci. 164, S115–S127.

Steppe, K., De Pauw, D.J.W., Lemeur, R., Vanrolleghem, P.A., 2006. A mathematical model linking tree sap flow dynamics to daily stem diameter fluctuations and radial stem growth. Tree Physiol. 26, 257–273.

Swidrak, I., Gruber, A., Kofler, W., Oberhuber, W., 2011. Effects of environmental conditions on onset of xylem growth in *Pinus sylvestris* under drought. Tree Physiol. 31, 483–493.

Taiz, L., Zeiger, E., 2006. Plant Physiology. Sinauer Associates, Sunderland, Massachusetts.

Turcotte, A., Morin, H., Krause, C., Deslauriers, A., Thibeault-Martel, M., 2009. The timing of spring rehydration and its relation with the onset of wood formation in black spruce. Agr. For. Meteorol. 149, 1403–1409.

Tyree, M.T., Zimmermann, M.H., 2002. Xylem Structure and the Ascent of Sap. Springer-Verlag, Heidelberg.

Vaganov, E.A., Hughes, M.K., Shashkin, A.V., 2006. Growth Dynamics of Conifer Tree Rings: An Image of Past and Future Environments. Springer-Verlag, New York.

Vicente-Serrano, S.M., Beguería, S., López-Moreno, J.I., 2010. A multiscalar drought index sensitive to global warming: the standardized precipitation evapotranspiration index. J. Clim. 23, 1696–1718.

Vieira, J., Rossi, S., Campelo, F., Freitas, H., Nabais, C., 2014. Xylogenesis of *Pinus pinaster* under a Mediterranean climate. Ann. For. Sci. 71, 71–80.

Wimmer, R., 2002. Wood anatomical features in tree rings as indicators of environmental change. Dendrochronologia 20, 21–36.

Wolkovich, E.M., Cook, B.I., Davies, T.J., 2014. Progress towards an interdisciplinary science of plant phenology: building predictions across space, time and species diversity. New Phytol. 201, 1156–1162.

Woodcock, D.W., 1989. Climate sensitivity of wood-anatomical features in a ring-porous oak (*Quercus macrocarpa*). Can. J. For. Res. 19, 639–644.

Xu, J., Lu, J., Bao, F., Evans, R., Downes, G.M., 2013. Climate response of cell characteristics in tree rings of *Picea crassifolia*. Holzforschung 67, 217–225.

Zimmermann, M.H., 1983. Xylem Structure and the Ascent of Sap. Springer-Verlag, Berlin.

Zweifel, R., Zeugin, F., Zimmermann, L., Newbery, D.M., 2006. Intraannual radial growth and water relations of trees – implications towards a growth mechanism. J. Exp. Bot. 57, 1445–1459.

Chapter 4

Abiotic Stresses on Secondary Xylem Formation

Jörg Fromm*, Silke Lautner**

*Institute for Wood Biology, University of Hamburg, Hamburg, Germany; **Faculty of
Wood Science and Technology, Eberswalde University for Sustainable Development,
Eberswalde, Germany

Chapter Outline

INTRODUCTION

The evolution of the secondary xylem is mainly driven by the adaptation of trees to alterations of environmental conditions. In particular, climate changes affect growth ring width and structure, which play an important role in dendrochronology. Climate change includes the effects of rising temperature and atmospheric CO_2 levels as well as frequently occurring drought stress in summer. With the ongoing global warming plants also shift their latitudinal and altitudinal ranges, for example, an upward altitudinal shift of 29 m per decade was detected in vascular plants over the last century in western Europe (Lenoir et al., 2008). Experimental studies of the impact of global warming on tree growth have shown that the length of the growing season and onset and termination dates of cambial activity are significantly affected (Gricar, 2007). Increasing temperatures correlate with changing transpiration rates and higher demands of conductive efficiency of xylem cells, therefore changes in vessel diameter and vessel frequency are often observed under abiotic stress conditions. For instance, during drought stress trees respond generally with a reduction in the size of earlywood vessels and an increase in vessel frequency (Lautner, 2013). Furthermore, the tropospheric

Secondary Xylem Biology. http://dx.doi.org/10.1016/B978-0-12-802185-9.00004-8

59

ozone concentration has been increasing for several decades (Ashmore, 2005) and is potentially one of the most harmful air pollutant for trees. Ozone affects crop yield and has been shown to cause reductions in growth and biomass of forest trees.

Apart from climate change parameters, the impact of salinity and nutrient deficiency are further important abiotic stresses that affect wood structure and physiology. Salinity globally appears to be a major environmental problem. Particularly cations like sodium affect nutrient uptake and the internal nutrient balance of the plant. Regarding trees, the question on how salt exposure interferes with nutrient uptake, growth, and wood production was studied on a molecular level in poplar, indicating that xylem differentiation was curtailed and the development of full-size vessels was impaired under salt stress (Escalante-Perez et al., 2009). Nutrient deficiency is another major abiotic stress that also affects wood formation. Mineral nutrients, such as nitrogen, phosphorus, and potassium, are primarily acquired by trees in the form of inorganic ions from the soil. After being taken up by the roots they are transported to various parts of the tree where they have a lot of biological functions. Moreover, characteristic deficiency symptoms occur when an essential element is supplied insufficiently. Regarding wood production, various changes such as reduction of cambial width, wood increment, and vessel size can result from inadequate nutrient supply.

EFFECTS OF NUTRIENT DEFICIENCY ON WOOD FORMATION

Macro- and micronutrients play significant roles in plant development and physiology. In many forests of the Northern Hemisphere N is the most limiting nutrient for tree growth (Rennenberg et al., 2009); however, in highly weathered tropical soils growth of trees is more limited by P (Vitousek et al., 2010) and K (Wright et al., 2011; Santiago et al., 2012) than by N. Remobilizations of N, P, and K within stem wood throughout wood ageing has been documented for various tree species such as chestnut (Colin-Belgrand et al., 1996) or Scots pine (Helmisaari and Siltala, 1989). In Eucalyptus, Laclau et al. (2001) studied the dynamics of nutrient translocation in stem wood across an age series.

Regarding nitrogen, it occurs primarily in amino acids, proteins, and nucleic acids and makes up around 4% of the plant's dry weight. In poplar, nitrogen fertilization increases photosynthate allocation in the shoots to the disadvantage of the roots (Pregitzer et al., 1990), causing an increase in crown growth and leaf number (Günthardt-Goerg et al., 1996; Ibrahim et al., 1997). Furthermore, a reduction in xylem width and an increase of the phloem width was observed at reduced N-supply in hybrid aspen (Puech et al., 2000).

Phosphorus is one of the key nutrients for various ecosystems, including forests, and its limitation has considerable impact on their primary productivity (Elser et al., 2007). Forest ecosystems often overcome soil P depletion by having developed forceful strategies in uptake, usage, storage, and remobilization of this element. There is evidence that trees draw phosphorus prior to shedding in autumn from their site of utilization, which is for the main part leaves, in order to deposit it in adjacent storage tissue and, hence, to have it short-dated allocable in spring. In previous investigations on 20-year-old beech trees, we were able to trace such autumnal translocation of P from leaves via the phloem tissue of the petiole into the stem tissue, where it was detected primarily in living cortex cells, in sieve elements of the phloem, as well as in the wood ray parenchyma (Eschrich et al., 1988). There is evidence that this translocation is mainly achieved by phloem transport. These results are in good agreement with assumptions on plant's recycling efficiency, retrieving essential components off the leaves, and storing them in twig tissue prior to remobilization in spring. In buds of deciduous trees Pi concentration in general decreased during the course of reactivation in spring, while the concentration of adenine nucleotides increased rapidly during swelling of buds in *Fagus*, *Quercus*, and *Fraxinus*, or during bud-break in *Acer* and *Alnus* (Fromm and Eschrich, 1986). Since changes of adenine nucleotide and Pi concentrations are correlated with energy-requiring processes, which are expected in buds during the course of reactivation, results indicate that P plays an important role in reactions during bud-break. Likewise, taking the below-ground P availability in the root tissue and the symbiotic activity of associated microorganisms into consideration would than allow to hypothesize on tree strategies of P cycling. During the process of wood formation phosphorus also is of importance. For example, an increase in P content was measured in the cambial zone of poplar during cambial reactivation (Arend and Fromm, 2000). On the one hand P is involved in energy-requiring processes within the cambium, for example, for ATP formation required for various processes. On the other hand, P is an important component of DNA. Also in the xylem developing zone P is essential for keeping up the characteristic cellular functions in wood forming processes and in the ray parenchyma cells. For example, P is required for ATP generation, which is consumed, for example, by the PM-H^+ATPase, as demonstrated by immunolocalization in poplar twigs (Arend et al., 2002).

In investigations on the effect of P on wood formation, we grew young poplars (*Populus tremula* × *Populus tremuloides*, clone T89) for 6 weeks in hydroponics supplied with Hoagland's nutrient solution modified in its phosphorus contents. Comparing the variations P deficiency, optimal P supply, and excess P supply, we found strong differences between the deficiency variation and the optimum and overnutrition variation concerning shoot-biomass production (Fig. 4.1a). In stem cross-sections, the wood increment and the vessel size showed a reduction under phosphorus starvation (0 mM P) in comparison to optimum P supply (1 mM P, Fig. 4.1b). Energy dispersive X-ray microanalysis of the root tissue revealed no detectable phosphorous content in roots grown in hydroponics under P starvation, neither in the central cylinder, nor in the cortex tissue, whereas rel. P content was detectable at similar range in the central cylinder of poplars grown under optimum and excess P supply. In the root cortex

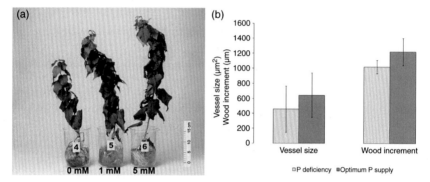

FIGURE 4.1   Poplar clones grown in hydroponics under different P supply (0, 1, and 5 mM). (a) After 6 weeks P deficiency variation showed clear biomass reduction. (b) Vessel size and wood increment decreased under P starvation (0 mM) compared to optimum P supply (1 mM).

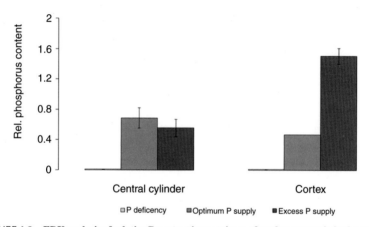

FIGURE 4.2   EDX analysis of relative P content in root tissue of poplars grown in hydroponics under different P supply (0 mM = deficiency, 1 mM = optimum, 5 mM = excess). No P was detectable under P deficiency, but balanced P content occurred in the central cylinder under optimum and excess P nutrition. P content increased in the cortex under excess P supply via the nutrient solution

P content increased under excess P supply in comparison to optimum P supply (Fig. 4.2). These results indicate a controlled P loading into the xylem transpiration stream, independently of excess P supply in the cortex. In addition, the observation that poplars grown under P starvation for several weeks still featured vital leaves, even though exhibiting reduced growth (Fig. 4.1a) along with no detectable P content in the roots, points to an intraplant relocation mechanism of already existing P in trees. However, grown in hydroponics, those poplars were not able to make use of mycorrhizal symbiosis. Mycorrhizal infections enhance root nutrient uptake and thereby tree growth by an increase of absorbing surface area, supporting mobilization of sparsely present nutrients and also greatly enhance uptake system efficiency, which might prove to be a crucial factor for trees growing in P-limited environments.

Potassium is important for osmoregulation, cell expansion, stomatal movements, enzyme activation, and floral induction (Fromm, 2010; Potchanasin et al., 2009). Moreover, numerous fertilization experiments have shown that potassium has an important effect on tree growth. For instance, after K^+ fertilization of eucalyptus trees the above-ground net primary production increased almost up to 100% over the first 36 months after planting (Laclau et al., 2009). In addition, a doubling of stem growth response to K^+ fertilizer was observed in *Pinus radiata* (Smethurst et al., 2007), while in *Picea abies* fertilization with K^+, Ca^{2+}, and Mg^{2+} causes 30% more biomass and increasing periclinal cell divisions in the cambium (Dünisch and Bauch, 1994). Poplar trees grown under potassium deficiency show leaf chlorosis after 5–7 weeks, which then develops into necrosis (Arend et al., 2004; Wind et al., 2004). Regarding wood formation, a high K^+ content could be measured in the poplar cambium in spring and summer, followed by a strong reduction in autumn and winter (Wind et al., 2004). Since these K^+ variations correlate with the radial width as well as with the osmotic potential of the cambial zone, results indicate that K^+ plays a key role in cell expansion. In line with these results is a high K^+ level in developing vessels in comparison to a low level in young fibers of poplar wood (Langer et al., 2002). This difference between vessels and fibers is most pronounced in trees grown under nonlimiting K^+ fertilization, while trees grown under K^+ deficiency do not show significant differences in K^+ level between various cell types (Langer et al., 2002). Moreover, with increasing K^+ supply, the vessel size significantly increased in poplar wood. However, the size of developing fibers was not affected by K^+ supply; therefore, the osmotic function for K^+ is mainly confined to vessel expansion (Langer et al., 2002). Furthermore, molecular and electrophysiological investigations indicate a strong involvement of specific K^+ channels in the regulation of wood formation. In particular the activity of two K^+ channels, PTORK and PTK2, is restricted to the period of wood formation in spring and summer in poplar, indicating essential roles in xylem development (Langer et al., 2002; Arend et al., 2005).

Calcium is generally required during cell division and as a second messenger for various responses to environmental signals (White and Broadley, 2003). In addition, calcium functions as a membrane stabilizer and plays a significant role in cell wall synthesis (Eklund and Eliasson, 1990), predominantly by linking pectin chains together. Within the process of wood formation calcium is also essential. In correlation to decreasing calcium levels in the phloem, the cambium, and the developing xylem of poplars grown under calcium deficiency, wood increment, vessel size, and fiber length are reduced (Lautner et al., 2007). Furthermore, the width of the cambial zone decreases under calcium deficiency and numerous small vacuoles occur within the cambial cells (Lautner et al., 2007). Regarding phloem physiology, poplars grown under calcium deficiency show a lower phloem loading rate in the leaves and a reduced phloem unloading within the stem (Schulte-Baukloh and Fromm, 1993).

In early spring a strong increase in the calcium level of cambial cells was detected in beech (Follet-Gueye et al., 1998) as well as poplar (Arend and Fromm, 2000). Similarly, calcium was also shown to increase in the apical meristem during bud break in spring, indicating a possible role in the induction of cell divisions (Lautner and Fromm, 2010). In addition, since calcium is known to activate enzymes, such as ATPases, amylases, and lipases (Bangerth, 1979), the transient meristematic increase in calcium might also be involved in the hydrolysis of starch and proteins. Another possible function of calcium within the process of wood formation might be a role in lignin polymerization within the cell wall (Westermark, 1982). Moreover, low calcium content led to lower lignin proportion in spruce wood, causing changes in wood hardness and elasticity (Wimmer and Lucas, 1997). Therefore, calcium and lignin content as well as the mechanical properties of wood seem to be closely related to each other. By using FTIR spectroscopy, direct evidence for a key role of calcium in lignification was given in poplar wood. Trees grown under calcium deficiency had decreased carbonyl as well as methoxyl groups from S-lignin (Lautner et al., 2007), causing a reduction of lignin concentration. To get a more profound understanding of the role of calcium in wood formation, future studies have to focus on calcium-mediated enzymes as well as molecular analyses of calcium channels in the cambial zone and developing xylem.

INFLUENCE OF DROUGHT STRESS

Xylem cell expansion is a turgor-driven process depending on cellular water uptake as well as on solute accumulation (Langer et al., 2002). Thus, drought stress can affect wood growth directly through its implications for the cambial meristem and the developing wood cells. In several investigations on trees exposed to drought stress it has been confirmed that turgor pressure is reduced in expanding cambial cell derivatives (e.g., Dünisch and Bauch, 1994; Abe et al., 2003) and that concentration of osmotically active solutes decreases in the cambial zone (Arend and Fromm, 2007). However, effects of drought stress on wood formation have shown a seasonal codependency as well as an inter- and intraspecific component. For example, in poplar it has been observed that under drought stress only one or two cambial cell derivatives occurred in early summer, reduced to none under drought stress in late summer (Arend and Fromm, 2007). In *Pinus halepensis*, closely dependent on cambial dimension, the number of tracheids decreased under drought conditions and remained at a very low level throughout the season when water stress was kept up, too (de Luis et al., 2011). Also, xylem element anatomy reveals distinct alterations when grown under drought stress conditions, especially in early-season. Fiber length as well as cross-sectional area was found to be reduced in poplar (Arend and Fromm, 2007). Likewise, vessel lumen area has also been found to be significantly reduced in early summer under drought stress conditions. This reduction in lumen area was found to be compensated for by an increased number of vessels, hence, keeping the overall vessel-area:fiber-area ratio similar to values found in control trees. Not only fiber length, but also vessel element length was found to decrease slightly under water deficiency, which has been reported, for example, for different oak species and olive (Garcia-Gonzalez and Eckstein, 2003; Eilmann et al., 2006; Gea-Izquierdo et al., 2012). In gymnosperm trees alterations in xylem element formations can also be detected under drought stress, as investigations on *Pinus* species have shown (Eilmann et al., 2011; Esteban et al., 2012). Here, an increase of tracheid lumen diameter reveals a tendency toward the formation of an optimized water conducting system along with restricted year ring formation. In contrast to these findings, other investigations show an increase of cell lumen area along with an increase of water supply (Sheriff and Whitehead, 1984; Sterck et al., 2008). Taking both sides into consideration, it seems that the physiological process of tracheid lumen formation under drought stress helps the tree balancing mechanical support requirements and water stress resistance while at the same time being also influenced by other environmental growth conditions as well as by species' provenances (Eilmann et al., 2011).

Apart from the cell shape of xylem elements, the cell wall composition is also altered under water deficiency. While the S_1 layer remains largely unaffected, in the S_2 variable lignification intensities can result in either increased or strongly decreased lignifications in a confined space (Donaldson, 2002). Cell wall lignification in tracheids' S_2 layer can even be reduced down to levels that cause their collapse, whereas the terminal S_3 layer does not seem to be greatly affected by drought-induced alterations in lignin deposition (Donaldson, 2002).

SALINITY-INDUCED CHANGES OF WOOD FORMATION

Salinity *a priori* reduces plant's ability to take up water. Hence, tree physiological responses to drought stress and salinity have much in common. Via osmotic adjustment and compartmentalization of minerals in the vacuole and apoplast region, plants promote dehydration tolerance, but with salinity, also secondary events can be observed, such as disruption of cell membrane integrity and altered cell metabolism. These effects also occur in the cambium of trees, which in turn produces xylem elements modified in anatomical and chemical aspects.

Under saline conditions, trees form a reduced cambial zone. Cambial cells appear rather disorganized, and cytoplasm of the cambial cells changes from being highly vacuolated in control trees toward exhibiting multiple smaller vacuoles, indicating a shift in the osmotic balance within the wood-forming region (Escalante-Perez et al., 2009). Lower concentrations of osmotically active solutes in the xylem-forming zone have an immanent effect on turgor of the developing xylem cells and, hence, also influences xylem element anatomy. Angiosperm trees therefore show decreased vessel diameter when exposed to salt stress (Baas et al., 1983; Escalante-Perez et al., 2009; Junghans et al., 2006), while at the same time vessel number per area increases. Thereby, the overall water conductivity remains more or less unaffected (Janz et al., 2012). To prevent against pending vessel collapse under osmotic stress, vessel cell walls are reinforced, as shown in poplar trees exposed to salinity (Junghans et al., 2006). Again, these anatomical features are subject to intraspecific variations, since within a tree genus, for example, *Populus*, species sensitive to salinity react significantly more pronounced than salt-tolerant species like *Populus euphratica*, which shows only minor changes in wood anatomy even under severe salt stress (Chen and Polle, 2010). In angiosperm trees, different to the vessel system, fiber and ray cell anatomy do not seem to be affected under salt stress conditions (Janz et al., 2012; Escalante-Perez et al., 2009). In gymnosperm trees, where water transport is ensured via the tracheids, reinforcement of the water-conducting xylem elements occurs under saline growth conditions, leading to thicker cell walls and, hence, increased wood density (Hacke et al., 2001). Also in tracheids, reduction in tracheid fiber length can be observed in various species under salinity (Khamis and Hammad, 2007).

HIGH AND LOW TEMPERATURE

In general, heat stress involves a multitude of cellular and metabolic processes in plants. A significant response to heat stress is a reduction in the synthesis of normal proteins and an accelerated synthesis of so-called heat-shock proteins. Such a response can already be detected when plants are exposed to temperatures of 5°C above their optimal growing conditions. Heat-shock proteins function as chaperones to facilitate folding of proteins. Regarding trees, Overdieck et al. (2007) found that changes in CO_2 levels caused wood anatomical changes and rising temperature caused increased growth of beech saplings. In addition, Thomas et al. (2007) observed a reduction in vessel lumen area in Eucalyptus at higher temperatures. Furthermore, in trees with thin bark, such as beech and maple, sun scald can be a serious problem causing large injuries at the nonshaded southern side. In contrast, species with thick bark, such as oak and pine, are not endangered by heat stress. However, following overheating of beech the cambium often dies back, the bark dries out, and wound wood is produced (Fig. 4.3a). Wounds can be many meters long and difficult to heal because of the possible attack of pathogens such as fungi and insects. However, the damage also depends on light intensity. In case of low light intensity trees are often able to protect themselves by producing a thicker bark.

In contrast to heat stress, low temperatures can damage tree stems in winter. Following a sudden drop in temperature below the freezing point, the outer zones of the stem are cooled down abruptly. Consequently, an uneven thermal cell contraction occurs between the surface and the center of the stem, often resulting in frost cracks (Fig. 4.3b). Following fiber orientation, they start at the base of the stem and extend into higher stem regions. Frost cracks appear particularly in oak, ash, elm, maple, and sweet chestnut and often extend radially from the stem surface to the pith. In order to prevent frost cracks these tree

FIGURE 4.3   (a) Following sun scald the cambium of beech often dies back and wound wood is produced (arrow). (b) Frost crack of an ash stem from a tree grown at valley side.

species should not be grown at sites where cold air suddenly occurs, such as gorges and valley sides. Following the crack the cambium starts to occlude it forming a frost scar, which is regarded as a serious defect in wood utilization.

EFFECTS OF RISING OZONE LEVELS ON DIAMETER GROWTH OF TREES

In recent years, high ozone levels resulting from traffic and industry have been generated in cities and even spread from urban to rural areas, becoming an important pollutant (Ashmore, 2005). Ozone concentration is steadily increasing (Jonson et al., 2006) and will most probably rise by a further 50% by the end of this century. The increased ozone concentration affects plant growth, may decrease forest productivity (Matyssek and Innes, 1999), and therefore cause considerable yield losses (Ashmore, 2005). It is known that ozone is taken up by the stomata and causes oxidative destruction of lipids and proteins of the plasma membrane. Numerous studies deal with the physiological and anatomical effects of ozone on young trees (e.g., Matyssek and Sandermann, 2003); however, a few field studies on adult trees also have shown interesting results. For instance, under humid conditions soil respiration rate was enhanced under elevated O_3 under beech and spruce, related to O_3-stimulated fine-root production only in beech (Nikolova et al., 2010). However, under drought, the stimulating effect of O_3 on soil respiration vanished under spruce, correlated with decreased fine-root production in spruce (Nikolova et al., 2010). Consequently, drought can override the stimulating O_3 influences on fine-root dynamics. Regarding wood increment, Wipfler et al. (2005) observed a decline in diameter growth at breast height of about 20% in ~50-year-old spruce trees and results on black cherry and yellow poplar stems showed ozone-induced radial growth losses of 8–12% and 30–43%, respectively (Somers et al., 1998). Also, Vollenweider et al. (2003) reported a reduction in stem growth of 28% in black cherry. Interestingly, long-term ozone fumigation from 2000 through 2007 induced a shift in resource allocation into height growth at the expense of diameter growth in Norway spruce and European beech (Pretzsch et al., 2010). This change in allometry leads to reduced stem stability in spruce. In addition, on the stand level, double ambient ozone caused a decrease of 10.2 m^3/ha/yr in European beech (Pretzsch et al., 2010).

CONCLUSIONS

Environmental factors, such as mineral deficiency, salinity, drought, high and low temperature, as well as ozone, significantly affect tree growth as well as wood formation. Many factors can act as a stressor when their dose is too high or too low and they also can function as a signal transducer in plant metabolism. They rarely act alone as a signal factor but are often cross-linked by various interactions. For instance, high temperatures are usually correlated with drought, which causes the formation of narrower vessels to reduce water conductivity as well as a reduction of tree-ring width. A similar response in vessel size and ring width is induced by salinity and mineral deficiency, indicating that different abiotic factors cause related responses in wood structure. Further studies on the effect of abiotic factors on the regulation of the cambium will provide further insight into the complex stress responses of trees.

REFERENCES

Abe, H., Nakai, T., Utsumi, Y., Kagawa, A., 2003. Temporal water deficit and wood formation in *Cryptomeria japonica*. Tree Physiol. 23, 859–863.

Arend, M., Fromm, J., 2000. Seasonal variation in the K, Ca and P content and distribution of plasma membrane H^+-ATPase in the cambium of *Populus trichocarpa*. In: Savidge, R., Barnett, J., Napier, R. (Eds.), Cell and Molecular Biology of Wood Formation. BIOS Scientific Publishers, Oxford, pp. 67–70.

Arend, M., Fromm, J., 2007. Seasonal change in the drought response of wood cell development in poplar. Tree Physiol. 27, 985–992.

Arend, M., Weisenseel, M.H., Brummer, M., Osswald, W., Fromm, J., 2002. Seasonal changes of plasma membrane H + -ATPase and endogenous ion current during cambial growth in poplar plants. Plant Physiol. 129, 1651–1663.

Arend, M., Monshausen, G., Wind, C., Weisenseel, M.H., Fromm, J., 2004. Effect of potassium deficiency on the plasma membrane H^+-ATPase of the wood ray parenchyma in poplar. Plant Cell Environ. 27, 1288–1296.

Arend, M., Stinzing, A., Wind, C., Langer, K., Latz, A., Ache, P., Fromm, J., Hedrich, R., 2005. Polar-localized poplar K^+ channel capable of controlling electrical properties of wood-forming cells. Planta 223, 140–148.

Ashmore, M.R., 2005. Assessing the future global impacts of ozone on vegetation. Plant Cell Environ. 28, 1–16.

Baas, P., Werker, E., Fahn, A., 1983. Some ecological trends in vessel characters. IAWA Bull. New Ser. 4, 141–159.

Bangerth, F., 1979. Calcium-related physiological disorders of plants. Annu. Rev. Phytopathol. 17, 97–122.

Chen, S., Polle, A., 2010. Salinity tolerance of *Populus*. Plant Biol. 12, 317–333.

Colin-Belgrand, M., Ranger, J., Bouchon, J., 1996. Internal nutrient translocation in chestnut tree stem wood: III. Dynamics across an age series of *Castanea sativa* (Miller). Ann. Bot. 78, 729–740.

de Luis, M., Novak, K., Raventos, J., Gricar, J., Prislan, P., Cufar, K., 2011. Cambial activity, wood formation and sapling survival of *Pinus halepensis* exposed to different irrigation regimes. For. Ecol. Manage. 262, 1630–1638.

Donaldson, L.A., 2002. Abnormal lignin distribution in wood from severely drought stressed *Pinus radiata* trees. IAWA J. 23, 161–178.

Dünisch, O., Bauch, J., 1994. Influence of mineral elements on wood formation of old growth spruce (*Picea abies* L. Karst.). Holzforschung 48, 5–14.

Eilmann, B., Weber, P., Rigling, A., Eckstein, D., 2006. Growth reactions of *Pinus sylvestris* L. and *Quercus pubescens* Willd. to drought years at a xeric site in Valais, Switzerland. Dendrochronology 23, 121–132.

Eilmann, B., Zweifel, R., Buchmann, N., Pannatier, E.G., Rigling, A., 2011. Drought alters timing, quantity, and quality of wood formation in Scots pine. J. Exp. Bot. 62, 2763–2771.

Eklund, L., Eliasson, L., 1990. Effects of calcium ion concentration on cell wall synthesis. J. Exp. Bot. 41, 863–867.

Elser, J.J., Bracken, M.E.S., Cleland, E.E., Gruner, D.S., Harpole, W.S., Hillebrand, H., Ngai, J.T., Seabloom, E.W., Shurin, J.B., Smith, J.E., 2007. Global analysis of nitrogen and phosphorus limitation of primary producers in freshwater, marine and terrestrial ecosystems. Ecol. Lett. 10, 1135–1142.

Escalante-Perez, M., Lautner, S., Nehls, U., Selle, A., Teuber, M., Schnitzler, J.-P., Teichmann, T., Fayyaz, P., Hartung, W., Polle, A., Fromm, J., Hedrich, R., Ache, P., 2009. Salt stress affects xylem differentiation of grey poplar (*Populus* × *canescens*). Planta 229, 299–309.

Eschrich, W., Fromm, J., Essiamah, S., 1988. Mineral partitioning in the phloem during autumn senescence of beech leaves. Trees 2, 73–83.

Esteban, L.G., Martin, J.A., de Palacios, P., Fernandez, F.G., 2012. Influence of region of provenance and climate factors on wood anatomical traits of *Pinus nigra* Arn. subsp salzmannii. Eur. J. For. Res. 131, 633–645.

Follet-Gueye, M.L., Verdus, M.C., Demarty, M., Thellier, M., Ripoll, C., 1998. Cambium, preactivation in beech correlates with a strong temporary increase of calcium in cambium and phloem but not in xylem cells. Cell Calcium 24, 205–211.

Fromm, J., 2010. Wood formation of trees in relation to potassium and calcium nutrition. Tree Physiol. 30, 1140–1147.

Fromm, J., Eschrich, W., 1986. Changes of adenine nucleotide and orthophosphate concentrations in buds of deciduous trees during spring reactivation. Trees 1, 42–46.

Garcia-Gonzalez, I., Eckstein, D., 2003. Climatic signal of earlywood vessels of oak on a maritime site. Tree Physiol. 23, 497–504.

Gea-Izquierdo, G., Fonti, P., Cherubini, P., Martin-Benito, D., Chaar, H., Canellas, I., 2012. Xylem hydraulic adjustment and growth response of *Quercus canariensis* Willd. to climatic variability. Tree Physiol. 32, 401–413.

Gricar, J., 2007. Xylo- and phloemogenesis in silver fir (*Abies alba* Mill.) and Norway spruce (*Picea abies* (L.) Karst.). Studia Forestalia Slovenica 132, 1–106.

Günthardt-Goerg, M.S., Schmutz, P., Matyssek, R., Bucher, J.B., Percy, K.E., Cox, R.M., Jensen, K.F., 1996. Leaf and stem structure of poplar (*Populus* × *euramericana*) as influenced by O_3, NO_2, their combination and different soil N supplies. Can. J. For. Res. 26, 649–657.

Hacke, U.G., Sperry, J.S., Pockman, W.T., Davis, S.D., McCulloch, K.A., 2001. Trends in wood density and structure are linked to prevention of xylem implosion by negative pressure. Oecologia 126, 457–461.

Helmisaari, H.S., Siltala, T., 1989. Variation in nutrient concentrations of *Pinus sylvestris* stems. Scan. J. For. Res. 4, 443–451.

Ibrahim, L., Proe, M.F., Cameron, A.D., 1997. Main effects of nitrogen supply and drought stress upon whole-plant carbon allocation in poplar. Can. J. For. Res. 27, 1413–1419.

Janz, D., Lautner, S., Wildhagen, H., Behnke, K., Schnitzler, J.P., Rennenberg, H., Fromm, J., Polle, A., 2012. Salt stress induces the formation of a novel type of pressure wood in two *Populus* species. New Phytol. 194, 129–141.

Jonson, J.E., Simpson, D., Fagerli, H., Solberg, S., 2006. Can we explain the trends in European ozone levels? Atmos. Chem. Phys. 6, 51–66.

Junghans, U., Polle, A., Duechting, P., Weiler, E., Kuhlman, B., Gruber, F., Teichmann, T., 2006. Adaptation to high salinity in poplar involves changes in xylem anatomy and auxin physiology. Plant Cell Environ. 29, 1519–1531.

Khamis, M.H., Hammad, H.H., 2007. Effect of irrigation by saline ground water on the growth of some conifer seedlings: mortality, growth, biomass and physical wood properties. Bull. Fac. Agric. Cairo Univ. 58, 36–45.

Laclau, J.-P., Bouillet, J.-P., Ranger, J., Joffre, R., Gouma, R., Saya, A., 2001. Dynamics of nutrient translocation in stem wood across an age series of a Eucalyptus. Ann. Bot. 88, 1079–1092.

Laclau, J.P., Almeida, J.C.R., Goncalves, J.L.M., Saint-Andre, L., Ventura, M., Ranger, J., Moreira, R.M., Nouvellon, Y., 2009. Influence of nitrogen and potassium fertilization on leaf lifespan and allocation of above-ground growth in Eucalyptus plantations. Tree Physiol. 29, 111–124.

Langer, K., Ache, P., Geiger, D., Stinzing, A., Arend, M., Wind, C., Regan, S., Fromm, J., Hedrich, R., 2002. Poplar potassium transporters capable of controlling K^+ homeostasis and K^+-dependent xylogenesis. Plant J. 32, 997–1009.

Lautner, S., 2013. Wood formation under drought stress and salinity. In: Fromm, J. (Ed.), Cellular Aspects of Wood Formation, Plant Cell Monographs 20. Springer Verlag, Berlin Heidelberg, pp. 187–202.

Lautner, S., Fromm, J., 2010. Calcium-dependent physiological processes in trees. Plant Biol. 12, 268–274.

Lautner, S., Ehlting, B., Windeisen, E., Rennenberg, H., Matyssek, R., Fromm, J., 2007. Calcium nutrition has a significant influence on wood formation in poplar. New Phytol. 174, 743–752.

Lenoir, J., Gegout, J.C., Marquet, P.A., de Ruffray, P., Brisse, H., 2008. A significant upward shift in plant species optimum elevation during the 20th century. Science 320, 1768–1771.

Matyssek, R., Innes, J.L., 1999. Ozone – a risk factor for trees and forests in Europe? Water Air Soil Pollut. 116, 199–226.

Matyssek, R., Sandermann, H., 2003. Impact of ozone on trees: an ecophysiological perspective. Progress Bot. 64, 349–404.

Nikolova, P.S., Andersen, C.P., Blaschke, H., Matyssek, R., Häberle, K.-H., 2010. Belowground effects of enhanced tropospheric ozone and drought in a beech/spruce forest (*Fagus sylvatica* L./*Picea abies* [L.] Karst). Environ. Pollut. 158, 1071–1078.

Overdieck, D., Ziche, D., Böttcher-Jungclaus, K., 2007. Temperature responses of growth and wood anatomy in European beech saplings grown in different carbon dioxide concentrations. Tree Physiol. 27, 261–268.

Potchanasin, P., Sringarm, K., Sruamsiri, P., Bangerth, K.F., 2009. Floral induction (FI) in longan (*Dimocarpus longan*, Lour.) trees: Part I. Low temperature and potassium chlorate effects on FI and hormonal changes exerted in terminal buds and sub-apical tissue. Sci. Hort. 122, 288–294.

Pregitzer, K.S., Dickmann, D.I., Hendrick, R., Nguyen, P.V., 1990. Whole-tree carbon and nitrogen partitioning in young hybrid poplars. Tree Physiol. 7, 79–93.

Pretzsch, H., Dieler, J., Matyssek, R., Wipfler, P., 2010. Tree and stand growth of mature Norway spruce and European beech under long-term ozone fumigation. Environ. Pollut. 158, 1061–1070.

Puech, L., Türk, S., Hodson, J., Fink, S., 2000. Wood formation in hybrid aspen (*Populus tremula* L. × *Populus tremuloides* Michx.) grown under different nitrogen regimes. In: Savidge, R., Barnett, J., Napier, R. (Eds.), Cell and Molecular Biology of Wood Formation. BIOS Scientific Publishers, Oxford, pp. 141–153.

Rennenberg, H., Dannemann, M., Gessler, A., Kreuzwieser, J., Simon, J., Papen, H., 2009. Nitrogen balance in forests: nutritional limitation of plants under climate stresses. Plant Biol. 11, S4–S23.

Santiago, L.S., Wright, S.J., Harms, K.E., Yawitt, J.B., Korine, C., Garcia, M.N., Turner, B.L., 2012. Tropical tree seedling growth responses to nitrogen, phosphorus and potassium addition. J. Ecol. 100, 309–316.

Schulte-Baukloh, C., Fromm, J., 1993. The effect of calcium starvation on assimilate partitioning and mineral distribution of the phloem. J. Exp. Bot. 44, 1703–1707.

Sheriff, D.W., Whitehead, D., 1984. Photosynthesis and wood structure in *Pinus radiata* D-Don during dehydration and immediately after rewatering. Plant Cell Environ. 7, 53–62.

Smethurst, P., Knowles, A., Churchill, K., Wilkinson, A., Lyons, A., 2007. Soil and foliar chemistry associated with potassium deficiency in *Pinus radiata*. Can. J. For. Res. 37, 1093–1105.

Somers, G.L., Chappelka, A.H., Rosseau, P., Renfo, J.R., 1998. Empirical evidence of growth decline related to visible ozone injury. For. Ecol. Manage. 104, 129–137.

Sterck, F.J., Zweifel, R., Sass-Klaassen, U., Chowdhury, Q., 2008. Persisting soil drought reduces leaf specific conductivity in Scots pine (*Pinus sylvestris*) and pubescent oak (*Quercus pubescens*). Tree Physiol. 28, 529–536.

Thomas, D.S., Montagu, K.D., Conroy, J.P., 2007. Temperature effects on wood anatomy, wood density, photosynthesis and biomass partitioning of *Eucalyptus grandis* seedlings. Tree Physiol. 27, 251–260.

Vitousek, P.M., Porder, S., Houlton, B.Z., Chadwick, O.A., 2010. Terrestrial P limitation: mechanisms, implications, and N–P interactions. Ecol. Appl. 20, 5–15.

Vollenweider, P., Woodcock, H., Keltry, M.J., Hofer, R.-M., 2003. Reduction of stem growth and site dependency of leaf injury in Massachusetts black cherries exhibiting ozone symptoms. Environ. Pollut. 125, 467–480.

Westermark, U., 1982. Calcium promoted phenolic coupling by superoxide radical – a possible lignification reaction in wood. Wood Sci. Technol. 16, 71–78.

White, P.J., Broadley, M.R., 2003. Calcium in plants. Ann. Bot. 92, 487–511.

Wimmer, R., Lucas, B.N., 1997. Comparing mechanical properties of secondary wall and cell corner middle lamella in spruce wood. IAWA J. 18, 77–78.

Wind, C., Arend, M., Fromm, F., 2004. Potassium-dependent cambial growth in poplar. Plant Biol. 6, 30–37.

Wipfler, P., Seifert, T., Heerdt, C., Werner, H., Pretzsch, H., 2005. Growth of adult Norway spruce (*Picea abies* [L.] Karst) and European beech (*Fagus sylvatica* L.) under free-air ozone fumigation. Plant Biol. 7, 611–618.

Wright, S.J., Yavitt, J.B., Wurzburger, N., Turner, B.L., Tanner, E.V.J., Sayer, E.J., Santiago, L.S., Kaspari, M., Hedin, L.O., Harms, K.E., Garcia, M.N., Corre, M.D., 2011. Potassium, phosphorus, or nitrogen limit root allocation, tree growth, or litter production in a lowland tropical forest. Ecology 92, 1616–1625.

Chapter 5

Flexure Wood: Mechanical Stress Induced Secondary Xylem Formation

Frank W. Telewski

Department of Plant Biology, W.J. Beal Botanical Garden, Michigan State University, East Lansing, MI, USA

Chapter Outline

INTRODUCTION

Mechanical support is one of three major functions of the secondary xylem, the conduction of water from roots to leaves and storage of photosynthate being the other two. All three functions impact the strength of the wood as it is formed by the tree. As has been reported in previous chapters of this book (Donaldson and Singh, 2016; Fromm and Kautner, 2016) and what will be the focus of this chapter, abiotic factors in the environment greatly influence the formation and thus the structure and mechanical strength of wood. Mechanical loading is the most common and prevalent abiotic factor influencing wood formation in trees and includes the influence of gravity on branches and displaced stems on self-loading of the vertical stem, and loading due to ice and snow accumulation (Niklas, 1998; Telewski, 2006). However, wind is a ubiquitous and persistent factor influencing tree growth, allometry, and wood formation and quality throughout its life history (Knight, 1803; Metzger, 1893; Grace, 1977; Coutts and Grace, 1995; Ennos, 1997; Niklas, 1998; Gardiner and Nicoll, 2006; Mitchell, 2013). The influence of wind on tree structure and wood quality was

Secondary Xylem Biology. http://dx.doi.org/10.1016/B978-0-12-802185-9.00005-X

first recorded by Theophrastus in 300 BC (Theophrastus, 1976; translation by Einarson and Link). He reported that trees growing in windy environments were shorter in height, had shorter internodes (more knots), and had closer grain and harder wood (greater wood density) compared to trees growing in more sheltered environments. The first recorded experiment to elucidate the influence of wind on tree growth was conducted by Knight (1803). This study was conducted by staying apple trees to prevent them from swaying in the wind and comparing the resulting growth to trees allowed to sway freely in the wind. Trees allowed to sway naturally in the wind were shorter and had thicker stems than the stayed trees. Metzger (1893) recognized wind as the most important factor in determining stem form and taper in trees and proposed his mechanistic theory of tree growth. Subsequent studies by Jacobs (1936, 1939, 1954), Fielding (1940, 1963, 1976), Larson (1965), and Grace (1977) supported Metzger's emphasis on wind as having the greatest influence on tree growth and form of all other environmental stimuli. In addition to influencing the overall morphology and allometry of tree stems and crowns, mechanical loading induced by wind also alters the structure and mechanical properties of wood, providing for increased functionality for the tree growing in windy environments. This modified wood was termed flexure wood by Telewski (1989) to differentiate it from compression wood and tension wood formed in response to gravity in displaced stems and branches.

While studying the influence of mechanical stimuli on plant growth, specifically the rubbing or brushing of plants, Jaffe (1973) coined the term thigmomorphogenesis to describe the altered pattern of plant growth in response to touch. Plants brushed to induce a mechanical stimulus were shorter in height and produced increased radial growth in their stems. The term thigmomorphogenesis has been universally applied to describe the influence of wind and other mechanical stimuli on plant growth. The focus of this chapter will be to report the influence of wind stress or the application of mechanical loading on wood formation and wood quality and distinguish flexure wood from reaction wood and wound wood.

REACTION WOOD

The formation and characteristics of reaction wood, compression wood in conifers and primitive nonporous angiosperms (wood lacking vessels), and tension wood in porous wood angiosperms (wood containing vessels) has been covered in the chapter on reaction wood in this volume by Donaldson and Singh (2016) and others elsewhere (Wilson and Archer, 1977; Timell, 1986a, 1986b, 1986c; Du and Yamamoto, 2007; Gardiner et al., 2014). The formation of reaction wood is most commonly associated with the displacement of a stem from the vertical position with regard to the gravitational vector or in the maintenance of plagiotropic branches in their horizontal orientation with respect to the vertical stem and gravity. In both instances, a strong case can be made for the role of mechanoperception in the initial sensing of organ orientation with regard to the pull of gravity, and the resulting formation of reaction wood (Telewski, 2006). The mechanoperception could involve the detection of sedimentation of amyloplasts triggering mechanosensitive ion channels upon impact or simply differential pressure placed on the cytoskeleton–plasma membrane–cell wall network of a cell or tissue reoriented within the gravitational field (Sievers et al., 1991; Baluška and Hasenstein, 1997; Yoder et al., 2001; Collings et al., 2001; Hoson et al., 2005; Telewski, 2006; Hashiguchi et al., 2012; Blancaflor, 2013). A component of mechanoperception in the formation of reaction wood can also be considered based upon studies in hypogravity where bending-induced mechanical strains induce compression wood formation in the stems of Douglas fir (*Pseudotsuga menziesii*) seedlings (Kwon et al., 2001) questioning the role of statoliths or gravity in compression wood formation. However, a direct link between mechanoperception and mechanosensitive ion channels and/or the cytoskeleton–plasma membrane–cell wall network in the gravitropic response is still inconclusive (Baldwin et al., 2013).

Beyond the influence of gravity on the induction and formation of reaction wood, several other environmental factors appear to induce compression wood formation, which do not involve graviperception as reviewed by Timell (1986a, 1986b, 1986c). For example, compression wood has also been observed to form in the concaved portion of leeward curved branches of trees growing in windy environments as part of the wind-blown or flag-form growth form (Hartig, 1901). The formation of the compression wood in these branches is perpendicular to the gravitational vector, but parallel to the curvature of the branch. In another example, a branch bent upward by the wind formed compression wood in the upper side of the bent branch. Although it is unclear as to the physiological regulation or function of this type of compression wood formation (Telewski, 2012), the formation of compression wood in the bent branches would appear to be consistent with that reported in bent Douglas fir seedlings growing under hypogravity conditions (Kwon et al., 2001); the mechanical bending of woody tissue can induce compression wood formation regardless of the influence of gravity or relative absence of gravity. Whereas the majority of gravitropic studies are conducted on roots and primary shoots (see reviews by Blancaflor (2013) and Baldwin et al. (2013)), there is a general dearth of literature on the role of graviperception and the role of statoliths in the formation of reaction wood. Given the important role reaction wood plays in determining wood quality, the lack of research or support for studies involved in developing a clearer understanding of it regulation and formation is shocking.

As Timell (1986c) concluded based on his comprehensive review of the literature, wind sway alone does not induce reaction wood formation in vertical stems, whereas wind-induced displacement of stems from the vertical position will result in reaction wood formation. The difference between wood formed in response to the dynamic motion of wind sway and wind-induced static displacement being the requirement of a presentation time (Timell, 1986b; Telewski, 1995) within the displaced stem as part of the perception pathway in the formation of reaction wood. In some studies, the presentation time has been interpreted as the time required for the sedimentation of statoliths in gravisensing cells. However, the role of a presentation time has not been addressed in wind-bent branches or in the bent stems growing under conditions of hypogravity. As will be discussed later, the wood formed in response to wind sway, termed flexure wood (Telewski, 1989), shares some anatomical and structural similarities to compression wood. The formation of flexure wood does not require a presentation time and the mechanoperception required to induce changes in cambial development is very rapid (Coutand, 2010).

WULSTHOLZ OR "BEADWOOD"

Flexure wood should not be confused with Wulstholz, a term coined by Trendelenburg (1940) to describe wood formed in trees in which wind loading exceeded the elastic limit of the trunk and resulted in compression slip plane failures (wounding) within the wood. The formation of Wulstholz, a type of wound wood, by the vascular cambium at the point of failure can be characterized as a type of wound response in which the newly formed wood creates a circumferential budge, or bead along the stem directly above the fracture. The newly formed wood is an attempt by the tree to stabilize the slip plane fracture. Koch et al. (2000) and Schmitt et al. (2007) published detailed characterizations of this type of wood. The newly formed wood must function to mechanically restabilize the weakened stem at the point of fracture in wind-exposed trees. In conifers, similar to flexure wood, Wulstholz is characterized by increases in xylem production, microfibril angle (MFA) and wood density (thicker S2 layer of tracheid cell walls and significantly smaller lumens), and a reduction in tracheid length compared to normal wood. Lignin content and hemicellulose content, especially mannose, are higher and the concentration of glucose is reduced when compared to normal wood. Biomechanically, compression strength and the elastic modulus are reduced (despite the increase in density). As a result of these chemical and anatomical changes, the tracheids of Wulstholz can be deformed and compressed greater than normal wood tracheids resisting fracture (Koch et al., 2000).

FLEXURE WOOD

As has been reported in other chapters in this book (Donaldson and Singh, 2016; Fromm and Kautner, 2016), the characteristics and quality of the wood, which differentiates from the vascular cambium, is determined by environmental cues together with the genetics of the tree. The prevailing environmental conditions will influence the characteristics of wood density, hydraulic conductivity, flexibility, strength, MFA and cell wall composition to meet the physiological and mechanical needs of the living tree. Flexure wood forms in trees in response

TABLE 5.1 Characteristics of Flexure Wood in Conifers and Dicotyledonous Angiosperms Compared to Normal Wood

Feature	Conifers	Angiosperms
Radial growth in the plain of flexure*	+	+
Spiral grain	+	+
Cell wall thickness	+	+
Resin ducts (nontraumatic)	+**	NA
Tracheid diameter	−	NA
Tracheid length	−	NA
Fiber length	NA	−
Vessel number	NA	−
Vessel diameter	NA	−
Wood density	+	0
Microfibril angle	+	+
Total lignin content	+?	0
Syringyl:guaiacyl	NA	+
Modulus of elasticity	−	−
Second moment of cross-sectional area	+	+
Flexural stiffness	+	+
Modulus of rupture	+	−

+, Increase in character; −, decrease; 0, no change; ?, uncertainty; NA, data not available, feature not present in taxonomic group.
*See text for exceptions and explanation.
**In the genus Pinus.

to dynamic bending or flexing of the woody tissues in which the external load does not exceed the elastic limit. By not exceeding the elastic limit, wounding does not occur in the wood or cambial region as is the case with the induction and formation of Wulstholz. A comparison list of the characteristics of flexure wood in conifers and dicot angiosperms is presented in Table 5.1 and discussed in detail in the subsequent section.

Radial Growth

Trees exposed to wind-induced sway, without displacement from the gravitational vector for the period of the presentation time will form flexure wood, not reaction wood – compression or tension wood (Timell, 1986c; Telewski, 1989, 1995). However, similar to the formation of reaction wood, flexure wood will form in the direction of the mechanical load; for reaction wood, the gravitational vector; for wind, the direction of flexure. This increase in cell production by the vascular cambium serves to increase the cross-sectional area in the direction of the mechanical load in both conifers (Jacobs, 1954; Larson, 1965; Bannon and Bindra, 1970; Reich and Ching, 1970; Burton and Smith, 1972; Carlton, 1976; Kellogg and Steucek, 1977, 1980; Telewski and Jaffe, 1981, 1986a, 1986b; Telewski, 1989, 1990; Dean, 1990; Valinger, 1992; Valinger et al., 1995; Stokes et al., 1995, 1997; Lundqvist and Valinger, 1996; Mickovski and Ennos, 2003; Watt et al., 2005; Meng et al., 2006; Lundström et al., 2008; Moore et al., 2014) and angiosperms (Knight, 1803; Neel and Harris, 1971; Harris et al., 1973; Wrigley and Smith, 1978; Lawton, 1982; Holbrook and Putz, 1989; Tateno, 1991; Telewski et al., 1997; Telewski and Pruyn, 1998; Pruyn et al., 2000; Quilhó et al., 2003; Liu et al., 2003; Kern et al, 2005; Leblanc-Fournier et al., 2008; Coutand et al., 2008, 2009, 2010; Reubens et al., 2009; Martin et al., 2010; Medhurst et al., 2011) (Fig. 5.1). However, it must be noted that in some angiosperms, notably *Juglans nigra*, *Acer saccharinum*, *Liquidambar styraciflua*, and *Cecropia schreberiana*, mechanical swaying or wind did not consistently induce an increase in radial growth and the production of more xylem (Heiligmann and Schneider, 1975; Ashby et al., 1979; Cordero, 1999).

These few conflicting results for angiosperms and the conflicting results of no observed increase in radial growth for a handful of conifer species, including *Pinus contorta* (Rees and Grace 1980a, 1980b), *P. menziesii* (Kellogg and Steucek, 1977, 1980), *Pinus taeda* (Telewski and Jaffe, 1981, 1986a), and *Pinus radiata* (Apiolaza et al., 2011), need to be addressed. In the case of the conifer species, all have been shown to increase radial growth in other mechanical stress studies; *Pi. contorta* (Meng et al. 2006, 2008), *P. menziesii* (Reich and Ching, 1970; Carlton, 1976; Mitchell, 2003), *Pi. taeda* (Burton and Smith, 1972; Telewski and Jaffe, 1981, 1986a; Telewski, 1990), and *Pi. radiata* (Jacobs, 1939, 1954; Watt et al. 2005). Similarly, for the angiosperm species, *J. nigra* (Leblanc-Fournier et al. 2008) and *L. styraciflua* (Neel and Harris, 1971; Holbrook and Putz, 1989) have both been shown to exhibit a thigmomorphogenetic response increasing radial growth in response to a mechanical stress. Of interest, *J. nigra* was used in one of the studies to characterize the early gene expression of a transcription factor associated with the thigmomorphogenetic response (Leblanc-Fournier et al., 2008). One possibility for the differences in growth response reported by the different researchers maybe related to the individual genetics within a species. This appears to be the case for both *Pi. taeda* and hybrid *Populus*. In a series of thigmomorphogenetic studies on different half-sib lines, Telewski reported variations in the growth and physiological responses (Telewski and Jaffe, 1981, 1986b, 1986c; Telewski, 1990). When exposed to mechanical sway, all half-sibs produced an elliptical cross-section, with the longest axis in the direction of the applied flexure. In the case on one of the half-sibs, the diameter of the longest axis was the same as the diameter of the control trees (Telewski and Jaffe, 1981). What caused the elliptical stem form was a growth reduction in the direction perpendicular to force application. When this study was replicated, the half-sib in question exhibited a

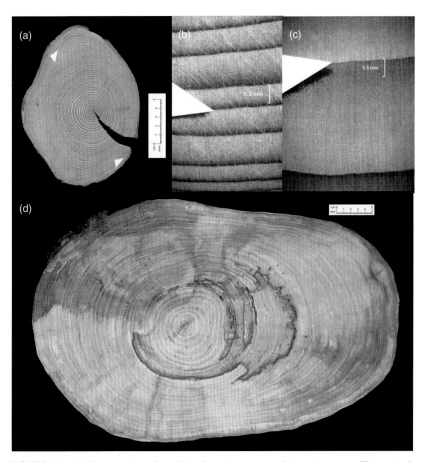

FIGURE 5.1   (a) Cross-section of an *Abies fraseri* tree exposed to a strong prevailing westerly wind on the summit of Roan Mountain, North Carolina, USA containing flexure wood. The tree was cut in 1982. Note the long axis of the elliptical cross-section corresponds to the prevailing wind direction. Compression wood is lacking in this sample. (b) Enlargement of the growth ring formed in 1976 on the windward side the *A. fraseri* sample shown in Fig. 5.1a. (c) Enlargement of the growth ring formed in 1976 on the leeward side of the *A. fraseri* sample shown in Fig. 5.1a. Note the much wider ring, with wider latewood band but an absence of compression wood. This portion of the ring is composed of flexure wood (Fig. 5.1b). (d) Cross-section of a *Tilia cordata* that was growing on the campus of Michigan State University next to a five-story building. The building redirected wind flow to move parallel to the building's face. The long axis of the elliptical cross-section containing flexure wood is parallel to the building and also the redirected bidirectional wind flow. Discoloration is due to the log being stored for a year after harvest. For a more detailed description of this phenomenon of wind effects on trees in urban environments, see Telewski et al. (1997).

similar growth response, produced fewer tracheids per radial file perpendicular to the applied force, and exhibited no change in the elastic modulus or flexural stiffness. However, similar to other reports, tracheid lengths were significantly reduced in response to flexing and the grain angle was significantly greater (Telewski and Jaffe, 1986a). In a study on wind-tolerant and -intolerant clones of hybrid *Populus*, Pruyn et al. (2000) reported that the intolerant clone could only alter stem allometry by decreasing height growth and increasing radial growth in response to flexing but not the biomechanical properties of the wood (no reduction in the elastic modulus), whereas the tolerant clone was able to alter both. In another study on hybrid *Populus*, Kern et al. (2005) reported some clones exhibited no significant change in stem cross-sectional area between control and mechanically flexed trees, whereas other clones exhibited a thigmomorphogenetic response in this morphological feature. These studies clearly indicate that the genetics of individual trees contributes significantly to how it will respond to wind and other mechanical loads. Such a genetic component in the response of trees to wind loading was also suggested by Telewski (2012) to the degree by which different conifer species respond to strong prevailing winds by the formation of a flagged crown. Trees with higher greenwood densities also have a higher propensity to produce a flagged crown compared to trees with lower greenwood densities. Aspects of genetic influence on the thigmomorphogenetic response and silvicultural practices will be addressed later.

Another factor that appears to influence tree response to wind and other mechanical loads is the type of mechanical stress being applied to the tree. For the duration of the nineteenth century, after Knight's publication (Knight, 1803) on wind effects in apple trees, until the end of the century, plant physiology was greatly influenced by the German plant physiologist, Julius Sachs, who formulated the mechanical growth theory (Sachs, 1873, 1882). This theory, once widely accepted by scholars of the day, was challenged as the nineteenth century drew to a close. One of the major challengers of the day was Frederick C. Newcombe. Newcombe (1895) published a critical review contesting the mechanical growth theory in which he states; "The formation of such growths as have been recounted in this paper is no longer to be explained as simple mechanics..." He also acknowledged that; "...the plant has the ability to respond to (mechanical) stress, but the notion stress is complex, and will doubtless by future research be subdivided." Distinguishing between the types of mechanical loads may certainly contribute to the type of response an individual tree may exhibit. This was clearly demonstrated by the work of Quirk and coworkers in the 1970s. In three separate studies they investigated the influence of torque, compression, and vibration on the growth and development of *Pinus resinosa*. Torque stress reduced height growth and increased radial growth (Quirk et al. 1975), whereas compression stress had no effect on height growth and decreased radial growth (Quirk and Freese, 1976b). Vibrational stress decreased height growth, but had no effect on radial growth (Quirk and Freese, 1976a).

In the case of *C. schreberiana,* a pioneer tropical rainforest species, the reported difference in response to wind of reduced radial growth may be a function of the ecological position the species hold in succession, and a function of its ontological stem anatomy. Stem stiffness can be accomplished in a number of ways, not just via an increase in material in the direction of loading, which appears to be the norm for the majority of species tested. Cordero (1999) explains that the seedlings of *C. schreberiana* have a very large pith filled with turgid parenchymatous cells that "could help to reduce local buckling because of its high stability" of the thin-walled, turgid cells. As the seedlings mature into later ontological stages, they increase the cross-sectional area of stiffer, denser xylem (fibers and vessels). The overall reduced growth rate in these seedlings is apparently due to a reduction in the maximum photosynthetic and respiration rates in the wind-exposed seedlings (Cordero, 1999).

Structure

Overall, the single characteristic at the macroscopic level to identify flexure wood is asymmetric radial growth resulting in an eccentric or elliptical stem cross-section, with the radius of greatest growth generally lacking features associated with reaction wood as described earlier (Fig. 5.1). The anatomical structural differences that characterize flexure wood at the macroscopic level are not as definitive as those associated with reaction wood (compression wood and tension wood). Whereas tension wood can usually be identified by the presence of gelatinous fibers and compression wood by the rounded tracheids in the transverse section, intercellular spaces, and greatly thickened S2 layer of the secondary cell wall, flexure wood, especially in transverse section, looks quite similar to normal wood. In conifers, it can take on the appearance of mild compression wood (Apiolaza et al., 2011). The differences in wood anatomy and structure of flexure wood are more subtle and impact the biomechanical properties via changes in microscopic properties of wood density, cellulose microfibrillar angle, and chemical composition.

In conifers, macroscopically, the newly formed tracheids formed under the influence of wind sway are smaller in size usually with thickened secondary cell walls resulting in an increase in wood density (Telewski 1989, 1990; Jungnikl et al., 2009). Flexure wood tracheids are also shorter in length (Telewski and Jaffe, 1986b; Telewski, 1989; Jungnikl et al., 2009) and were reported to occasionally lack an S3 wall layer with occasional checking in the S2 layer of some tracheids, but lacked intercellular air spaces (Apiolaza et al., 2011). The stems of trees with flexure wood were also reported to have an increase in grain angle of spiral grain in conifers (Telewski and Jaffe, 1986a, 1986b; Kubler, 1991; Skatter and Kucera, 1997; Eklund and Sall, 2000; Fonweban et al., 2013) and angiosperms (Richter, 2006). Schulgasser and Witztum (2007) discussed that the influence of wind on the formation of spiral grain in trees improving the mechanical strength of the trunk under a torsional load was consistent with their model for a mechanism of spiral grain formation. However, a study on spiral grain in *Pinus longaeva* reports no correlation between wind exposure and the occurrence of spiral grain (Wing et al., 2014).

At the microscopic level the angle of the cellulose microfibrils increased (Cave and Walker, 1994; Telewski, 1989; Jungnikl et al., 2009), although Brennan et al. (2012) failed to find a significant increase in MFA in tracheid cell walls in the core wood of rocked (to simulate sway induced flexure wood) or displaced (to induce compression wood) seedlings of *Pi. radiata*. However, the authors noted that had outerwood been included in the study, "differences in cellulose microfibril angles across analogous phenotypes would likely have been observed."

Within members of the genus *Pinus*, flexure wood is also characterized by an increase in the number of resin ducts (Telewski and Jaffe, 1986b; Watt et al., 2009a; Apiolaza et al., 2011; Jones et al., 2013) and extractable fats, waxes, and resins (Telewski and Jaffe, 1981; Brennan et al., 2012). Resin ducts are a normal feature of the wood of the genus *Pinus* and the increase in resin duct number was not of the traumatic resin duct type. Traumatic resin ducts were not observed in the flexure of *Abies* (Telewski, 1989). Moore et al. (2014) reported no significant difference in the occurrence of resin features in the wood of free-swaying verses guyed 13-year-old *Pi. radiata* trees.

In porous wood angiosperms, mechanical flexing reduces the diameter and frequency of vessels and length of fibers in the xylem of *L. styraciflua* (Neel and Harris, 1971), and reduction in vessel lumen area, vessel diameter, and vessel frequency in hybrid *Populus* (Kern et al., 2005). These reports are consistent with the findings of Christensen-Dalsgaard et al. (2007) who reported that wood formed under the greatest mechanical loads in six tropical tree species had the smallest vessels and lowest vessel frequency. In *Quercus suber*, saplings exposed to wind had more fibers, less axial parenchyma, more rays, and increased vessel frequency than saplings sheltered from the wind (Quilhó et al., 2003). When stems of *L. styraciflua* were guyed to prevent wind sway, no difference in wood density was observed when compared to free-swaying trees (Holbrook and Putz, 1989). Seedlings of *C. schreberiana* grown exposed to wind were characterized with lower stem density and elastic modulus when compared to seedlings grown protected from the wind (Cordero 1999). Similar to flexure wood in conifers, flexure wood in angiosperms is also characterized by an increase in the cellulose MFA (Telewski, Mansfield and Koehler, unpublished data).

Chemical Composition

In addition to anatomical changes characteristic of flexure wood induced by wind and other mechanical loads, alterations in cell wall chemistry have also been reported. In conifers, lignin is mostly composed of the monolignol coniferyl alcohol. Total insoluble lignin was reported to significantly increase in one half-sib line of *Pi. taeda* in response to flexing (Telewski and Jaffe, 1981). However, Brennan et al. (2012) reported no significant change in the composition or amount of cell-wall polymers between normal wood and flexure wood, including total lignin content. Subtle differences were observed in the core wood polysaccharide anomeric region spectra for normal wood, compression wood, opposite wood, and flexure wood.

In angiosperms, the lignin polymer is composed primarily of two monolignols – syringyl alcohol and coniferyl alcohol – which give rise to the guaiacyl unit. Mechanical flexing induced flexure wood did not have a higher total lignin content compared to normal wood in hybrid *Populus*. However, it did increased the syringyl monolignol content (Koehler and Telewski, 2006; Kohler et al., 2006).

Biomechanics

Changes in the cellular structure and chemical composition of the cell wall will alter the biomechanical properties of the wood as well as increases in the amount of wood produced in the direction of mechanical loading. The stiffness of a stem is determined by the product of two parameters: the elastic modulus (*E*) times the second moment of cross-sectional area (*I*).

$$EI = E \times I \tag{5.1}$$

An increase in *E* will result in more flexible wood, whereas an increase in *I* will result in a stiffer stem. *I* is determined by

$$I = \frac{1}{4}\pi r^4 \tag{5.2}$$

where *r* is the radius of the circular cross-section. For a stem of elliptical cross-section, the equation for *I* for loading in the direction of the long axis of the ellipse becomes

$$I = \frac{1}{4}\pi a^3 b \tag{5.3}$$

where *a* is the radius of the long axis of the ellipse. Therefore, elliptical stems are stiffer in the direction of the long axis of the ellipse, or in the case of trees exposed to a strong bi- or unidirectional wind, the direction of the prevailing wind direction. This can clearly be observed in the stems of trees growing in urban environments next to multistory buildings, where the long axis of the elliptical stem is parallel to the face of the building, where wind is restricted to movement up or down the street (Fig. 5.1d, Telewski et al., 1997). This pattern of tree growth conserves the use of photosynthate, applying it to growth where increased mechanical stiffness is required, resulting in the elliptical cross-sectional growth form.

Although the wind-induced increase in radial growth in the direction of mechanical loading serves to increase the flexural stiffness of the entire stem, on a per unit volume basis of the wood, the elastic modulus decreases for both conifers and angiosperms, which increases flexibility (Telewski and Jaffe, 1986a,b; Telewski, 1989; Telewski and Pruyn, 1998; Pruyn et al., 2000; Anten et al., 2005; Koehler and Telewski, 2006; Martin et al., 2010;

Apiolaza et al., 2011). An exception to this was reported by Holbrook and Putz (1989) when they observed no significant decrease in E when comparing free swaying stems of *L. styraciflua* to guyed stems to prevent wind induced flexing. The decrease in E is never sufficient to override the increase in I, hence the product EI results in a stiffer stem. Ecologically, this is a significant adaptation, which provides for a stem that is less likely to be deflected by the wind, yet due to a lower value of E, is more pliable, able to absorb bending energy (Telewski, 1989, 1995, 2012; Pruyn et al., 2000).

Overall stem strength in trees exposed to flexing, as determined by the modulus of rupture, is higher in conifer stems (Telewski and Jaffe, 1986b), but was reported to be decreased in hybrid *Populus* (Kern et al., 2005).

EARLY GENETIC EVALUATION FOR WOOD QUALITY AND WIND FIRMNESS

Selecting for wind firmness is a desirable trait, especially in regions prone to wind throw and wind snap. This has proven to be especially true for *Pi. radiata* plantations in New Zealand (Waghorn et al., 2007; Watt et al., 2009b; Waghorn and Watt, 2013) and hybrid poplar fiber farming in the Pacific Northwest (Harrington and DeBell, 1996). In *Pi. radiata* breeding programs, the modulus of elasticity (E) is considered more important than strength (modulus of rupture) as a predictor of wood quality (Watt et al., 2009b); however, Watt et al. (2009b) acknowledged the importance of wind on altering E in trees and the need to better account for its potential impact on modeling wood quality. In a follow up study to Harrington and DeBell (1996), Pruyn et al. (2000) reported that the wind firm clone of hybrid poplar exhibited a greater thigmomorphogenetic response to stem flexing. The reports of increased wood density and decrease in E in flexure wood are counter to most standards for determining wood quality in which higher density equates with higher E. Selecting for wind firmness may result in a reduction in wood quality due to the lower values of E in flexure wood, especially in coniferous species. A more integrative approach in the selection for wood quality taking into consideration wood biomechanics and its ecological function (Fournier et al., 2013) may yield an improved end product for tree selection for use in the forest products industry.

FOREST PRODUCTS CONTAINING FLEXURE WOOD

Unfortunately, from a wood products perspective, the studies reviewed here focus on the mechanical properties of flexure wood in the greenwood or living state. To the best of the author's knowledge, no studies have been published on the mechanical properties of dried wood containing flexure wood as there have been for reaction wood. It would appear that the decrease in E in flexure wood, despite an increase in wood density (especially in coniferous species), would result in a more flexible timber, a less desirable trait. Additionally, the occurrence of spiral grain, a feature of flexure wood, is known to decrease wood quality in dried lumber by reducing dimensional stability and strength, and can result in warping and twisting (Fonweban et al., 2013). Therefore, in terms of forest products, especially for the production of dimensional lumber, wood derived from windy environments are more flexible and likely to bend under lower loads than wood derived from trees growing in sheltered environments when applied as a structural element in a building or other wood structure. Nothing is currently known about the brittle nature of flexure wood or the strength (modulus of rupture) in the dried condition. This is an area ripe for research, especially if breeding programs begin to select for wind firmness as a favorable trait. Once again, it is critical to consider the whole tree in terms of functional biomechanics (Fournier et al., 2013) to better understand the full and proper utilization of wood derived from trees growing in windy environments. This is a good example that what is good for the living tree is not necessarily good for forest products and vice versa.

CONCLUSIONS

There is still much that needs to be determined with regard to the structure and function of flexure wood in the greenwood condition and especially in the dried wood condition. For example, the reported decrease in E is not consistent with an increase, or at least no decrease in wood density (Telewski, 1989). In the forest products industry, wood density has consistently been used as a proxy for wood stiffness and mechanical strength. The decrease in E may be attributed to the increase in MFA in conifers and angiosperms exposed to flexing. But this is not the first reported case where wood mechanical properties do not correlate with wood density. Compression wood, which has a higher density than normal wood, is more brittle in the dried state than normal wood. Yet, it has a very dynamic biomechanical function in the living tree, in the green state. In a study on branch junctions, Jungnikl et al. (2009) reported that wood around branch junctions exposed to wind had wood of higher density and MFA, whereas the wood at branch bases was characterized with lower density and higher MFA. They suggested that the wood around the branch junction was optimized for toughness whereas the wood of branch bases was optimized for flexibility and deformability. Another contradiction occurs in the flexure wood of angiosperms. Increasing the syringyl content of the lignin polymer of hybrid *Populus* via genetic overexpression resulted in an increase in E; however, under conditions of flexure, syringyl content increased while E declined (Kohler et al., 2006; Koehler and Telewski, 2006).

In comparing flexure wood to reaction wood (compression wood and tension wood) some interesting similarities and contradictions occur. In conifers, flexure wood has many characteristics of compression wood and could be classified as incipient compression wood (Brennan et al., 2012). However, flexure wood in ring-porous angiosperms is not that similar to tension wood and is functionally similar to flexure wood in conifers. Although flexure wood and tension wood share the features of reduced vessel area and a few gelatinous fibers, flexure wood in angiosperms is characterized by an increase in MFA (decreased in tension wood), no significant change in lignin content (decreased in tension wood), no change or a possible slight decrease in glucose (cellulose) content (increased in tension wood), and an increase in the syringyl mono-lignol content of lignin (no change in tension wood). In some ways, flexure wood in angiosperms more closely resembles compression wood than tension wood. This could be, in large part, due to the mechanical requirement of flex-ure wood functioning in a dynamic state of mechanical loading, compared to the mechanical requirement of compression or tension wood functioning under static loading. Wind-induced flexing induces an alternating tensional and compressional load on the stem. Since wood is weaker under compres-sional loading than tensional loading, the ability of the vascular cambium to differentiate cells more suited to withstand compressional loading in conifers and angiosperms under conditions of alternating compressional and tensional loading is likely a selective advantage to maintain a vertical and intact stem in regions of wind loading (Telewski, 1989). Finally, the key distinguishing factor between flexure wood and reaction wood is the mode of mechanoper-ception, with reaction wood induction occurring after a requisite period of a presentation time has been met, whereas the formation of flexure wood as part of the thigmomorphogenetic response is nearly instantaneous after the first flexure is initiated.

REFERENCES

Anten, N.P.R., Casado-Garcia, R., Nagashima, H., 2005. Effects of mechanical stress and plant density on mechanical properties, growth, and lifetime reproduction of Tobacco plants. Am. Nat. 166, 650–660.

Apiolaza, L.A., Butterfield, B., Chauhan, S.S., Walker, C.E., 2011. Characterization of mechani-cally perturbed young stems: can it be used for wood quality screening? Ann. For. Sci. 68, 407–414.

Ashby, W.C., Kolar, C.A., Hendricks, T.R., Phares, R.E., 1979. Effects of shaking and shading on growth of three hardwood species. Forest Sci. 25, 212–216.

Baldwin, K.L., Strohm, A.K., Masson, P.H., 2013. Gravity sensing and signal transduction in vas-cular plant primary roots. Am. J. Bot. 100, 126–142.

Baluška, F., Hasenstein, K.H., 1997. Root cytoskeleton: its role in perception of and response to gravity. Planta 203, S69–S78.

Bannon, M.W., Bindra, M., 1970. The influence of wind on ring-width and cell length in conifer stems. Can. J. Bot. 48, 255–259.

Blancaflor, E.B., 2013. Regulation of plant gravity sensing and signaling by the actin cytoskeleton. Am. J. Bot. 100, 143–152.

Brennan, M., McLean, J.P., Altaner, C.M., Ralph, J., Harris, P.J., 2012. Cellulose microfibril angles and cell-wall polymers in different wood types of *Pinus radiata*. Cellulose 19, 1385–1404.

Burton, J.D., Smith, D.M., 1972. Guying to prevent wind sway influences loblolly pine growth and wood properties. U.S. For. Serv. Res. Pap. SO-80, Atlanta, Georgia, USA.

Carlton, R.R., 1976. Influences of the duration of periodic sway on the stem form development of four-year-old Douglas-fir (*Pseudotsuga menziesii*). Unpublished Master's Thesis, Oregon State University, USA.

Cave, I.D., Walker, J.C.F., 1994. Stiffness of wood in fast-grown plantation softwoods: the influence of microfibril angle. Forest. Prod. J. 44, 43–48.

Christensen-Dalsgaard, K.K., Fournier, M., Ennos, A.R., Barfod, A.S., 2007. Changes in vessel anatomy in response to mechanical loading in six species of tropical trees. New Phytol. 176, 610–622.

Collings, D.A., Zsuppan, G., Allen, N.S., Blancaflor, E.B., 2001. Demonstration of prominent actin filaments in the root columella. Planta 212, 392–403.

Cordero, R.A., 1999. Ecophysiology of *Cercopia schreberiana* saplings in two wind regimes in an elfin cloud forest: growth, gas exchange, architecture and stem biomechanics. Tree Physiol. 19, 153–163.

Coutand, C., 2010. Mechanosensing and thigmomorphogenesis, a physiological and biomechanical point of view. Plant Sci. 179, 168–182.

Coutand, C., Dupraz, C., Jaouen, G., Ploquin, S., Adam, B., 2008. Mechanical stimuli regulate the allocation of biomass in trees: demonstration with young *Prunus avium* trees. Ann. Bot. 101, 1421–1432.

Coutand, C., Martin, L., Leblanc-Fournier, N., Decourteix, M., Julien, J.-L., Moulia, B., 2009. Strain mechanosensing quantitatively controls diameter growth and PtaZFP2 gene expression in poplar. Plant Physiol. 151, 223–232.

Coutand, C., Chevol, M., Lacointe, A., Rowe, N., Scotti, I., 2010. Mechanosensing of stem bending and its interspecific variability in five neotropical rainforest species. Ann. Bot. 105, 341–347.

Coutts, M.P., Grace, J., 1995. Wind and Trees. Cambridge University Press, Cambridge, UK.

Dean, T.J., 1990. Effect of growth rate and wind sway on the relation between mechanical and water-flow properties in slash pine seedlings. Can. J. For. Res. 21, 1501–1506.

Donaldson, L.A., Singh, A.P., 2016. Reaction wood. In: Kim, Y.S. (Ed.), Secondary Xylem Biology, Elsevier, Oxford, UK. pp. 93–110.

Du, S., Yamamoto, F., 2007. An overview of the biology of reaction wood formation. J. Int. Plant Biol. 49, 131–143.

Eklund, L., Sall, H., 2000. The influence of wind on spiral grain formation in conifer trees. Trees 14, 324–328.

Ennos, A.R., 1997. Wind as an ecological factor. Trends Ecol. Evol. 12, 108–111.

Fielding, J.M., 1940. Leans in Monterey pine (*Pinus radiata*) plantations. Aust. For. 5, 21–25.

Fielding, J.M., 1963. Aspects of the nature and characteristics of Monterey pine (*P. radiata* D. Don.). PhD Thesis, University of Queensland, Australia. pp. 184.

Fielding, J.M., 1976. Some characteristics of the crown and stem of *Pinus radiata*. Bull. For. Timb. Bur. Aust. 43, 32.

Fonweban, J., Mavrou, I., Gardiner, B., Macdonald, E., 2013. Modelling the effect of spacing and site exposure on spiral grain angle on Sitka spruce (*Picea sitchensis* (Bong.) Carr.) in Northern Britain. Forestry 86, 331–342.

Fournier, M., Dlouhá, J., Jaouen, G., Almeras, T., 2013. Integrative biomechanics for tree ecology: beyond wood density and strength. J. Exp. Bot. 64, 4793–4815.

Fromm, J., Kautner, S., 2016. Abiotic stresses on secondary xylem formation. In: Kim, Y.S. (Ed.), Secondary Xylem Biology, Elsevier, Oxford, UK. pp. 59–71.

Gardiner, B., Nicoll, B., 2006. Trees as engineering structures: standing up to the wind. J. Biomech. 39 (Suppl. 1), S352.

Gardiner, B., Barnett, J., Saranpaa, P., Gril, J., 2014. Biology of Reaction Wood. Springer-Verlag, Berlin, Germany. p. 274.

Grace, J., 1977. Plant Response to Wind. Academic Press, London, UK. p. 204.

Harrington, C.A., DeBell, D.S., 1996. Above- and below-ground characteristics associated with wind toppling in a young *Populus* plantation. Trees 11, 109–118.

Harris, R.W., Leiser, A.T., Neel, P.L., Long, D., Stice, N.W., Maire, R.G., 1973. Trunk development of young trees. Calif. Agric. 27, 7–9.

Hartig, R., 1901. Holzuntersuchungen Altes und Neues. Julius Springer, Berlin.

Hashiguchi, Y., Tasaka, M., Morita, M.T., 2012. Mechanism of higher plant gravity sensing. Am. J. Bot. 100, 91–100.

Heiligmann, R., Schneider, G., 1975. Black walnut seedlings growth in wind protected microenvironments. For. Sci. 21, 293–297.

Holbrook, N.M., Putz, F.E., 1989. Influence of neighbors on tree form: effects of lateral shade and prevention of sway on the allometry of *Liquidambar styraciflua* (Sweet Gum). Am. J. Bot. 76, 1740–1749.

Hoson, T., Saito, Y., Soga, K., Wakabaysashi, K., 2005. Signal perception, transduction, and response in gravity resistance. Another graviresponse in plants. Adv. Space. Res. 36, 1196–1202.

Jacobs, M.R., 1936. The effect of wind on trees. Aust. For. 1, 25–32.

Jacobs, M.R., 1939. A study of the effect of sway on trees. Aust. Com. For. Bur. Bul., No.26.17 pp.

Jacobs, M.R., 1954. The effect of wind sway on the form and development of *Pinus radiata*. Aust. J. Bot. 2, 35–51.

Jaffe, M.J., 1973. Thigmomorphogenesis: the response of plant growth and development to mechanical stimulation, with special reference to *Bryonia dioica*. Planta 144, 143–157.

Jones, T.G., Downes, G.M., Watt, M.S., Kimberley, M.O., Culvenor, D.S., Ottenschlaeger, M., Estcourt, G., Xue, J., 2013. Effect of stem bending and soil moisture on the incidence of resin pockets in *radiata* pine. N. Z. J. For. Sci. 43, 10.

Jungnikl, K., Goebbels, J., Burgert, I., Fratzl, P., 2009. The role of material properties for the mechanical adaptation at branch junctions. Trees 23, 605–610.

Kellogg, R.M., Steucek, G.L., 1977. Motion induced growth effects in Douglas-fir. Can. J. For. Res. 7, 94–99.

Kellogg, R.M., Steucek, G.L., 1980. Mechanical stimulation and xylem production in Douglas-fir. For. Sci. 26, 643–651.

Kern, K.A., Ewers, F.W., Telewski, F.W., Koehler, L., 2005. Mechanical perturbation affects conductivity, mechanical properties and aboveground biomass of hybrid poplars. Tree Physiol. 25, 1243–1251.

Knight, T., 1803. Account of some experiments on the descent of sap in trees. Philos. Trans. R. Soc. London 93, 277–289.

Koch, G., Bauch, J., Puls, J., Schwab, E., 2000. Biological, chemical and mechanical characteristics of 'Wulstholz' as response to mechanical stress in living trees of *Picea abies* [L.] Karst. Holsforschung 54, 137–143.

Koehler, L., Telewski, F.W., 2006. Biomechanics and transgenic wood. Am. J. Bot. 93, 1273–1278.

Kohler, L., Ewers, F.W., Telewski, F.W., 2006. Optimizing for multiple functions: mechanical and structural contributions of cellulose microfibrils and lignin in strengthen tissues. In: Stokke, D.D., Groom, L.H. (Eds.), Characterization of the Cellulosic Cell Wall. Blackwell Publishing, Ames, Iowa, USA. pp. 20–29.

Kubler, H., 1991. Function of spiral grain in trees. Trees 5, 125–135.

Kwon, M., Bedgar, D.L., Piastuch, W., Davin, L.B., Lewis, N.G., 2001. Induced compression wood formation in Douglas fir (*Pseudotsuga menziesii*) in microgravity. Phytochemistry 57, 847–857.

Larson, P.R., 1965. Stem form of young *Larix* as influenced by wind and pruning. For. Sci. 4, 412–424.

Lawton, R.O., 1982. Wind stress and elfin stature in a montane rain forest tree: an adaptive explanation. Am. J. Bot. 69, 1224–1230.

Leblanc-Fournier, N., Coutand, C., Crouzet, J., Brunel, N., Lenne, C., Moulia, B., Julien, J.-L., 2008. *Jr-ZFP2*, encoding a Cys2/His2-type transcription factor, is involved in the early stages of the mechano-perception pathway and specifically expressed in mechanically stimulated tissues in woody plants. Plant Cell Environ. 31, 715–726.

Liu, X., Silins, U., Lieffers, V.J., Man, R., 2003. Stem hydraulic properties and growth in lodgepole pine stands following thinning and sway treatment. Can. J. For. Res. 33, 1295–1303.

Lundqvist, L., Valinger, E., 1996. Stem diameter growth of Scots pine trees after increased mechanical load in the crown during dormancy and (or) growth. Ann. Bot. 77, 59–62.

Lundström, T., Jonas, T., Volkwein, A., 2008. Analyzing the mechanical performance and growth adaptation of Norway spruce using a non-linear finite-element model and experimental data. J. Exp. Bot. 59, 2513–2528.

Martin, L., Leblanc-Fournier, N., Julien, J.L., Moulia, B., Coutand, C., 2010. Acclimation kinetics of physiological and molecular responses of plants to multiple mechanical loadings. J. Exp. Bot. 61, 2403–2412.

Medhurst, J., Ottenschlaeger, M., Wood, M., Harwood, C., Beadle, C., Valencia, J.C., 2011. Stem eccentricity, crown dry mass distribution, and longitudinal growth strain of plantation-grown *Eucalyptus nitens* after thinning. Can. J. For. Res. 11, 2209–2218.

Meng, S.X., Lieffers, V.J., Reid, D.E.B., Rudnicki, M., Silins, U., Jin, M., 2006. Reducing stem bending increases the height growth of tall pines. J. Exp. Bot. 57, 3175–3182.

Meng, S.X., Huang, H., Lieffers, V.J., Nunifu, T., Yang, Y., 2008. Wind speed and crown class influence the height–diameter relationship of lodgepole pine: nonlinear mixed effects modeling. For. Ecol. Man. 256, 570–577.

Metzger, C., 1893. Der Wind als maßgebender Faktor für das Wachsthum der Bäume. Mündener Forstl. Hefte 3, 35–86.

Mickovski, S.B., Ennos, R.A., 2003. The effect of unidirectional stem flexing on shoot and root morphology and architecture in young *Pinus sylvestris* trees. Can. J. For. Res. 33, 2202–2209.

Mitchell, S.J., 2003. Effects of mechanical stimulus, shade, and nitrogen fertilization on morphology and bending resistance in Douglas-fir seedlings. Can. J. For. Res. 33, 1602–1609.

Mitchell, S.J., 2013. Wind as a natural disturbance agent in forests: a synthesis. Forestry 86, 173-157.

Moore, J.R., Crown, D.J., Lee, J.R., McKinley, R.B., Brownlie, R.K., Jones, T.G., Downes, G.M., 2014. The influence of stem guying on radial growth, stem form and internal resin features in *radiata* pine. Trees 28, 1197–1207.

Neel, P.R., Harris, L.W., 1971. Motion-induced inhibition of elongation and induction of dormancy in *Liquidambar*. Science 173, 58–59.

Newcombe, F.C., 1895. The regulatory formation of mechanical tissue. Bot. Gaz. 20, 411–448.

Niklas, K.J., 1998. The influence of gravity and wind on land plant evolution. Rev. Paleobot. Palynol. 102, 1–14.

Pruyn, M.L., Ewers, B., Telewski, F.W., 2000. Thigmomorphogenesis: changes in the morphology and mechanical properties of two *Populus* hybrids in response to mechanical perturbation. Tree Physiol. 20, 535–540.

Quilhó, T., Lopes, F., Pereira, H., 2003. The effect of tree shelter on the stem anatomy of cork oak (*Quercus suber*) plants. IAWA J. 24, 385–395.

Quirk, J.T., Freese, F., 1976a. Effects of mechanical stress on growth and anatomical structure of red pine (*Pinus resinosa* Ait.): compression stress. Can. J. For. Res. 6, 195–202.

Quirk, J.T., Freese, F., 1976b. Effects of mechanical stress on growth and anatomical structure of red pine (*Pinus resinosa* Ait): stem vibration. Can. J. For. Res. 6, 375–381.

Quirk, J.T., Smith, D.M., Freese, F., 1975. Effects of mechanical stress on growth and anatomical structure of red pine (*Pinus resinosa* Ait.): torque stress. Can. J. For. Res 5, 691–699.

Rees, D.J., Grace, J., 1980a. The effect of wind on extension growth of *Pinus contorta* Douglas. Forestry 53, 145–153.

Rees, D.J., Grace, J., 1980b. The effect of shaking on extension growth of *Pinus contorta* Douglas. Forestry 53, 154–166.

Reich, F.P., Ching, K.K., 1970. Influence of bending stress on wood formation of young Douglas fir. Holzforschung 24, 68–70.

Reubens, B., Pannemans, B., Danjon, F., DeProft, M., DeBaets, S., DeBaerdemaeker, J., Poesen, J., Muys, B., 2009. The effect of mechanical stimulation on root and shoot development of young containerized *Quercus robur* and *Robinia pseudoacacia* trees. Trees 23, 1213–1228.

Richter, J., 2006. Influence of site factors on the occurrence of spiral grain in European beech. Forst Holz. 61, 470–472.

Sachs, J., 1873. Lehrbuch der Botanik, nach dem Gegenwärtigen Stand der Wissenschaft. Wilhelm Engelmann, Leipzig, Germany.

Sachs, J., 1882. Vorlesungen über Pflanzen-Physiologie. Wilhelm Engelmann, Leipzig, Germany.

Schmitt, U., Frankenstein, C., Koch, G., 2007. Mechanical stress as stimulus for structurally and chemically altered walls in wood xylem cells. In: Schmitt, U., Singh, A.P., Harris, P.J., The Plant Cell Wall – Recent Advances and New Perspectives, pp. 119–128. Proceedings of the Second New Zealand–German Workshop on Plant Cell Walls, Hamburg, Germany, October 4–6, 2006.

Schulgasser, K., Witztum, A., 2007. The mechanism of spiral grain formation in trees. Wood Sci. Technol. 41, 133–156.

Sievers, A., Buchen, B., Volkmann, D., Heinowicz, Z., 1991. Role of cytoskeleton in gravity perception. In: Lloyd, C.W. (Ed.), The Cytoskeletal Basis of Plant Growth and Form. Academic Press, London, UK. pp. 169–182.

Skatter, S., Kucera, B., 1997. Spiral grain – an adaptation of trees to withstand stem breakage caused by wind-induced torsion. Holz. Roh. Werkst 55 (4), 207–213.

Stokes, A., Fitter, A.H., Coutts, M.P., 1995. Responses of young trees to wind and shading-effects on root architecture. J. Exp. Bot. 46, 1139–1146.

Stokes, A., Nicoll, B.C., Coutts, M.P., Fitter, A.H., 1997. Responses of young Sitka spruce clones to mechanical perturbation and nutrition: effects on biomass allocation, root development, and resistance to bending. Can. J. For. Res. 27, 1049–1057.

Tateno, M., 1991. Increase in lodging safety factor of thigmomorphogenetically dwarfed shoots of mulberry. Physiol. Plant 81, 239–243.

Telewski, F.W., 1989. Structure and function of flexure wood in *Abies fraseri*. Tree Physiol. 5, 113–122.

Telewski, F.W., 1990. Growth, wood density and ethylene production in response to mechanical perturbation in *Pinus taeda*. Can. J. For. Res. 20, 1277–1283.

Telewski, F.W., 1995. Wind induced physiological and developmental responses in trees. In: Coutts, M.P., Grace, J. (Eds.), Wind and Trees. Cambridge University Press, Cambridge, UK. pp. 237–263.

Telewski, F.W., 2006. A unified hypothesis of mechanoperception in plants. Am. J. Bot. 93, 1306–1316.

Telewski, F.W., 2012. Is windswept tree growth negative Thigmotropism? Plant Sci. 184, 20–28.

Telewski, F.W., Jaffe, M.J., 1981. Thigmomorphogenesis: changes in the morphology and chemical composition induced by mechanical perturbation in six month old *Pinus taeda* seedlings. Can. J. For. Res. 11, 380–387.

Telewski, F.W., Jaffe, M.J., 1986a. Thigmomorphogenesis: field and laboratory studies of *Abies fraseri* in response to wind or mechanical perturbation. Physiol. Planta 66, 211–218.

Telewski, F.W., Jaffe, M.J., 1986b. Thigmomorphogenesis: anatomical, morphological and mechanical analysis of genetically different sibs of *Pinus taeda* in response to mechanical perturbation. Physiol. Planta 66, 219–226.

Telewski, F.W., Jaffe, M.J., 1986c. Thigmomorphogenesis: on the physiological response of *Pinus taeda* and *Abies fraseri* to mechanical perturbation: with emphasis on the role of ethylene. Physiol. Planta 66, 227–233.

Telewski, F.W., Pruyn, M.L., 1998. Thigmomorphogenesis: a dose response to flexing in *Ulmus americana* L. seedlings. Tree Physiol. 18, 65–68.

Telewski, F.W., Gardner, B.A., White, G., Plovanich-Jones, A., 1997. Wind flow around multistorey buildings and its influence on tree growth. Plant Biomechanics, Conference Proceedings. The University of Reading, UK, vol. I. 185–192, September 7–12, 1997.

Theophrastus, 1976. In: Einarson, B., Link, G.K.K. (Eds.), *De causisplantarum*: In Three Volumes, Trans. Harvard Univ. Press, Cambridge, MA, USA.

Timell, T.E., 1986a. Compression Wood in Gymnosperms, vol. 1. Springer-Verlag, Berlin, Germany.

Timell, T.E., 1986b. Compression Wood in Gymnosperms, vol. 2. Springer-Verlag, Berlin, Germany.

Timell, T.E., 1986c. Compression Wood in Gymnosperms, vol. 3. Springer-Verlag, Berlin, Germany.

Trendelenburg, R., 1940. Über Faserstauchungen in Hoh und ihre Überwallung durch den Baum. Holz. Roh. Werkst 3, 209–221.

Valinger, E., 1992. Effects of wind sway on stem form and crown development of Scots pine (*Pinus sylvestris* L.). Aust. For. 55, 15–21.

Valinger, E., Lundqvist, L., Sundberg, B., 1995. Mechanical bending stress applied during dormancy and (or) growth stimulates stem diameter growth of Scots pine seedlings. Can. J. For. Res. 25, 886–890.

Waghorn, M.J., Watt, M.S., 2013. Stand variation in *Pinus radiata* and its relationship with allometric scaling and critical buckling height. Ann. Bot. 111, 675–680.

Waghorn, M.J., Watt, M.S., Mason, E.G., 2007. Influence of tree morphology, genetics, and initial stand density on outerwood modulus of elasticity of 17-year-old *Pinus radiata*. For. Ecol. Manage. 244, 86–92.

Watt, M.S., Moore, J.R., McKinlay, B., 2005. The influence of wind on branch characteristics of *Pinus radiata*. Trees 19, 58–65.

Watt, M.S., Downes, G.M., Jones, T., Ottenschlaeger, M., Leckie, A.C., Smaill, S.J., Kimberley, M.O., Brownlie, R., 2009a. Effect of stem guying on the incidence of resin pockets. For. Ecol. Manage. 258, 1913–1917.

Watt, M.S., Clinton, P.C., Parfitt, R.L., Ross, C., Coker, G., 2009b. Modelling the influence of site and weed competition on juvenile modulus of elasticity in *Pinus radiata* across broad environmental gradients. For. Eco. Manage. 258, 1479–1488.

Wilson, B.F., Archer, R.R., 1977. Reaction wood-induction and mechanical action. Ann. Rev. Plant Phys. 28, 23–43.

Wing, M.R., Knowles, A.J., Melbostad, S.R., Jones, A.K., 2014. Spiral grain in bristlecone pines (*Pinus longaeva*) exhibits no correlation with environmental factors. Trees 28, 487–491.

Wrigley, M.P., Smith, G.S., 1978. Staking and pruning effects on trunk and root development of four ornamental trees. N. Z. J. Exp. Agric. 6, 309–311.

Yoder, T.L., Zheng, H.-Q., Todd, P., Staehelin, L.A., 2001. Amyloplast sedimentation dynamics in maize columella cells support a new model for the gravity-sensing apparatus of roots. Plant Physiol. 125, 1045–1060.

Chapter 6

Reaction Wood

Lloyd A. Donaldson, Adya P. Singh

Manufacturing and Bioproducts Group, Scion (New Zealand Forest Research Institute), Rotorua, New Zealand

Chapter Outline

INTRODUCTION

Reaction wood is formed by leaning stems and branches in woody plants including trees and shrubs. Reaction wood forms in response to gravity and is thus part of the gravitropic response of the tree. There are two types of reaction wood: (1) compression wood, which occurs in conifers (softwoods), and (2) tension wood, which occurs in flowering trees (hardwoods). The function of reaction wood is to overcome stem lean thus ensuring the tree grows upright. In a leaning stem the lower side experiences compressive strain. Compression wood overcomes this negative strain by generating compressive forces. The upper side experiences tensile strain. Tension wood overcomes this positive strain by generating tensile forces. In tree-like plants that do not form secondary xylem, such as palms and bamboo, reaction wood does not occur and stems grow vertically due to the response of the shoot tip to light (phototropism). Reaction wood is primarily of interest because it is considered a defect in sawn timber due to its unusual properties and hence has been studied extensively. For detailed reviews the reader is referred to the three-volume monograph on compression wood by Timell (1986) and the review by Donaldson and Singh (2013). For tension wood, useful reviews include Du and Yamamoto (2007), and a very comprehensive review on gelatinous fibers in xylem and other tissues by Mellerowicz

Secondary Xylem Biology. http://dx.doi.org/10.1016/B978-0-12-802185-9.00006-1

and Gorshkova (2012). The review by Felton and Sundberg (2013) includes a detailed discussion on hormone signaling during tension wood formation.

COMPRESSION WOOD

Formation

Compression wood is formed on the lower side of the stem in leaning trees or branches (Fig. 6.1). Its function is to correct the leaning stem back to the vertical position by generating compression stress, which pushes the stem back to the upright position. In branches, compression wood overcomes the strain induced by gravity thus maintaining the position of the branch. Studies in zero-gravity environments have indicated that the strain caused by stem bending induces compression wood (Kwon et al., 2001). However, direct sensing of stem position by cambial cells and by shoot tips also plays a role under the influence of gravity. Plant hormones including ethylene and auxin are involved in signaling roles during compression wood formation (Du and Yamamoto, 2007). The exact mechanism of gravity perception in conifers is unknown but may involve sedimentation of amyloplasts (Nakamura et al., 2001) or mechanosensitive ion channels (Hoson et al., 2005).

Compression wood is easy to recognize on the cut end of a freshly felled log. Stems containing compression wood show eccentric growth, that is, they grow more wood on the compression (lower) side than on the opposite (upper) side,

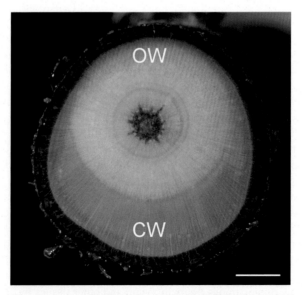

FIGURE 6.1   Compression wood (CW) and opposite wood (OW) in a young pine stem. Scale bar = 2 mm.

so the stem will tend to be oval in shape, and the center of the stem marked by the pith will be closer to the upper side of the stem, a condition known as eccentric pith (Fig. 6.1). Compression wood typically occurs as wide crescent-shaped bands of darker brown wood that occur either within growth rings or as broad areas covering several growth rings. Compression wood may sometimes look like latewood, the darker colored outer part of growth rings. Compression wood can occur in a range of types from mild to severe, depending on how much or for how long the stem is leaning, the growing conditions, and the tree species. Mild compression wood may often have the same appearance as normal wood and detecting it often requires microscopic examination (Yumoto et al., 1983).

Compression wood can be compared to opposite wood (wood from the same growth ring but formed on the opposite side of the stem to the compression wood) or to normal wood (wood formed by a tree that does not contain any compression wood, or from a growth ring that does not contain compression wood). Opposite wood is generally very similar to normal wood in terms of chemistry and anatomy (Timell, 1986).

Compression wood is typical of leaning stems but it can also occur in apparently straight stems when the lean has been corrected, where the stem lean was transient as in the case of wind, or as a result of stimulated growth caused by fertilizer treatment or silviculture for example. Compression wood is often associated with stem wood just below branches and is commonly found on windy sites where it may occur as fine bands within growth rings, resulting from variable weather conditions (Mayr et al., 2005). In seedlings or young trees compression wood may occur as randomly distributed arcs, sometimes in a spiral pattern, formed as stem lean is corrected in response to wind from varying directions, or from bending as a result of rainfall, or irrigation if growing in a nursery for example. Compression wood is found in all conifers, and also occurs in Yew (*Taxus*) and *Ginkgo*, although it may be less distinctive in these trees (Timell, 1983).

Structure

In addition to differences in color and basic density, compression wood has a different microscopic structure (Fig. 6.2) and chemical composition (Table 6.1) compared to normal or opposite wood (Timell, 1982). The characteristics of compression wood may vary with severity but typical features include the following:

- Macroscopic features
  - Increased basic density due to thicker cell walls and smaller diameter tracheids.
  - Darker color due to thicker cell walls and increased lignification.
- Microscopic features
  - Shorter tracheids with thicker cell walls and a rounded shape in transverse view.

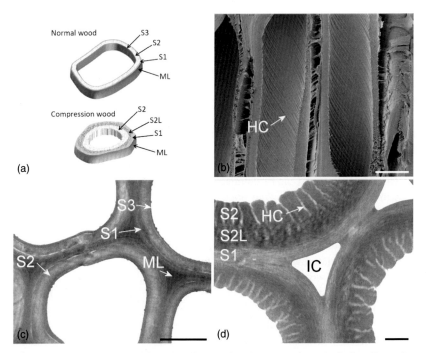

FIGURE 6.2   Ultrastructural features of normal and compression wood of radiata pine. (a) Cell wall layers for normal and compression wood tracheids. (b) Tangential longitudinal scanning electron microscopy (SEM) image of compression wood showing helical cavities on the lumen surface. HC, helical cavities. Scale bar = 20 μm. (c) Transverse image of normal wood (transmission electron microscopy (TEM), permanganate stained). Scale bar = 5 μm. (d) Transverse image of compression wood (TEM, permanganate stained). ML, middle lamella; IC, intercellular space. Scale bar = 2 μm.

TABLE 6.1 Composition of Compression (Pine) and Tension Wood (Poplar) (%)

Composition	MFA	Cellulose	Mannan	Xylan	Galactan	Lignin
Compression wood	>30°	38	8	6	9	36
Opposite wood	<30°	51	13	5	2	27
Tension wood	<5°	54		30		14
Opposite wood	<20°	43		36		19

Donaldson et al., 2004; Donaldson, 2008; Nanayakkara et al., 2009; Mellerowicz and Gorshkova, 2012.

- Tracheids with intercellular spaces at the cell corners.
- A thicker S1 layer in the secondary wall.
- A secondary wall divided into an outer S2L region, which is enriched in lignin and β(1,4)-galactan, and a less lignified inner S2 region containing helical cavities, with no S3 layer.
- High microfibril angle in the secondary cell wall.
- Distorted bordered and cross-field pits.
- Chemical composition
  - Increased lignin content and an increased proportion of *p*-hydroxyphenyl lignin monomer units compared to guaiacyl units.
  - Reduced cellulose, mannan, and xylan content.
  - Presence of significant amounts of β(1,4)-galactan.

In mild compression wood, not all of the mentioned features may be present. Mild compression wood is characterized by an S2L layer that may be restricted to the secondary wall near the corners of the cell resulting in lesser increases in lignin content and β(1,4)-galactan. Intercellular spaces, helical cavities, and a rounded cell shape are typically absent from less severe compression wood (Yumoto et al., 1983; Donaldson et al., 1999; Nanayakkara et al., 2009).

Detecting β(1,4)-galactan requires either chemical analysis (Nanayakkara et al., 2009) or immunocytochemistry (Donaldson and Knox, 2012). However, the increased lignification, which characterizes the S2L layer, can easily be detected by fluorescence microscopy and it is possible to screen samples qualitatively using lignin autofluorescence without any need for staining (Donaldson et al., 2004). Mild compression wood may look like normal wood macroscopically so screening with fluorescence is the most reliable way to detect this type of wood. Very few studies have compared wood properties of mild and severe compression wood, partly because the presence of mild forms has not been effectively detected (Burgert et al., 2004). Confusion may arise where the term "mild compression wood" refers to small amounts of compression wood, rather than compression wood of reduced severity in terms of its chemistry and anatomy. The amount of compression wood can usually be estimated visually or by image analysis (Duncker and Spiecker, 2009).

Severe compression wood will typically have a highly lignified S2L layer that extends around the circumference of the secondary wall (Timell, 1982) as well as helical cavities that follow the microfibril orientation (Fujita et al., 1973). Intercellular spaces are characteristic of the more severe forms of compression wood (Fig. 6.2).

Compression wood has high microfibril angles compared to normal or opposite wood. However, in juvenile wood where the microfibril angle is already high, compression and opposite wood may have similar angles (Donaldson, 2008).

Apart from qualitative detection of mild versus severe compression wood, some studies have attempted to develop quantitative methods for measuring compression wood severity. Two approaches have been used: (1) Chemical analysis, including infrared spectroscopy, can be used to detect increases in lignification and the amount of β(1,4)-galactan, which can be related to severity assessed by microscopy (Nanayakkara et al., 2009; McLean et al., 2014). (2) Fluorescence spectroscopy can also be used to measure changes in the fluorescence emission of compression wood compared to normal wood and the ratio of two fluorescence peaks at blue and green wavelengths has been used as a quantitative measure (Donaldson et al., 2010). Such quantitative measurements have only recently been applied to understanding the relationship between compression wood severity and wood properties, or for screening clones for variation in compression wood severity related to genotype (McLean et al., 2014; Nanayakkara et al., 2014).

Galactan in compression wood is spatially associated with increased lignification and corresponding relative decreases in mannan and xylan (Donaldson and Knox, 2012). Other polysaccharides may show specific changes in compression wood. For example, an α(1–5)-arabinan is associated with intercellular spaces in compression wood of Sitka spruce while β(1–3)-glucan (callose/laricinan) is associated with helical cavities in compression wood of larch and spruce (Altaner et al., 2007).

Genetics

Compression wood is obviously strongly affected by the environment, but there are also indications that it is influenced by genotype although such studies have not been definitive with heritabilities covering a wide range (Apiolaza et al., 2011). Results with clonal trials have been variable but there are indications that different clones form different amounts of compression wood (Nanayakkara et al., 2014). There is often only a weak relationship between stem lean, growth rate, and compression wood severity (Lachenbruch et al., 2010), and there is a real need to compare the response of clones and families to induced stem lean in terms of detailed microscopic and chemical analysis (Nanayakkara et al., 2014).

Because compression wood is readily inducible by simply leaning the stems of small trees, it lends itself to molecular studies to identify differential gene expression associated with compression wood formation. Unfortunately, while many genes are differentially expressed between normal and compression wood, relatively few of these genes can be characterized in terms of function (Yamashita et al., 2008). Many genes involved in compression wood formation are associated with lignification and polysaccharide metabolism including regulatory proteins and transcription factors (Zhang et al., 2000; Bedon et al., 2007; Mast et al., 2010).

Mechanism of Stress Generation

Compression wood formation generates forces that result in correction of stem lean. The exact mechanism is as yet unresolved but changes in chemical

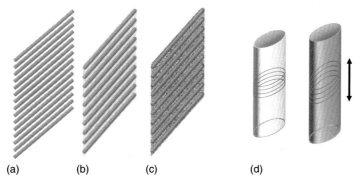

(a)   (b)   (c)   (d)

FIGURE 6.3 **Mechanism of stress generation in compression wood.** The secondary cell wall with a high microfibril angle (a) undergoes swelling as a result of galactan deposition (b), which slightly reduces the microfibril angle, generating a longitudinal compression stress which is fixed in place by lignification (c, d).

composition and structure are clearly involved (Fig. 6.3). Currently, there are two competing theories of stress generation: one depending on matrix swelling (Okuyama et al., 1998) and the other depending on cellulose contraction (Bamber, 2001). Both of these mechanisms will be affected by the increased microfibril angle that typically occurs in compression wood (Burgert et al., 2004; Joffre et al., 2014). Matrix swelling is a strong contender with lignin and galactan involved in this process (Brennan et al., 2012). As $\beta(1,4)$-galactan is deposited into the outer secondary wall, it generates longitudinal compression stress by undergoing expansion between cellulose microfibrils. Stress is then fixed in the cell wall by lignification, which might also contribute to further swelling. Cellulose contraction may also contribute but the exact roles of these various mechanisms remain unclear. Helical cavities are formed in a cytologically controlled way during wall deposition but probably facilitate longitudinal expansion of the secondary wall during matrix swelling as well as dimensional changes during wetting/drying cycles of timber in service.

Wood Properties

Compression wood is darker in color and higher in density compared to normal wood of comparable age. Increased lignin and reduced cellulose content of compression wood results in increased chemical consumption and lower yields during pulping. Increased hardness and brittleness cause difficulties with sawing, drilling, and nailing. However, the most significant property difference in terms of wood utilization is an increase in longitudinal shrinkage in compression wood relative to normal or opposite wood (Leonardon et al., 2010). This difference is due to changes in two cell wall properties, the increased $\beta(1,4)$-galactan content, which makes compression wood more hygroscopic, and the high microfibril angle, which acts to translate matrix swelling/shrinkage resulting from β(1,4)-galactan into longitudinal changes in dimensions with variation in moisture content (Gindl, 2002; Burgert et al., 2004; Yeh et al., 2005; Joffre et al., 2014). High microfibril angles contribute to a reduction in stiffness but this is offset to some extent by increased wood density. Because compression wood is often distributed unevenly, a piece of timber with compression wood on one side and normal wood on the other will have differential shrinkage and will hence tend to warp or bend. Compression wood will thus be associated with distortion of timber during drying or in service.

TENSION WOOD

Formation

Like compression wood, tension wood in hardwood trees forms as a result of stem lean, but unlike compression wood, tension wood forms on the upper side of the leaning stem with corresponding stem eccentricity also on the upper side (Fig. 6.4). Tension wood acts by pulling the stem into the erect position and thus generates tensile stress rather than the compression stress that occurs in compression wood. Not all hardwoods form tension wood; notable exceptions are the vessel-less angiosperms, such as *Pseudowintera* and *Sarcandra* (Aiso et al., 2014), which form compression wood-like reaction wood, and the shrub *Buxus*, which has a highly lignified S2L layer similar to that found in conifer compression wood (Yoshizawa et al., 1993). Some hardwood species have a low propensity for forming reaction wood. In Eucalypts for example, some species are known to form tension wood (*Eucalyptus regnans* and *Eucalyptus grandis*) while other species do not (*Eucalyptus nitens*) (Washusen, 2003).

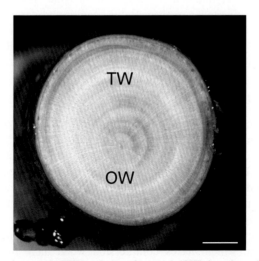

FIGURE 6.4   Tension wood (TW) and opposite wood (OW) in a branch of birch. Scale bar = 2 mm.

Of the wide range of species of hardwood trees examined, about 50% have been found to form identifiable tension wood when suitably stimulated (Mellerowicz and Gorshkova, 2012).

Tension wood is not always visually detectable on logs or sawn timber but in at least some species it can be seen as a dark-colored or furry/rough region of wood sometimes in patches or in arcs (Coutand et al., 2004). Tension wood may occur in milder forms as scattered patches that consist of only a few or even single affected fibers. As with compression wood, comparison can be made to opposite wood or normal wood (Mellerowicz and Gorshkova, 2012).

Stems require a bending stimulus of 24–48 h before tension wood is formed (Jourez and Avella-Shaw, 2003). As with softwoods the exact mechanism of gravity perception is not understood in hardwood trees but because single tension wood fibers can occur in isolation there must be a mechanism localized to individual developing fiber cells.

Plant hormones including ethylene are involved in tension wood induction (Andersson-Gunnerås et al., 2003). Ethylene precursors (ACC – 1-aminocyclopropane-1-carboxylic acid) are reduced on the upper side of leaning stems indicating increased conversion to ethylene corresponding to increased enzyme (ACC oxidase) activity also on the upper side of the stem. ACC oxidase activity is important in controlling asymmetric ethylene production associated with tension wood formation. Tension wood formation may involve auxin and gibberellin although evidence for direct effects of auxin on tension wood formation are sometimes conflicting suggesting that interactions between plant hormones may be important. Exact mechanisms have yet to be elucidated (Du and Yamamoto, 2007).

Structure

Typical tension wood is characterized by the presence of gelatinous fibers or G-fibers (Fig. 6.5). Gelatinous fibers typically have an additional cell wall layer inside the secondary wall called a gelatinous or G-layer, which is unlignified, consisting largely of cellulose (Pilate et al., 2004; Hedenström et al., 2009; Mellerowicz and Gorshkova, 2012) but with some matrix polysaccharides, including pectins and xyloglucan, as well as arabinogalactan proteins (AGP) (Nishikubo et al., 2007; Bowling and Vaughn, 2008). In some species, the lumen surface of the G-layer may be lined with a weakly lignified zone (Gierlinger and Schwanninger, 2006; Donaldson, 2013). Because the G-layer is always preceded by at least one secondary (S) layer, the G-layer is considered to be a tertiary cell wall layer to distinguish it from the secondary cell wall. This layer is often poorly attached to the lignified secondary wall and hence typically forms a gelatinous mass in the lumen as a result of cutting of transverse sections for microscopy (Clair et al., 2005), or as a result of high shrinkage during drying. The G-layer increases in thickness toward the middle of the fiber and is thus thinner at the tips of fibers (Okumura et al., 1977). G-layers

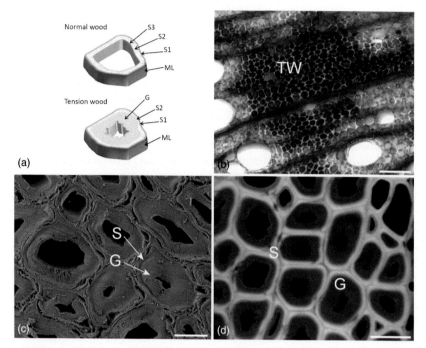

FIGURE 6.5   Ultrastructural features of normal and tension wood cell walls in *E. grandis*. (a) Cell wall layers for normal and tension wood fibers. (b) Tension wood containing gelatinous fibers stained with chlorazol black. Scale bar = 100 μm. (c) Transverse SEM image of tension wood showing thick gelatinous wall (G) and thin secondary wall (S). Scale bar = 5 μm. (d) Transverse section of tension wood stained with safranine and imaged by confocal fluorescence showing a bright lignified secondary wall (S) and a dark unlignified G-layer. Scale bar = 20 μm.

may sometimes be polylaminate (Ruelle et al., 2007b). Gelatinous fibers can be identified by specific stains such a chlorazol black (Fig. 6.5), safranine/astra blue, toluidine blue, or zinc-chlorine-iodide (Vazquez-Cooz and Meyer, 2004; Dogu and Grabner, 2010). Gelatinous fibers can also be identified by the lack of fluorescence due to the absence of lignin (Vazquez-Cooz and Meyer, 2004). It is the gelatinous layer that produces the characteristic furry appearance of tension wood surfaces on sawn timber because the G-layer gets torn out of the fiber during sawing (Coutand et al., 2004).

The composition of the G-layer has been studied by immunohistochemistry (Bowling and Vaughn, 2008; Sandquist et al., 2010; Kim and Daniel, 2012) and microspectrometry (Gierlinger and Schwanninger, 2006). These studies indicate a high content of crystalline cellulose as well as smaller amounts of matrix polysaccharides. In some cases small amounts of lignin or lignin-like materials may be present throughout the G-layer or on its lumen surface. The lignin composition in the lignified secondary wall of gelatinous fibers may be altered compared to normal fibers with a mixed syringyl/guaiacyl lignin in the secondary wall of G-fibers compared to mainly syringyl lignin in normal fibers of poplar (Olsson et al., 2011; Donaldson, 2013). During wood formation, lignification of the S wall is still taking place when the G-layer begins to form (Yoshinaga et al. 2012).

Proteins involved in wall formation, such as lignin biosynthesis enzymes, xyloglucan endotransglucosylase, and fasciclin-like arabinogalactan protein, have been detected in the G-layer. AGPs are characteristic of G-layers (Lafarguette et al., 2004; Kaku et al., 2009).

Cellulose microfibrils in the G-layer are oriented close to the fiber axis and are aggregated in to branched macrofibrils or bundles of microfibrils, forming a very porous network held together by small amounts of matrix polysaccharides (Daniel et al., 2006; Lehringer et al., 2009; Kim and Daniel, 2012). The dominant matrix polysaccharide in the G-layer is xyloglucan as compared to xylan in the secondary cell wall (Table 6.1). Mannans may also be present in the G-layer but at much lower levels than in the S-layers of fibers. Pectins, especially the rhamnogalacturonan-1 type, as well as $\beta(1,4)$-galactan, have been demonstrated immunologically in G-layers of some species (Arend, 2008; Bowling and Vaughn, 2008). The galactan is localized to the S–G boundary suggesting some involvement in interlayer adhesion. It is interesting that increased galactan content is a characteristic of tension and compression wood.

Although typical tension wood is characterized by the presence of G-fibers, it is generally accepted that not all tension wood contains such fibers. A number of studies have demonstrated the absence of G-fibers in leaning stems, although relatively few such studies have measured the stresses present in these stems and confirmed the presence of tensile stress (Clair et al., 2006b; Ruelle et al. 2007a).

Genetics

Because tension wood is readily inducible by stem leaning, studies of gene expression have been carried out to study tension wood formation. Changes in gene expression are associated with cellulose biosynthesis, pectin, and xylan degradation/modification, mannan biosynthesis, and lignification (Decou et al., 2009). Transcription of genes associated with fasciclin-like arabinogalactan proteins is significantly increased in tension wood formation (Déjardin et al., 2004; Djerbi et al., 2004; Lafarguette et al., 2004; Andersson-Gunnerås et al., 2006). Fasciclin may be associated with β-tubulin and hence microfibril orientation (Lafarguette et al., 2004).

Cellulose biosynthesis is enhanced during tension wood formation, involving sucrose synthase, endoglucanase, and three cellulose synthase genes associated with secondary cell wall formation (Déjardin et al., 2004; Bhandari et al., 2006; Lu et al., 2008). These enzymes are coregulated with increased expression on the upper tension side of the stem and suppression on the lower compression side. Cellulose synthase (CesA) expression in developing tension wood may increase to six times the levels found in opposite or normal wood from 6 h to 1 week after induction (Paux et al., 2005).

Genes involved in lignin biosynthesis may also show differential expression in tension wood but trends are less well resolved compared to cellulose synthase perhaps because lignin biosynthesis is a multigene pathway (Paux et al., 2005). Lignin biosynthesis, including the overall phenylpropanoid pathway and genes for laccase involved in monolignol polymerization, are downregulated in bent stems of *Liriodendron* (yellow poplar) (Jin et al., 2011).

MicroRNAs (miRNAs) are a group of small noncoding RNAs playing vital roles in plant development and growth. Some of these miRNAs are associated with tension wood formation where they are up- or downregulated (Lu et al., 2008).

There are few if any quantitative genetics studies of tension wood heritability. The fact that different species of the same genus, *Eucalyptus* for example, seem to have different propensity toward tension wood formation suggests that there is an element of genetic control.

Mechanism

Induction of tension wood by leaning the stem results in significant tensile stress on the upper side of the stem within 2–3 weeks of tilting (Plomion et al., 2003; Yoshida et al., 2003). Shrinkage in the G-layer of up to 4.5% is the likely source of this tensile stress and shrinkage due to residual strain may be up to 10 times that of normal wood (Clair and Thibaut, 2001; Clair et al., 2006). Generation of tensile stress could result from either contraction of vertically oriented cellulose microfibrils, or by lateral swelling of the cellulose network within the G-layer (Goswami et al., 2008; Clair et al., 2011). Xyloglucan-endotransglycosylase enzyme (XET) is localized to the G-layer and the boundary between G- and S-layers during fiber development and in mature wood, and is thought to play an essential role in repairing xyloglucan cross-linkages that break during shrinkage of the G-layer, thus contributing to tension stress generation (Nishikubo et al., 2007). Suppression of xyloglucan formation using fungal xyloglucanase by genetic modification removes the stress generation capability of tension wood (Baba et al., 2009).

Although the exact mechanism of stress generation in tension wood is not fully understood, a number of theories have been proposed including the G-layer swelling (or pressure) hypothesis (Goswami et al., 2008; Burgert and Fratzl, 2009), and the G-layer tension hypothesis (Clair et al., 2006) (Fig. 6.6):

- G-layer swelling may exert a lateral pressure on the surrounding S-layers, which have a relatively high microfibril angle relative to the G-layer. The resulting circumferential enlargement of the S-wall may generate longitudinal shrinkage and hence tensile stress.

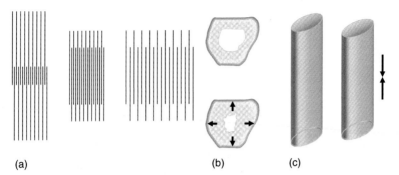

(a)   (b)   (c)

FIGURE 6.6 Mechanism of stress generation in tension wood. The G-layer with a low microfibril angle (a) undergoes shrinkage by a mechanism involving contraction of microfibrils and/or matrix swelling involving xyloglucan and XET, resulting in longitudinal shrinkage and lateral swelling of the G-layer (b) and subsequent generation of tensile stress in the adjacent S-wall (c).

- G-layer shrinkage resulting from changes in cellulose molecular organization may directly generate tensile stress, possibly by interactions between cellulose and matrix polysaccharides. However, given the weak adhesion of the G-layer to the S-wall, exactly how this strain is transmitted to the S-wall remains unclear. It seems likely that strain is transmitted to the S-wall by shear rather than at the fiber tips.

As there is experimental evidence for both behaviors *in situ*, it seems likely that longitudinal shrinkage of the G-layer contributes to G-layer swelling and hence to the production of tensile stress. The absence of lignification would certainly facilitate such behavior as would the very high porosity of the G-layer (Clair et al., 2008). AGPs and pectins may be responsible for generating the swelling (Bowling and Vaughn, 2008).

Properties

Tension wood has high longitudinal shrinkage due to high residual strain, which may lead to timber distortion and in this respect is similar to compression wood (Jourez et al., 2001; Washusen and Ilic, 2001; Clair et al., 2003, 2008; Ruelle et al., 2007a). However, the presence of tension wood also results in difficulties with sawing and surface finishing. Tension wood has a tendency to fibrillate so that planned surfaces lack smoothness (woolly wood) (Coutand et al., 2004). The energy required for sawing is increased by cell wall fibrillation. Both these properties can be attributed to the G-layer of individual fibers, which shows poor adhesion to the S-wall and hence is easily torn out during sawing (Yamamoto et al., 2005). Tension wood also has an increased Young's modulus associated with increased stiffness (Coutand et al., 2004; Yamamoto et al., 2005). Residual stress associated with tension wood may contribute to the formation of shakes and end checks in hardwood logs after felling (Castéra et al., 1994).

The high cellulose content of tension wood makes it a suitable feedstock for biofuel applications, especially in species that can be grown as coppice. Because the G-layer is unlignified and porous, accessibility to cell-wall-degrading enzymes is relatively high without problematic pretreatments (Brereton et al., 2011; Muñoz et al., 2011).

REFERENCES

Aiso, H., Ishiguri, F., Takashima, Y., Iizuya, K., Yokota, S., 2014. Reaction wood anatomy in a vessel-less angiosperm *Sarcandra glabra*. IAWA J. 35, 116–126.

Altaner, C., Knox, J.P., Jarvis, M.C., 2007. *In situ* detection of cell wall polysaccharides in Sitka spruce (*Picea sitchensis* [Bong.] Carrière) wood tissue. Bioresources 2, 284–295.

Andersson-Gunnerås, S., Hellgren, J.M., Björklund, S., Regan, S., Moritz, T., Sundberg, B., 2003. Asymmetric expression of a poplar ACC oxidase controls ethylene production during gravitational induction of tension wood. Plant J. 34, 339–349.

Andersson-Gunnerås, S., Mellerowicz, E.J., Love, J., Segeman, B., Ohmiya, Y., Coutinho, P.M., Nilsson, P., Henrissat, B., Moritz, T., Sundberg, B., 2006. Biosynthesis of cellulose-enriched tension wood in *Populus*: global analysis of transcripts and metabolites identifies biochemical and developmental regulators in secondary wall biosynthesis. Plant J. 45, 144–165.

Apiolaza, L., Chauhan, S.S., Walker, J.C.F., 2011. Genetic control of very early compression and opposite wood in *Pinus radiata* and its implications for selection. Tree Genet. Genom. 7, 563–571.

Arend, M., 2008. Immunolocalisation of (1,4)-β-galactan in tension wood fibers of poplar. Tree Physiol. 28, 1263–1267.

Baba, K., Park, Y.-W., Kaku, T., Kaida, R., Takeuchi, M., Yoshida, M., Hosoo, Y., Ojio, Y., Okuyama, T., Taniguchi, T., Ohmiya, Y., Kondo, T., Shani, Z., Shoseyov, O., Awano, T., Serada, S., Norioka, N., Norioka, S., Hayashi, T., 2009. Xyloglucan for generating tensile stress to bend tree stem. Mol. Plant 2, 893–903.

Bamber, R.K., 2001. A general theory for the origin of growth stresses in reaction wood: how trees stay upright. IAWA J. 22, 205–212.

Bedon, F., Grima-Pettenati, J., Mackay, J., 2007. Conifer R2R3-MYB transcription factors: sequence analyses and gene expression in wood-forming tissues of white spruce (*Picea glauca*). BMC Plant Biol. 7, 17.

Bhandari, S., Fujino, T., Thammanagowda, S., Zhang, D., Xu, F., Joshi, C.P., 2006. Xylem-specific and tension stress-responsive co-expression of KORRIGAN endoglucanase and three secondary wall-associated cellulose synthase genes in aspen trees. Planta 224, 828–837.

Bowling, A.J., Vaughn, K.C., 2008. Immunocytochemical characterization of tension wood: gelatinous fibres contain more than just cellulose. Am. J. Bot. 95, 655–663.

Brennan, M., McLean, J.P., Altaner, C.M., Ralph, J., Harris, P.J., 2012. Cellulose microfibril angles and cell-wall polymers in different wood types of *Pinus radiata*. Cellulose 19, 1385–1404.

Brereton, N.J.B., Pitre, F.E., Ray, M.J., Karp, A., Murphy, R.J., 2011. Investigation of tension wood formation and 2,6-dichlorbenzonitrile application in short rotation coppice willow composition and enzymatic saccharification. Biotechnol. Biofuels 4, 13.

Burgert, I., Fratzl, P., 2009. Plants control the properties and actuation of their organs through the orientation of cellulose fibrils in their cell walls. Integr. Comp. Biol. 49, 69–79.

Burgert, I., Frühmann, K., Keckes, J., Fratzl, P., Stanzl-Tschegg, S., 2004. Structure-function relationships of four compression wood types: micromechanical properties at the tissue and fibre level. Trees 18, 480–485.

Castéra, P., Nepveu, G., Mahé, F., Valentin, G., 1994. A study on growth stresses, tension wood distribution and other related wood defects in poplar (*Populus euramericana* cv I214): end splits, specific gravity and pulp yield. Ann. Sci. For. 51, 301–313.

Clair, B., Thibaut, B., 2001. Shrinkage of the gelatinous layer of poplar and beech tension wood. IAWA J. 22, 121–131.

Clair, B., Jaouen, G., Beauchêne, J., Fournier, M., 2003. Mapping radial, tangential and longitudinal shrinkages and relation to tension wood in discs of the tropical tree *Symphonia globulifera*. Holzforschung 57, 665–671.

Clair, B., Thibaut, B., Sugiyama, J., 2005. On the detachment of gelatinous layer in tension wood fibre. J. Wood Sci. 51, 218–221.

Clair, B., Alméras, T., Yamamoto, H., Okuyama, T., Sugiyama, J., 2006. Mechanical behaviour of cellulose microfibrils in tension wood in relation with maturation stress generation. Biophysical J. 91, 1128–1135.

Clair, B., Ruelle, J., Beauchêne, J., Prévost, M.F., Fournier, M., 2006b. Tension wood and opposite wood in 21 tropical rain forest species. 1. Occurrence and efficiency of G-layer. IAWA J. 27, 329–338.

Clair, B., Gril, J., Di Renzo, F., Yamamoto, H., Quignard, F., 2008. Characterisation of a gel in the cell wall to elucidate the paradoxical shrinkage of tension wood. Biomacromolecules 9, 494–498.

Clair, B., Alméras, T., Pilate, G., Jullien, D., Sugiyama, J., Riekel, C., 2011. Maturation stress generation in Poplar tension wood studied by synchrotron radiation microdiffraction. Plant Physiol. 155, 562–570.

Coutand, C., Jeronimidis, G., Chanson, B., Loup, C., 2004. Comparison of mechanical properties of tension and opposite wood in *Populus*. Wood Sci. Technol. 38, 11–24.

Daniel, G., Filonova, L., Kallas, A.M., Teeri, T.T., 2006. Morphological and chemical characterisation of the G-layer in tension wood fibres of *Populus tremula* and *Betula verrucosa*: labelling with cellulose-binding module CBM1$_{HjCel7A}$ and fluorescence and FE-SEM microscopy. Holzforschung 60, 618–624.

Decou, R., Lhernould, S., Laurans, F., Sulpice, E., Leplé, J.-C., Déjardin, A., Pilate, G., Costa, G., 2009. Cloning and expression analysis of a wood-associated xylosidase gene (PtaBXL1) in poplar tension wood. Phytochemistry 70, 163–172.

Déjardin, A., Leplé, J.-C., Lesage-Descauses, M.-C., Costa, G., Pilate, G., 2004. Expressed sequence tags from poplar wood tissues: a comparative analysis from multiple libraries. Plant Biol. 6, 55–64.

Djerbi, S., Aspeborg, H., Nilsson, P., Sundberg, B., Mellerowicz, E., Blomqvist, K., Teeri, T.T., 2004. Identification and expression analysis of genes encoding putative cellulose synthases (CesA) in the hybrid aspen, *Populus tremula* (L.) × *P. tremuloides* (Michx.). Cellulose 11, 301–312.

Dogu, A.D., Grabner, M., 2010. A staining method for determining severity of tension wood. Turk. J. Agric. For. 34, 381–392.

Donaldson, L.A., 2008. Microfibril angle: measurement, variation and relationships. IAWA J. 29, 345–386.

Donaldson, L.A., 2013. Softwood and hardwood lignin fluorescence spectra of wood cell walls in different mounting media. IAWA J. 34, 3–19.

Donaldson, L.A., Knox, J.P., 2012. Localisation of cell wall polysaccharides in normal and compression wood of radiata pine – Relationships with lignification and microfibril orientation. Plant Physiol. 158, 642–653.

Donaldson, L.A., Singh, A.P., 2013. Structure and formation of compression wood. In: Fromm, J. (Ed.), Cellular Aspects of Wood Formation Plant Cell Monographs. Springer, Heidelberg, pp. 225–256.

Donaldson, L.A., Singh, A.P., Yoshinaga, A., Takabe, K., 1999. Lignin distribution in mild compression wood of *Pinus radiata* D. Don. Can. J. Bot. 77, 41–50.

Donaldson, L.A., Grace, J.C., Downes, G., 2004. Within tree variation in anatomical properties of compression wood in radiata pine. IAWA J. 25, 253–271.

Donaldson, L.A., Radotić, K., Kalauzi, A., Djikanović, D., Jeremić, M., 2010. Quantification of compression wood severity in tracheids of *Pinus radiata* D. Don using confocal fluorescence imaging and spectral deconvolution. J. Struct. Biol. 169, 106–115.

Du, S., Yamamoto, F., 2007. An overview of the biology of reaction wood formation. J. Integrative Plant Biol. 49, 131–143.

Duncker, P., Spiecker, H., 2009. Detection and classification of Norway spruce compression wood in reflected light by means of hyperspectral image analysis. IAWA J. 30, 59–70.

Felton, J., Sundberg, B., 2013. Biology, chemistry and structure of tension wood. In: Fromm, J. (Ed.), Cellular Aspects of Wood Formation Plant Cell Monographs. Springer, Heidelberg, pp. 203–224.

Fujita, M., Saiki, H., Harada, H., 1973. The secondary wall formation of compression wood tracheids. On the helical ridges and cavities. Bull. Kyoto Univ. For. 45, 192-163.

Gierlinger, N., Schwanninger, M., 2006. Chemical imaging of poplar wood cell walls by confocal Raman microscopy. Plant Physiol. 140, 1246–1254.

Gindl, W., 2002. Comparing mechanical properties of normal and compression wood in Norway spruce: the role of lignin in compression parallel to the grain. Holzforschung 56, 395–401.

Goswami, L., Dunlop, J.W.C., Jungnikl, K., Eder, M., Gierlinger, N., Coutand, C., Jeronimidis, G., Fratzl, P., Burgert, I., 2008. Stress generation in the tension wood of poplar is based on lateral swelling power of the G-layer. Plant J. 56, 531–538.

Hedenström, M., Wiklund-Lindström, S., Öman, T., Lu, F., Gerber, L., Schatz, P., Sundberg, B., Ralph, J., 2009. Identification of lignin and polysaccharide modifications in *Populus* wood by chemometric analysis of 2D NMR spectra from dissolved cell walls. Mol. Plant 2, 933–942.

Hoson, T., Saito, Y., Soga, K., Wakabayashi, K., 2005. Signal perception, transduction, and response in gravity resistance. Another graviresponse in plants. Adv. Space Res. 36, 1196–1202.

Jin, H., Do, J., Moon, D., Noh, E.W., Kim, W., Kwon, M., 2011. EST analysis of functional genes associated with cell wall biosynthesis and modification in the secondary xylem of the yellow poplar (*Liriodendron tulipifera*) stem during early stage of tension wood formation. Planta 234, 959–977.

Joffre, T., Neagu, R.C., Bardage, S.L., Gamstedt, E.K., 2014. Modelling of the hygroelastic behaviour of normal and compression wood tracheids. J. Struct. Biol. 185, 89–98.

Jourez, B., Avella-Shaw, T., 2003. Effect of gravitational stimulus duration on tension wood formation in young stems of poplar (*Populus euramericana* cv 'Ghoy'). Ann. For. Sci. 60, 31–41.

Jourez, B., Riboux, A., Leclerq, A., 2001. Comparison of basic density and longitudinal shrinkage in tension wood and opposite wood in young stems of *Populus euramericana* cv. Ghoy when subjected to a gravitational stimulus. Can. J. For. Res. 31, 1676–1683.

Kaku, T., Serada, S., Baba, K., Tanaka, F., Hayashi, T., 2009. Proteomic analysis of the G-layer in poplar tension wood. J. Wood Sci. 55, 250–257.

Kim, J.S., Daniel, G., 2012. Distribution of glucomannans and xylans in poplar xylem and their changes under tension stress. Planta 236, 35–50.

Kwon, M., Bedgar, D.L., Piastuch, W., Davin, L.B., Lewis, N.G., 2001. Induced compression wood formation in Douglas fir (*Pseudotsuga menziesii*) in microgravity. Phytochemistry 57, 847–857.

Lachenbruch, B., Droppelmann, F., Balocchi, C., Peredo, M., Perez, E., 2010. Stem form and compression wood formation in young *Pinus radiata* trees. Can. J. Forest Res. 40, 26–36.

Lafarguette, F., Leplé, J.-C., Déjardin, A., Laurans, F., Costa, G., Lesage-Descauses, M.-C., Pilate, G., 2004. Poplar genes encoding fasciclin-like arabinogalactan proteins are highly expressed in tension wood. New Phytol. 164, 107–121.

Lehringer, C., Daniel, G., Schmitt, U., 2009. TEM/FE-SEM studies on tension wood fibres of *Acer* spp., *Fagus sylvatica* L. and *Quercus robur* L. Wood Sci. Technol. 43, 691–702.

Leonardon, M., Altaner, C.M., Vihermaa, L., Jarvis, M.C., 2010. Wood shrinkage: influence of anatomy, cell wall architecture, chemical composition and cambial age. Eur. J. Wood Prod. 68, 87–94.

Lu, S., Li, L., Yi, X., Joshi, C.P., Chiang, V.L., 2008. Differential expression of three eucalyptus secondary cell wall-related cellulose synthase genes in response to tension stress. J. Exp. Bot. 59, 681–695.

Mast, S., Peng, L., Jordan, T.W., Flint, H., Phillips, L., Donaldson, L., Strabala, T.J., Wagner, A.W., 2010. Proteomic analysis of membrane preparations from developing *Pinus radiata* compression wood. Tree Physiol. 30, 1456–1468.

Mayr, S., Bardage, S., Brändström, J., 2005. Hydraulic and anatomical properties of light bands in Norway spruce compression wood. Tree Physiol. 26, 17–23.

McLean, J.P., Jin, G., Brennan, M., Nieuwoudt, M.K., Harris, P.J., 2014. Using NIR and ATR-FTIR spectroscopy to rapidly detect compression wood in *Pinus radiata*. Can. J. For. Res. 44, 820–830.

Mellerowicz, E.J., Gorshkova, T.A., 2012. Tensional stress generation in gelatinous fibres: a review and possible mechanism based on cell-wall structure and composition. J. Exp. Bot. 63, 551–565.

Muñoz, C., Baeza, J., Freer, J., Mendonça, R.T., 2011. Bioethanol production from tension and opposite wood of *Eucalyptus globulus* using organosolv pretreatment and simultaneous saccharification and fermentation. J. Ind. Microbiol. Biotechnol. 38, 1861–1866.

Nakamura, T., Negishi, Y., Funada, R., Yamada, M., 2001. Sedimentation amyloplasts in starch sheath cells of woody stems of Japanese cherry. Adv. Space Res. 27, 957–960.

Nanayakkara, B., Manley-Harris, M., Suckling, I.D., Donaldson, L.A., 2009. Quantitative chemical indicators to assess the gradation of compression wood. Holzforschung 63, 431–439.

Nanayakkara, B., Lagane, F., Hodgkiss, P., Dibley, M., Smaill, S., Riddell, M., Harrington, J., Cown, D., 2014. Effects of induced drought and tilting on biomass allocation, wood properties, compression wood formation and chemical composition of young *Pinus radiata* genotypes (clones). Holzforschung 68, 455–465.

Nishikubo, N., Awano, T., Banasiak, A., Bourquin, V., Ibatulin, F., Funada, R., Brumer, H., Teeri, T.T., Hayashi, T., Sundberg, B., Mellerowicz, E.J., 2007. Xyloglucan endo-transglycosylase (XET) functions in gelatinous layers of tension wood fibers in Poplar – A glimpse into the mechanism of the balancing act of trees. Plant Cell Physiol. 48, 843–855.

Okumura, S., Harada, H., Saiki, H., 1977. Thickness variation of the G layer along a mature and a differentiating tension wood fiber in *Populus euramericana*. Wood Sci. Technol. 11, 23–32.

Okuyama, T., Takeda, H., Yamamoto, H., Yoshida, M., 1998. Relation between growth stress and lignin concentration in the cell wall: ultraviolet microscopic spectral analysis. J. Wood Sci. 44, 83–89.

Olsson, A.-M., Bjurhager, I., Gerber, L., Sundberg, B., Salmén, L., 2011. Ultrastructural organisation of cell wall polymers in normal and tension wood of aspen revealed by polarisation FTIR microspectroscopy. Planta 233, 1277–1286.

Paux, E., Carocha, V., Marques, C., Mendes de Sousa, A., Borralho, N., Sivadon, P., Grima-Pettenati, J., 2005. Transcript profiling of *Eucalyptus* xylem genes during tension wood formation. New Phytol. 167, 89–100.

Pilate, G., Chabbert, B., Cathala, B., Yoshinaga, A., Leplé, J.-C., Laurans, F., Lapierre, C., Ruel, K., 2004. Lignification and tension wood. CR Biol. 327, 889–901.

Plomion, C., Pionneau, C., Baillères, H., 2003. Analysis of protein expression along the normal to tension wood gradient in *Eucalyptus gunnii*. Holzforschung 57, 353–358.

Ruelle, J., Beauchene, J., Thibaut, A., Thibaut, B., 2007a. Comparison of physical and mechanical properties of tension and opposite wood from ten tropical rainforest trees from different species. Ann. For. Sci. 64, 503–510.

Ruelle, J., Yoshida, M., Clair, B., Thibaut, B., 2007b. Peculiar tension wood structure in *Laetia procera* (Poepp.) Eichl. (Flacourtiaceae). Trees 21, 345–355.

Sandquist, D., Filonova, L., von Schantz, L., Ohlin, M., Daniel, G., 2010. Microdistribution of xyloglucan in differentiating poplar cells. BioResources 5, 796–807.

Timell, T.E., 1982. Recent progress in the chemistry and topochemistry of compression wood. Wood Sci. Technol. 16, 83–122.

Timell, T.E., 1983. Origin and evolution of compression wood. Holzforschung 37, 1–10.

Timell, T.E., 1986. Compression Wood in Gymnosperms. Springer-Verlag, Berlin.

Vazquez-Cooz, I., Meyer, R.W., 2004. Occurrence and lignification of libriform fibers in normal and tension wood of red and sugar maple. Wood Fibre Sci. 36, 56–70.

Washusen, R., 2003. The relationship between longitudinal growth strain, tree form and tension wood at the stem periphery of ten- to eleven-year-old *Eucalyptus globulus* Labill. Holzforschung 57, 308–316.

Washusen, R., Ilic, J., 2001. Relationship between transverse shrinkage and tension wood from three provenances of *Eucalyptus globulus* Labill. Holz Roh- Werkst. 59, 85–93.

Yamamoto, H., Abe, K., Arakawa, Y., Okuyama, T., Grill, J., 2005. Role of the gelatinous layer (G-layer) on the origin of the physical properties of the tension wood of *Acer sieboldianum*. J. Wood Sci. 51, 222–233.

Yamashita, S., Yoshida, M., Yamamoto, H., Okuyama, T., 2008. Screening genes that change expression during compression wood formation in *Chamaecyparis obtusa*. Tree Physiol. 28, 1331–1340.

Yeh, T.-F., Goldfarb, B., Chang, H.-M., Peszlen, I., Braun, J.L., Kadla, J.F., 2005. Comparison of morphological and chemical properties between juvenile wood and compression wood of loblolly pine. Holzforschung 59, 669–674.

Yoshida, M., Ikawa, M., Kaneda, K., Okuyama, T., 2003. Stem tangential strain on the tension wood side of *Fagus crenata* saplings. J. Wood Sci. 49, 475–478.

Yoshinaga, A., Kusumoto, H., Laurans, F., Pilate, G., Takabe, K., 2012. Lignification in poplar tension wood lignified cell wall layers. Tree Physiol. 32, 1129–1136.

Yoshizawa, N., Satoh, M., Yokota, S., Idei, T., 1993. Formation and structure of reaction wood in *Buxus microphylla* var. *insularis* Nakai. Wood Sci. Technol. 27, 1–10.

Yumoto, M., Ishida, S., Fukazawa, K., 1983. Studies on the formation and structure of the compression wood cells induced by artificial inclination in young trees of *Picea glauca* IV. Gradation of the severity of compression wood tracheids. Res. Bull. Coll. Exp. For. Hokkaido Univ. 39, 409–454.

Zhang, Y., Sederoff, R.R., Allona, I., 2000. Differential expression of genes encoding cell wall proteins in vascular tissues from vertical and bent loblolly pine trees. Tree Physiol. 20, 457–466.

Part II

Function and Pathogen Resistance of Secondary Xylem

Chapter 7

Bordered Pit Structure and Cavitation Resistance in Woody Plants

Yuzou Sano
Laboratory of Woody Plant Biology, Research Faculty of Agriculture, Hokkaido University, Sapporo, Japan

Chapter Outline

INTRODUCTION

A pit is a small hole in the secondary wall of a plant cell (Fig. 7.1). The essential components of a pit are the pit cavity and the pit membrane. The pit cavity is an empty space in the secondary wall and is continuous with the cell lumen. It is formed as a result of the localized absence of deposition of the secondary wall during the thickening of the cell wall (Fig. 7.2). The pit membrane is the outer layer of the cell wall, which partitions the pit cavity from the adjacent cell. Pits can be divided into simple pits and bordered pits on the basis of the morphology of the pit cavity. A simple pit is a pit in which the width of the pit cavity

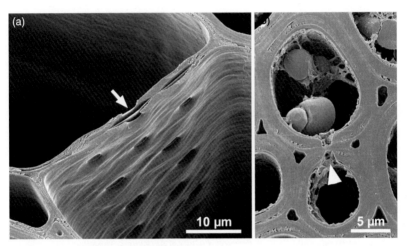

FIGURE 7.1 Scanning electron micrographs of pits. (a) Bordered pits between vessel elements in *Populus sieboldii*. Arrow: bordered pit pair. (b) Simple pits between ray parenchyma cells in *Q. crispula*. Arrowhead: simple pit pair.

is consistent or just slightly changes toward the cell lumen (Fig. 7.1b). On the other hand, a bordered pit is a pit in which the pit cavity becomes noticeably narrower toward the cell lumen (Fig. 7.1a). The constricted opening between pit cavity and cell lumen is called pit aperture, and the overhanging secondary wall that covers the pit cavity is called pit border. In the case of bordered pits, the pit cavity covered by the dome-like pit border is also called pit chamber (Panshin and de Zeeuw, 1980; Evert, 2006).

In general, a pit of one cell is located exactly opposite a pit of an adjacent cell wall. Such a complementary pair of pits is called a pit pair (Fig. 7.2). In common walls between water-conducting elements, such as tracheids and

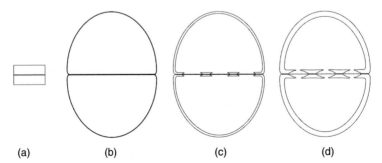

(a)　　　　　(b)　　　　　(c)　　　　　(d)

FIGURE 7.2 Schematic drawing of bordered pit formation between adjacent vessel elements viewed in transverse section. (a) Future vessel elements at undifferentiating stage in cambial zone. (b) Forming vessel elements that reached full size. (c) Early stage of secondary wall deposition. (d) Mature vessel elements. Certain cell wall components that were present during the formation are selectively removed by enzymes at the final stage, and the pit membranes become porous enough to permit water flow.

vessel elements, bordered pit pairs are formed. Some of the cell wall components that temporally existed during the process of cell wall formation are removed from the pit membranes at the final stage of cell differentiation (Fig. 7.2) (Thomas, 1968; O'Brien, 1970; Imamura and Harada, 1973; Kim et al., 2011; Kim and Daniel, 2013). Consequently, the pit membranes are porous enough to allow sufficient water flow from one element to the adjacent element. The pit borders also have vital roles. This structure insures mechanical strength while contributing to expose a larger permeable area of the pit membrane (Zimmerman, 1983).

Bordered pit pairs play an important role in water conduction in living trees and in the penetration of liquid into timber. Intensive efforts have been made to clarify their structures. Recently, attention has been especially focused on how the water flow and progression of cavitation are regulated at bordered pit membranes, and new findings have been gained to deepen our understanding on the mechanisms. In this chapter, I will review the recent findings in addition to providing a brief overview of our knowledge on the basic structure of pits and its association with cavitation resistance.

STRUCTURE OF BORDERED PIT MEMBRANES

Bordered Pit Membranes in Conifers

Torus (pl. tori) and Margo

Bordered pits are formed in (longitudinal) tracheids, ray tracheids, and strand tracheids in conifers. In general, bordered pits between tracheids are present almost exclusively in radial walls of tracheids except for the growth ring boundary and the outermost part of the growth ring. The size of the pits and their pit membrane porosity differ between earlywood and latewood (Fig. 7.3a). The intertracheary pit membranes in earlywood are characterized by clear distinction between the torus and margo (Fig. 7.3a, b). A torus is the thickened central region of the pit membrane. A margo is the porous outer region of the pit membrane. The periphery of the pit membrane is also thickened and called rim or annulus (Fig. 7.3c). In margo, wider strands of cellulose microfibrils that radiate from the torus to annulus suspend the torus. Finer microfibril strands that bridge between the radiating wider strands are deposited. In water-conducting sapwood, water flows from a tracheid to an adjacent tracheid through the porous margo. In general, the flow resistance per pit area between tracheids in conifers is approximately 60 times lower than that between vessels in angiosperms because of the porous nature of the margo (Pittermann et al., 2005). A recent computational model analysis demonstrated that wider openings located in the inner region of the margo contribute to the flow through an intertracheary pit for the most part (Schulte, 2012).

Classification

The intertracheary pit membranes between tracheids have been classified into several types based on structural characters (e.g., Harada, 1964; Liese, 1965;

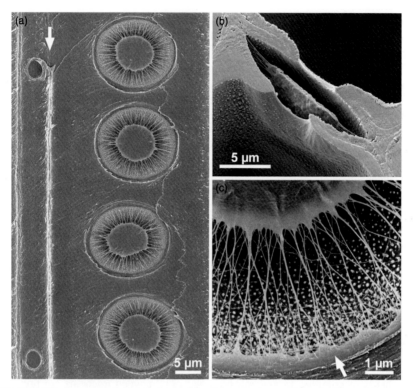

FIGURE 7.3 **Scanning electron micrographs of intertracheary pits in conifer.** (a) Pit membranes in *Abies sachalinensis*. Arrow: growth ring boundary. (b) Pit pair in *Pinus densiflora*. (c) Portion of pit membrane in *A. sachalinensis*. Arrow: "annulus" or "rim."

Bauch et al., 1972; Fujikawa and Ishida, 1972). Although the proposals of the classifications are not consistent among the researchers, it seems reasonable to roughly divide the pit membranes into two types. One type is characterized by the deposition of amorphous materials on tori. The outline of the torus is distinct in this type (Figs 7.3c and 7.4a). This type is found in all species of Pinaceae and some taxa of Cupressaceae (e.g., *Juniperus*, *Widdringtonia*), Cephalotaxaceae (e.g., *Cephalotaxus*), and Podocarpaceae (e.g., *Nageia*). In another type, no deposition of amorphous materials occurs on the tori, and the texture of microfibrils, which are continuous from the margo region, are apparent on torus surfaces without any extraction treatments (Fig. 7.4b). The intertracheary pit membranes of conifers have been further classified into several types based on morphological features such as density of microfibrilar strands in margo and the thickness of tori (e.g., Liese, 1965; Bauch et al., 1972). However, such characters are variable, and it is well known that there are considerable differences in the margo porosity and outline of tori between earlywood and latewood. A study showed that width and density of microfibrilar strands in margo differ between

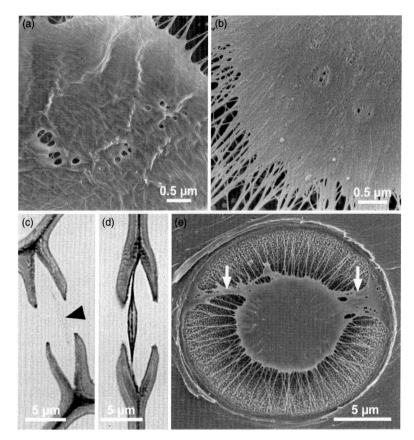

FIGURE 7.4   Scanning electron and ultraviolet micrographs of intertracheary pits in conifer.
(a, b) Portions of pit membranes of *A. sachalinensis* and *Cryptomeria japonica*, respectively.
(c, d) Pit pairs of outer sapwood and heartwood of *C. japonica*, respectively. Arrowhead: unlignified torus. (e) Pit membrane with torus extensions (arrows) in *A. sachalinensis*.

trees grown in bright open field and those grown in dark understory (Schoonmaker et al., 2010).

Water Flow Through a Pit Pair

A torus is much denser than the net-like margo, and it has often been generalized in literatures that tori are impermeable. However, some reports demonstrated that minute pores are apparent in tori with deposition of amorphous
materials (Thomas, 1969; Fujikawa and Ishida, 1972; Sano et al., 1999; Jansen
et al., 2012). It is considered that such minute pores are remnants of plasmo-
desmata, which are present during cell wall formation (e.g., Thomas, 1969).
Minute openings have also been shown to exist in shallow depressions that are
present in tori without deposition of amorphous materials (Fig. 7.4b) (Sano and

Nakada, 1998). These findings suggest that cell wall components are less densely packed in the inside of the torus, as compared with normal lignified cell walls. It is not surprising that the tori are sparsely packed with cell wall components because enzymatic digestion of certain matrix components occurs in the pit membranes at the final stage of cell wall formation (Thomas, 1968; O'Brien, 1970; Imamura and Harada, 1973; Kim et al., 2011; Kim and Daniel, 2013). Indeed, aromatic compounds impregnate into the torus during the course of heartwood formation (Fig. 7.4c, d) (Sano and Nakada, 1998).

Torus Extension

In some species, partial deposition of amorphous materials that are continuous between torus and annulus are commonly present in the margo region (Fig. 7.4e). This type of bridging by amorphous materials in margo regions is called torus extension or extended torus (e.g., Sano et al., 1999 and literatures cited therein; Pittermann et al., 2010; Schoonmaker et al., 2010; Plavcová et al., 2013). It seems likely that torus extension is exclusively present in tori with deposition of amorphous materials. The presence or absence of torus extension is used as a character to identify conifer wood (IAWA Committee, 2004).

Pit Membranes Mediated with Ray Tracheids and in Latewood

Intertracheary pit membranes between tracheids and ray tracheids are similar to those between (longitudinal) tracheids, but the strands of microfibrils in margo between tracheids and ray tracheids are apparently much denser than those between tracheids (e.g., Thomas, 1969). Intertracheary pit membranes in latewood differ in size and in the density of microfibrilar strands (or porosity) in margo, from those in earlywood (Fig. 7.3b). The distinction between torus and margo is often unclear due to the presence of amorphous materials in the entire region of the pit membranes.

Bordered Pit Membranes in Angiosperms

Layered Structure

Studies by the pioneers of transmission electron microscopic studies of wood structures showed consistently that the bordered pit membranes are homogeneous structures with a primary wall texture, without distinction between torus and margo, consisting of several layers and lacking visible openings that can be detected by transmission electron microscopy (e.g., Côté, 1958; Harada et al., 1958; Schmid and Machado, 1968). As a consequence, it has often been generalized in literatures that bordered pit membranes consist of two primary wall layers with a middle lamella between them. However, wide variations in the basic structures have been revealed to date.

The presence or absence of the middle lamella is recognizable by checking the sections. Figure 7.5a shows an ultrathin section of intervessel pit. It is apparent that the primary walls are quite thin and adhere to each other except

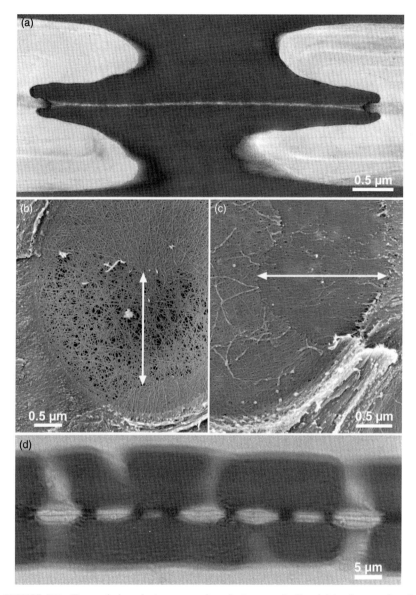

FIGURE 7.5   Transmission electron, scanning electron, and ultraviolet micrographs of bordered pits in angiosperms. (a) An intervessel pit pair in *P. sargentii*, stained with KMnO4. (b, c) Portions of intertracheary pit membrane in *Q. crispula* and intervessel pit membrane in *F. mandshurica* var. *japonica*, respectively. Double sided arrows: portions from which primary walls peeled away. (d) Intervessel pits in current year growth ring of *F. mandshurica* var. *japonica*.

for the peripheral region, and that the middle lamella is absent. Such layered structures are also confirmable by fractography (Schmid and Machado, 1968; Sano, 2005). Figure 7.5b,c shows portions of intervessel pit membranes of which primary walls are partly peeled away. The middle lamella is invisible on the peeled part in the case of *Quercus crispula*, whereas a layer whose texture is obviously different from the true primary wall is visible beneath the primary walls in *Fraxinus mandshurica* var. *japonica*. Although records concerning the layered structures of intervessel pit membranes are limited, it seems likely that the presence of the middle lamella in intervessel pit membranes is not so common among angiospermous woods. Even if the middle lamella is present between primary walls, the pit membranes are not lignified in the outer layer of sapwood, differently from the middle lamella in nonpit regions, which is highly lignified (Fig. 7.5d) (Sano and Fukazawa, 1994).

Torus and Pseudotorus

It was believed that bordered pit membranes are exclusively homogeneous without distinction between torus and margo in angiospermous woods until Ohtani and Ishida (1978) first reported the presence of torus-bearing pit membranes in species belonging to the genera *Osmanthus* and *Daphne*. After their discovery, torus-bearing pit membranes have been recorded for several taxa (Fig. 7.6a, b). According to Dute and Elder (2011), torus-bearing pit membranes have been recorded for over 80 species within 13 genera representing 5 families (Cannabaceae, Oleaceae, Rosaceae, Thymelaceae, and Ulmaceae). My colleagues and I further found that torus-bearing pit membranes are present in bordered pits between wood fibers (tracheids) in *Schisandra chinensis*, Schisandraceae (Sano et al., 2013). Such homoplastic occurrences in a wide range of angiosperms suggest that torus-bearing pit membranes are more common than has been acknowledged.

The nature of tori in angiosperms is more complicated than those in conifers. Timing of torus thickenings during differentiation differs among species (Dute et al., 2008 and literature cited therein). In addition, chemical composition of tori seems to differ among taxa (Coleman et al., 2004).

In some taxa, anomalous thickenings that are similar to but different from true tori are present in bordered pit membranes in tracheary elements. Since Parameswaran and Liese (1973) first recorded that the torus-like thickenings exist in *Ribes sanguineum* (Grossulariaceae), similar thickenings and related structures have been also found in species belonging to Elaeagnaceae, Ericaseae, Oleaceae, Rhamnaceae, and Rosaceae (Rabaey et al., 2008 and literatures cited therein). The term "pseudotorus" has been used to refer collectively to such structures (Rabaey et al., 2006, 2008; Jansen et al., 2007).

The "pseudotori" seem to be classified into a few types. The pseudotori found in *Prunus sargentii* are unique. They look like rounded plates that partly overhang pit membranes from pit membrane annulus (Fig. 7.6c–e). The pit membranes are partly missing or extraporous beneath the plate (Fig. 7.6d).

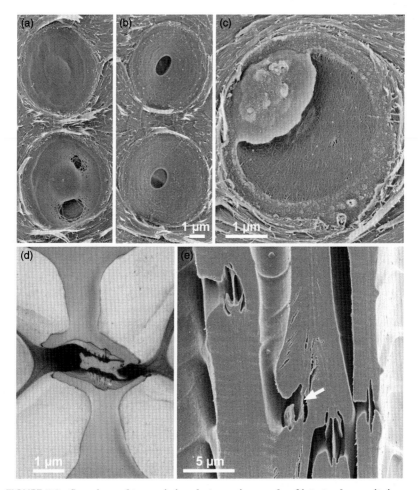

FIGURE 7.6 **Scanning and transmission electron micrographs of intertracheary pits in angiosperms.** (a, b) A complimentary pair of fractured plain between tracheary elements of *Cercocarpus* sp. (c–e) Interfiber pits with pseudotori in *P. sargentii*. (d) Stained with PATAg. Arrow: pit membrane aspirated to pseudotorus.

This type of pseudotori might close such extraporous zone in the pit membranes when pit aspiration occurs (Fig. 7.6e, arrow). If so, it is more appropriate to refer to the plate-like thickenings as "pseudopit borders."

Porosity of Intervascular Pit Membranes

In contrast to early electron microscopic studies that failed to visualize openings in bordered pit membranes (e.g., Côté, 1958; Harada et al., 1958; Schmid and Machado, 1968), Bonner and Thomas (1972) clearly demonstrated the porous nature of intervascular pit membranes of *Liriodendron tulipifera*.

Then, intervascular pit membranes with visible micropores have been recorded for various taxa (e.g., Wheeler, 1982; Meylan and Butterfield, 1982; Sano, 2005). Some studies have quantitatively shown the interspecific differences in the porosity of intervessel pit membranes, and their association with cavitation resistance (e.g., Sperry and Tyree, 1988; Choat et al., 2003; Jansen et al., 2009). The porosity of intervessel pit membranes considerably differ not only among species but also among individual pits within a single tree (Sano, 2004, 2005). According to my observations, localized extraporous regions with single to a few wide pores up to several hundred nanometers are present in a few to 20% of homogeneous intervascular pit membranes in some taxa (e.g., *Acer, Betula, Cercocarpus, Fagus, Osmanthus, Quercus, Salix, Tilia, Trochodendron,* and *Ulmus*) (Fig. 7.7a), whereas such small extraporous regions are never seen in some species (e.g., *Fraxinus* spp., *Amborella trichopoda, Robinia pseudoacacia,* and *Juglans mandshurica*). Such extraporous regions tend to appear near the periphery of pit membrane (Sano, 2004, 2005; Hacke et al., 2007). We are able to check if such extraporous regions are artificially induced by partial removal of the surface layer of the pit membrane when wood pieces were split during sample preparation for SEM, by observing complementary pairs of the split planes. For example, Fig. 7.7b,c shows a complementary pair of fractured plains, which are arranged like an opened book. Fragments of intertracheary pit membranes that correspond to holes in pit membranes are seen on pit borders. This is an indication that holes seen in the pit membranes (Fig. 7.7b) are artificially induced. In Fig. 7.6a,b, by contrast, no fragments of pit membranes that correspond to the extraporous parts are seen on the opposite side in the similar pair. In this case, it is certain that the extraporous regions in the pit membranes are not induced by splitting, although we cannot exclude the possibility of artifacts caused by other phenomena. In any instance, it is likely that the structure and the physical properties differ between intervessel pit membranes with and without such extraporous regions, and that the differences affect the conductivity and cavitation resistance.

Matrix Substances in Pit Membranes

Zwieniecki et al. (2001) demonstrated that the flow rate through stem segments was reversibly enhanced by changing the liquids from deionized water to dissociating solutes, such as KCl and NaCl. They hypothesized that such reversible enhancement of liquid flow was caused by the reversible shrinkage of the pectin that covered the cellulose microfibrils in the intervessel pit membranes. However, recent transmission electron microscopic studies using immunolabeling techniques indicated consistently that pectin is absent from intervessel pit membranes except for the thickened part at the periphery, namely, the pit membrane annulus (Plavcová and Hacke, 2011; Plavcová et al., 2011; Kim et al., 2011; Kim and Daniel, 2013) although there are evidences indicating that pectin surely exists in the tori of conifers (e.g., Thomas, 1975). The presence of unknown extraneous materials that cover intervessel pits has been also found in some

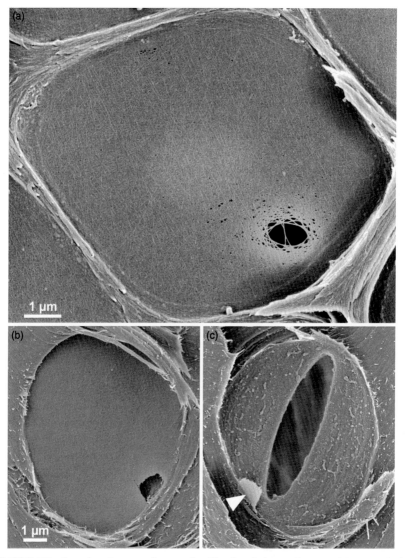

FIGURE 7.7 Scanning electron micrographs of bordered pit membranes. (a) Intervessel pit membrane with extraporous region in outer sapwood in *Pop. sieboldii*. (b, c) A complimentary pair of fractured plain between tracheids of *Am. trichopoda*. Arrowhead: fragment of pit membrane.

taxa (Sano, 2005). Moreover, analysis of intervessel pit membrane surfaces by atomic force microscopy has shown that the surface texture apparently changes from smooth to rough when the hydrated pit membranes were dried (Pesacreta et al., 2005). Further studies are needed to clarify the chemical components consisting the pit membranes and their real configuration in the living trees.

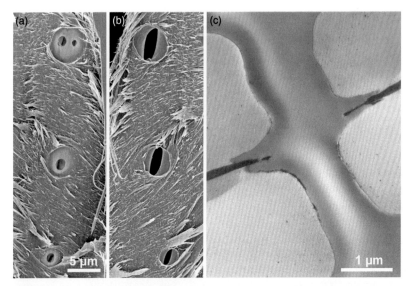

FIGURE 7.8   Scanning and transmission electron micrographs of intertracheary pits in angiosperms. (a, b) A complimentary pair of fractured plain between fiber-tracheids of *Tilia japonica*. (c) Interfiber pit with perforated pit membrane in *P. sargentii*. Stained with PATAg.

Perforated Pit Membranes

A perforated pit membrane is a vestigial pit membrane of which the central part is largely missing or very sparse (Fig. 7.8). The presence of perforated pit membranes between imperforate tracheary elements has been demonstrated in a wide range of angiosperms from basal to derived groups (Thomas, 1976; Sano and Fukuzawa, 1994; Sano and Jansen, 2006; Sano et al., 2011, 2013). According to the studies discussed earlier, the perforated pit membranes tend to occur in smaller interfiber pits of which pit borders develop minutely. The occurrence of perforated interfiber pit membranes is closely associated with the pitting between the fiber cells and vessel elements in woods with such direct contacts. Pits are absent or very rare, or blind pits are present on either side of the common walls between vessel elements and wood fibers with perforated interfiber pit membranes. By contrast, pit pairs are commonly present in vessel elements and wood fibers with homogeneous interfiber pit membranes (Sano et al., 2008). The type of interfiber pit membranes are closely associated with the specialization and the conductive nature of the fiber cells (Sano et al., 2011).

CAVITATION RESISTANCE

Air Seeding

Bordered pit pairs work not only as a pathway for water flow between cells but also as a barrier to prevent the spread of cavitation and embolism in a conductive

Transpiration

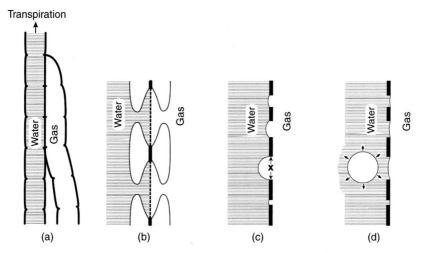

(a)   (b)   (c)   (d)

FIGURE 7.9   Schematic drawing of "air seeding" proposed by Zimmerman (1983). (a) A water-filled (functioning) vessel and a gas-filled (dysfunctional) vessel that are partitioned by bordered pit pairs. (b) An enlargement of the intervessel pit pairs. (c) Just before air seeding. The meniscus is not small enough to pass through the widest pore (x) in the pit membranes. (d) Immediately after air seeding. The air bubble grows because negative pressure continues to generate by transpiration.

system of living trees. Micropores in bordered pit membranes are so small for the surface tension of water that a gas in a cavitated cell does not easily enter an adjacent water-filled cell via pit pairs between the cells. The gas entry (air seeding) from a cavitated cell to an adjacent water-filled cell occurs when the pressure difference between the cells exceeds a certain limit and the diameter of meniscus in the micropore in the pit membrane becomes smaller than the diameter of the micropores (Fig. 7.9). The pressure that allows air seeding (ΔP) across bordered pit membranes is given by the following equation (Zimmerman, 1983; Sperry and Tyree, 1988; Choat et al., 2008):

$$\Delta P = \frac{4\tau\cos\theta}{D} \tag{7.1}$$

where D is the diameter of the widest micropore in pit membranes (μm), τ is the surface tension of water (sap in conductive cells) (0.072 N/m at 20°C), and θ is the contact angle at the surface of the pit membrane (°). θ is usually assumed to be 0° because pit membranes in conductive sapwood are not lignified and hydrophilic. In living trees, air seeding occurs from a cavitated cell to a water-filled conductive cell when negative pressure increases by transpiration and the pressure difference between the cells exceeds ΔP. After an air seeding occurs, the air bubble will grow and finally embolize the cell lumen because a strong negative pressure is continuously generated due to transpiration.

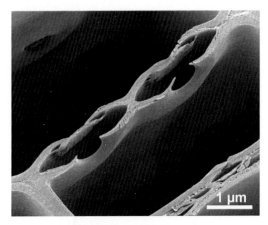

FIGURE 7.10 Scanning electron micrograph of aspirated intertracheary pit membranes in *Larix kaempferi.*

Pit Structure and Cavitation Resistance in Conifers

When bordered pit membranes are flexible enough, they are aspirated to a pit border of water-filled cells (e.g., Zimmerman, 1983). In the case of aspirated intertracheary pits in conifers, one side of pit apertures is tightly sealed by the tori (Fig. 7.10), and air-seeding pressure greatly increases. Regarding this phenomenon, studies in conifers found that torus overlap (O) or the ratio of torus to pit aperture diameter is positively related with the resistance to cavitation or water stress (Domec et al., 2008; Hacke and Jansen, 2009; Delzon et al., 2010; Pittermann et al., 2010). The torus overlap is defined as Eq. (7.2) by Hacke et al. (2004) and as Eq. (7.3) by Delzon et al. (2010):

$$O = \frac{D_t - D_a}{D_m - D_a} \qquad (7.2)$$

$$O = \frac{D_t - D_a}{D_t} \qquad (7.3)$$

where D_t is the diameter of the torus (μm), D_a is the diameter of the pit aperture (μm), and D_m is the diameter of the pit membrane (μm). In both cases, the larger the value is, the wider the contact area of the torus with pit aperture is when pit aspiration occurs, as well as the ratio of torus to pit aperture. By contrast, Jansen et al. (2012) reported that there was no significant relationship between O and the cavitation resistance as a result of examinations on 33 conifer species representing 19 genera from 5 families. Alternatively, they found that cavitation resistance significantly differed between species with punctured tori, as shown in Fig. 7.4a, and species without punctured tori.

Attempts have been made to clarify the relationship between other micromorphological characters, such as torus extension and torus thickness, and cavitation resistance (Delzon et al., 2010; Pittermann et al., 2010; Schoonmaker et al., 2010; Jansen et al., 2012). However, the results are inconsistent among studies.

Pit Structure and Cavitation Resistance in Angiosperms

According to the air-seeding theory, porosity of bordered pit membranes is a determinant of cavitation resistance. Therefore, attempts have been made to confirm the coincidence between the micropore size of the pit membrane measured explicitly and those calculated according to the capillary Eq. (7.1) and stem vulnerability measurements for homogeneous intervessel pit membranes in angiosperms. Some studies reported that real pore size roughly corresponded with theoretical pore size predicted based on air-seeding pressure (Sperry and Tyree, 1988; Jarbeau et al., 1995). However, studies showed that the pore diameter, as directly observed by scanning electron microscopy, is much smaller than that based on stem vulnerability measurements (e.g., Choat et al., 2003; Jansen et al., 2009; Lens et al., 2011). A possible explanation for this discrepancy is that the largest class pores that allow air seeding are too rare to detect by scanning electron microscopy (e.g., Choat et al., 2003, 2008; Jansen et al., 2009). This view was supported by empirical evidence that vulnerability to embolism was correlated with the area of pit membrane per vessel (Wheeler et al., 2005).

In contrast to the studies that failed to detect wider pores in intervessel pit membranes, the author found that extraporous regions with openings up to several hundred nanometers in width commonly present in some species with intervascular pit membranes without an apparent intercellular layer (Fig. 7.7a) (Sano 2004, 2005). However, it is unlikely that air seeding occurs through the extraporous regions because such regions tend to appear in the peripheral parts of individual pit membranes and this type of two-layered pit membranes are easily aspirated.

Recently, analyses have been attempted to link the cavitation resistance and bordered pit characters. It has been consistently shown that there is a significant positive correlation between the thickness of intervessel pit membrane and cavitation resistance (Jansen et al., 2009; Lens et al., 2011; Scholz et al., 2013). Lens et al. (2011) further indicated that the depth of the pit chamber (distance from pit membrane surface and inner edge of pit aperture) was negatively correlated with cavitation resistance. These studies also indicated that the thickness of pit membranes is closely linked to their porosity, and that the pit chamber depth affects the porosity of pit membranes when the pit membranes are stretched during pit aspiration. Therefore, it is certain that porosity is a key factor in cavitation resistance.

CONCLUSIONS

The structure and function of bordered pits in the xylem have been fairly clarified to date. However, recent studies have still failed to obtain clear evidences showing how the bordered pits regulate water flow and the progression of cavitation in the xylem (Plavcová et al., 2013; Rockwell et al., 2014). Further careful studies linking the structural analysis using adequate methods and physiological methods are required to develop our understanding of the mechanisms of how bordered pits regulate water flow and cavitation.

REFERENCES

Bauch, J., Liese, W., Schultze, R., 1972. The morphological variability of the bordered pit membranes in gymnosperms. Wood Sci. Technol. 6, 162–184.

Bonner, L.D., Thomas, R.J., 1972. The ultrastructure of passageways in vessel of yellow poplar (*Liriodendron tulipifera* L.). Part 1: Vessel pitting. Wood Sci. Technol. 6, 196–203.

Choat, B., Ball, M., Luly, J., Holtum, J., 2003. Pit membrane porosity and water stress-induced cavitation in four co-existing dry rainforest tree species. Plant Physiol. 131, 41–48.

Choat, B., Cobb, A., Jansen, S., 2008. Structure and function of bordered pits: new discovery and impacts on whole plant hydraulic function. New Phytol. 177, 608–626.

Coleman, C.M., Prather, B.L., Valente, M.J., Dute, R.R., Miller, M.E., 2004. Torus lignification in hardwoods. IAWA J. 25, 435–447.

Côté, W.A., 1958. Electron microscope studies of pit membrane structure, implications in seasoning and preservation of wood. Forest Prod. J. 8, 296–301.

Delzon, S., Douthe, C., Sala, A., Cochard, H., 2010. Mechanism of water-stress induced cavitation in conifers: bordered pit structure and function support the hypothesis of seal capillary-seeding. Plant Cell Environ. 33, 2101–2111.

Domec, J.-C., Lachenbruch, B., Meinzer, F.C., Woodruff, D.R., Warren, J.M., et al., 2008. Maximum height in a conifer is associated with conflicting requirements for xylem design. Proc. Natl. Acad. Sci. USA 105, 12069–12074.

Dute, R.R., Elder, T., 2011. Atomic force microscopy of torus-bearing pit membranes. IAWA J. 32, 415–430.

Dute, R., Hagler, L., Black, A., 2008. Comparative development of intertracheary pit membranes in *Abies firma* and *Metasequoia glyptostroboides*. IAWA J. 29, 277–289.

Evert, R.F., 2006. Esau's Plant Anatomy. Wiley-Liss, Hoboken.

Fujikawa, S., Ishida, S., 1972. Study on the pit of wood cells using scanning electron microscopy. III. Structural variation of bordered pit membrane on the radial wall between tracheids in Pinaceae species. Mokuzai Gakkaishi 18, 477–483.

Hacke, U.G., Jansen, S., 2009. Embolism resistance of three boreal conifer species varies with pit structure. New Phytol. 182, 675–686.

Hacke, U.G., Sperry, J.S., Pittermann, J., 2004. Analysis of circular bordered pit function – II. Gymnosperm tracheids with torus-margo pit membranes. Am. J. Bot. 91, 386–400.

Hacke, U.G., Sperry, J.S., Feild, T.S., Sano, Y., Sikkema, E.H., Pittermann, J., 2007. Water transport in vesselless angiosperms: conducting efficiency and cavitation safety. Int. J. Plant Sci. 168, 1113–1126.

Harada, H., 1964. Further observation on the pit structure of wood. Mokuzai Gakaishi 10, 221–225.

Harada, H., Miyazaki, Y., Wakashima, T., 1958. Electron microscopic investigation on the cell wall structure of wood. Bull. Forest Exp. Sta. Meguro 104, 1–115.

IAWA Committee, 2004. IAWA list of microscopic features for softwood identification. IAWA J. 25, 1–70.

Imamura, Y., Harada, H., 1973. Electron microscopic study of the development of the bordered pit in coniferous tracheids. Wood Sci. Technol. 7, 189–205.

Jansen, S., Sano, Y., Choat, B., Rabaey, D., Lens, F., et al., 2007. Pit membranes in tracheary elements of Rosaceae and related families. Am. J. Bot. 94, 503–514.

Jansen, S., Choat, B., Pletsers, A., 2009. Morphological variation of intervessel pit membranes and implications to xylem function in angiosperms. Am. J. Bot. 96, 409–419.

Jansen, S., Lamy, J.B., Burlett, R., Cochard, H., Gasson, P., et al., 2012. Plasmodesmatal pores in the torus of bordered pit membranes affect cavitation resistance of conifer xylem. Plant Cell Environ. 35, 1109–1120.

Jarbeau, J.A., Ewers, F.W., Davis, S.D., 1995. The mechanism of water-stress-induced embolism in two species of chaparral shrubs. Plant Cell Environ. 18, 189–196.

Kim, J.S., Daniel, G., 2013. Developmental localization of homogalacturonan and xyloglucan epitopes in pit membranes varies between pit types in two poplar species. IAWA J. 34, 245–262.

Kim, J.S., Awano, T., Yoshinaga, A., Takabe, K., 2011. Temporal and spatial diversities of the immunolabeling of mannan and xylan polysaccharides in differentiating earlywood ray cells and pits of *Cryptomeria japonica*. Planta 233, 109–122.

Lens, F., Sperry, J.S., Christman, M.A., Choat, B., Rabaey, D., Jansen, S., 2011. Testing hypotheses that link wood anatomy to cavitation resistance and hydraulic conductivity in the genus *Acer*. New Phytol. 190, 709–723.

Liese, W., 1965. The fine structure of bordered pits in soft wood. In: Côté, W.A. (Ed.), Cellular Ultrastructure of Woody Plants. Syracuse Univ. Press, Syracuse, pp. 271–290.

Meylan, B.A., Butterfield, B.G., 1982. Pit membrane structure in the vessel-less wood of *Pseudowintera dandy* (Winteraceae). IAWA Bull. New Ser. 3, 167–175.

O'Brien, T.P., 1970. Further observations on hydrolysis of cell wall in xylem. Protoplasma 69, 1–14.

Ohtani, J., Ishida, S., 1978. Pit membrane with torus in dicotyledonous woods. Mokuzai Gakkaishi 24, 673–675.

Panshin, A.J., de Zeeuw, C., 1980. Textbook of Wood Technology, fourth ed., McGraw-Hill, New York.

Parameswaran, N., Liese, W., 1973. Anomalous structures in the bordered pits of fibre-tracheids of *Ribes sanguineum*. Wood Fiber 5, 76–79.

Pesacreta, T.C., Groom, L.H., Rials, T.G., 2005. Atomic force microscopy of the intervessel pit membrane in the stem of *Sapium sebiferum* (Euphorbiaceae). IAWA J. 26, 397–426.

Pittermann, J., Sperry, J.S., Hacke, U.G., Wheeler, J.K., Sikkema, E.H., 2005. The torus-margo pit valve makes conifers hydraulically competitive with angiosperms. Science 310, 1924.

Pittermann, J., Choat, B., Jansen, S., Stuart, S., Lynn, L., Dawson, T., 2010. The relationships between cavitation safety and hydraulic efficiency in the pit membranes of conifers belonging to the Cupressaceae: the evolution of form and function. Plant Physiol. 153, 1919–1931.

Plavcová, L., Hacke, U.G., 2011. Heterogeneous distribution of pectin epitopes and calcium in different pit types of four angiosperm species. New Phytol. 192, 885–897.

Plavcová, L., Hacke, U.G., Sperry, J.S., 2011. Linking irradiance-induced changes in pit membrane ultrastructure with xylem vulnerability to cavitation. Plant Cell Environ. 34, 501–513.

Plavcová, L., Jansen, S., Klepsch, M., Hacke, U.G., 2013. Nobody's perfect: can irregularities in pit structure influence vulnerability to cavitation? Front. Plant Sci. 4, 453.

Rabaey, D., Lens, F., Smets, E., Jansen, S., 2006. Micromorphology of pit membranes in tracheary elements of Ericales: new records of tori or pseudo-tori. Ann. Bot. 98, 943–951.

Rabaey, D., Huysmans, S., Lens, F., Smets, E., Jansen, S., 2008. Micromorphology and systematic distribution of pit membrane thickenings in Oleaceae: tori and pseudo-tori. IAWA J. 29, 409–424.

Rockwell, F.E., Wheeler, J.K., Holbrook, N.M., 2014. Cavitation and its discontents: opportunities for resolving current controversies. Plant Physiol. 164, 1649–1660.

Sano, Y., 2004. Intervascular pitting across the annual ring boundary in *Betula platyphylla* var. *japonica* and *Fraxinus manshurica* var. *japonica*. IAWA J. 25, 129–140.

Sano, Y., 2005. Inter- and intraspecific structural variations among inter- vascular pit membranes, as revealed by field-emission scanning electron microscopy. Am. J. Bot. 92, 1077–1084.

Sano, Y., Fukazawa, K., 1994. Structural variations and secondary changes in pit membranes in *Fraxinus mandshurica* var. *japonica*. IAWA J. 15, 283–291.

Sano, Y., Jansen, S., 2006. Perforated pit membranes in imperforate tracheary elements of some angiosperms. Ann. Bot. 97, 1045–1053.

Sano, Y., Nakada, R., 1998. Time course of the secondary deposition of incrusting materials on bordered pit membranes in *Cryptomeria japonica*. IAWA J. 19, 285–299.

Sano, Y., Kawakami, Y., Ohtani, J., 1999. Variation in the structure of intertracheary pit membranes in *Abies sachalinensis*, as observed by field-emission scanning electron microscopy. IAWA J. 20, 375–388.

Sano, Y., Ohta, T., Jansen, S., 2008. The distribution and structure of pits between vessels and imperforate tracheary elements in angiosperm woods. IAWA J. 29, 1–15.

Sano, Y., Morris, H., Shimada, H., Ronse De Craene, L.P., Jansen, S., 2011. Anatomical features associated with water transport in imperforate tracheary elements of vessel-bearing angiosperms. Ann. Bot. 107, 953–964.

Sano, Y., Utsumi, Y., Nakada, R., 2013. Homoplastic occurrence of perforated pit membranes and torus-bearing pit membranes in ancestral angiosperms as observed by field-emission scanning electron microscopy. J. Wood Sci. 59, 95–103.

Schmid, R., Machado, R.D., 1968. Pit membranes in hardwoods – fine structure and development. Protoplasma 66, 185–204.

Scholz, A., Rabaey, D., Stein, A., Cochard, H., Smets, E., Jansen, S., 2013. The evolution and function of vessel and pit characters with respect to cavitation resistance across 10 *Prunus* species. Tree Physiol. 33, 684–694.

Schoonmaker, A.L., Hacke, U.G., Landhausser, S.M., Lieffers, V.J., Tyree, M.T., 2010. Hydraulic acclimation to shading in boreal conifers of varying shade tolerance. Plant Cell Environ. 33, 382–393.

Schulte, P.J., 2012. Computational fluid dynamics models of conifer bordered pits show how pit structure affects flow. New Phytol. 193, 721–729.

Sperry, J.S., Tyree, M.T., 1988. Mechanism of water stress-induced xylem embolism. Plant Physiol. 88, 581–587.

Thomas, R.J., 1968. The development and ultrastructure of the bordered pit membrane in the southern yellow pines. Holzforschung 22, 38–44.

Thomas, R.J., 1969. The ultrastructure of southern pine bordered pit membranes as revealed by specialized drying techniques. Wood Fiber 1, 110–123.

Thomas, R.J., 1975. The effect of polyphenol extraction on enzyme degradation of bordered pit tori. Wood Fiber 7, 207–215.

Thomas, R.J., 1976. Anatomical features affecting liquid penetrability in three hardwood species. Wood Fiber 7, 256–263.

Wheeler, E.A., 1982. Ultrastructural characteristics of red maple (*Acer rubrum* L.). Wood Fiber 14, 43–53.

Wheeler, J.K., Sperry, J.S., Hacke, U.G., Hoang, N., 2005. Inter-vessel pitting and cavitation in woody Rosaceae and other vesselled plants: a basis for a safety versus efficiency trade-off in xylem transport. Plant Cell Environ. 28, 800–812.

Zimmerman, M.H., 1983. Xylem Structure and the Ascent of Sap. Springer-Verlag, New York.

Zwieniecki, M.A., Melcher, P.J., Holbrook, N.M., 2001. Hydrogel control of xylem hydraulic resistance in plants. Science 291, 1059–1062.

Chapter 8

Fungal Degradation of Wood Cell Walls

Geoffrey Daniel

Department of Forest Products/Wood Science, Swedish University of Agricultural Sciences, Uppsala, Sweden

Chapter Outline

GENERAL BACKGROUND

Wood (lignocellulose) is colonized and degraded by a wide variety of biological agents including fungi, bacteria, and insects if suitable environmental conditions are available (Blanchette et al., 1990; Eriksson et al., 1990; Eaton and Hale, 1993; Daniel and Nilsson, 1998; Daniel, 2003; Goodell et al., 2003). These organisms can cause very different types of decay with destruction having a negative impact on construction/archeological wood leading to a reduction

Secondary Xylem Biology. http://dx.doi.org/10.1016/B978-0-12-802185-9.00008-5

in service life/historical value but also positive impact by contributing benefi-
cially in forest ecosystems by recycling CO_2 to the atmosphere. Wood in aerobic
terrestrial and aquatic environments can be colonized rapidly by fungi and bacteria and decay of wood structural and nonstructural components often initiate
quickly.

Fungi can cause a variety of decay types in wood commonly classified as
mold, blue stain (sapstain), white rot, brown rot, and soft rot. Rate of colonization, decay, and removal of wood structural components varies greatly between
rot types, fungal species, and even strains. A very important aspect discovered
very early (Hartig, 1874) however, was that a particular fungus causes a specific type of decay and that the decay will always be the same no matter what
wood type (i.e., coniferous or angiosperm wood). Thus, the presence of fungal
fruit bodies (e.g., basiocarps) can allow identification of the main types of decay
present no matter if it is growing on a standing tree in a forest, stored lumber, or
house facade. Variations in the morphology of decay patterns in wood and wood
cell walls reflect therefore rather local variations in the microdistribution of
wood chemical components (i.e., lignin, cellulose, hemicelluloses), presence of
extractives, or effect of wood treatments (preservatives, chemical modification).

Our understanding of the morphological aspects of wood colonization and
attack of wood cell walls by fungi is quite well developed. However, our understanding of molecular, biochemical, and chemical aspects of fungal decay
are still evolving and considerable details are lacking. Substantial reductions
in decay of wood in service would be possible if our current knowledge and
understanding of decay patterns were used in practice.

Wood Structure and Importance of Wood Chemistry on Decay

The structural features of wood vary greatly among the many different types
of wood species found on earth. Typical morphological features of softwoods
(e.g., *Picea abies*) and hardwoods (e.g., *Betula verrucosa*) are shown in Fig. 8.1.
In general terms, softwoods have a much simpler construction, evolved much
earlier (ca. 200 million years ago), and are composed primarily of one cell type
(i.e., tracheids; 90–95%), which has both water transport and support functions
in the living tree. In contrast, hardwoods are much more complicated, evolve
much later (ca. 100 million years ago) and are composed of a variety of different cell types (i.e., fibers, fiber tracheids, vessels, parenchyma cells) of which
vessels are the most characteristic and in which support (fibers) and water transport functions (vessels) are separated (Jane, 1956; Kollmann and Cote, 1968;
Panshin and Zeeuw, 1970; Carlquist, 2001; Daniel, 2009). All wood cells are
composed of different ratios of cellulose, lignin, and hemicelluloses. In broad
terms, coniferous wood (i.e., gymnosperm) has higher lignin content and lignin
composed of guaiacyl monomers while hardwoods (i.e., angiosperms) have in
general (at least in temperate woods) a lower lignin content composed of syringyl and guaiacyl monomers. Large variations in the ratio of syringyl:guaiacyl

monomers exist and in some tropical hardwoods not only high lignin levels are known (Nilsson et al., 1988; Lhate et al., 2010) but also wood species containing high guaiacyl lignin (Nilsson et al., 1988). An additional major difference between soft- and hardwoods is the type and concentration of hemicelluloses present with glucuronoxylan characteristic for hardwoods and galactoglucomannans for softwoods. A further difference that can strongly affect fungal decay is the microdistribution of the different chemical components at the wood cell wall level, particularly in hardwoods. Differences can exist in wood chemistry across growth rings and within different cell types and even across wood cell walls and between the compound middle lamellae separating wood cells (Saka et al., 1982) (Fig. 8.1). A final difference that can affect wood decay generally and at the wood cell wall level is the presence of extractives with a large variety found in tropical hardwoods, which can provide increased durability. Extractive type and location is of great importance and wood species with high extractive levels present in the cell walls (e.g., greenheart; *Ocotea rodiei*) showing greater durability than when extractives are only present in the cell lumina. The chemistry of wood cell walls has a major influence on how rapid cell walls are degraded by different fungi and correspondingly this is also very important for understanding for development of protective agents against wood decay fungi.

Almost all fungi including those that cause molds and blue stain can degrade exposed cellulose and hemicelluloses (i.e., polysaccharides). However, in lignified cell walls, the cellulose is generally surrounded by the hemicelluloses and in turn by lignin (Fig. 8.1). Essentially for fungi to degrade lignified cell walls, they must have an ability to overcome the "lignin barrier." Fungi causing "true decay" have developed various strategies for overcoming the lignin barrier (shown later) while lower fungi, mold, and blue stain have a very poor ability to do this and thus they are either restricted to degrading soluble sugars or storage materials where present or attack nonlignified wood cells (e.g., epithelial, parenchyma cells) or specialized wood cells where the cellulose is exposed and the lignin very low/absent (e.g., tension wood fibers in hardwoods) (Blanchette et al., 1994; Encinas and Daniel, 1995).

Fundamentally, lignin provides the natural durability of wood with wood species with high lignin and guaiacyl type lignin monomers normally showing greatest resistance against wood decay. There is a further relationship between lignin type, level (percent), and retention of various metals (e.g., copper) used in earlier wood protection strategies (Butcher and Nilsson, 1982). With modern chemical wood modification strategies the relationship with lignin is, however, less tenuous and best protection may be given by reaction with cellulose (e.g., acetylation). Despite variations in wood chemistry and knowledge of wood species with high durability, no wood species is currently known that is completely resistant against wood decay by fungi. Wood exposed in oxygen-free environments (i.e., fungal wood decay is an aerobic process) or maintained dry can last for thousands of years without decay assuming no insects are present (Nilsson et al., 1990).

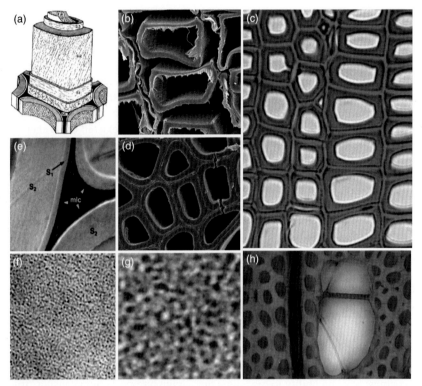

FIGURE 8.1   (a) Schematic diagram (Côté, 1977) and light microscopy (LM), scanning electron microscopy (SEM), and transmission electron microscopy (TEM) images of wood cell structure. The S2 layer dominates and is important for strength of the cells, which is regulated by the orientation of the cellulose microfibrils (MFA). S1 and S3 layers have a flat MFA orientation and the primary wall a random distribution; (b) spruce tracheids after delignification with absence of lignin in middle lamellae; (c) LM of spruce tracheids; (d) SEM of birch fibers; (e–g) TEM of birch S2 layer at increasing magnifications; (h) LM of birch fibers and vessels; hyphae may be restricted to the cell lumen of wood cells (e.g., molds, blue stain), produce cavities in the S2 and S1 layers (soft rot Type I, white rot), cause cell wall thinning initiated from the cell lumen (e.g., soft rot Type II; simultaneous white rot); preferentially remove lignin and hemicelluloses starting from the cell lumen (preferential white rot); depolymerize cellulose, hemicelluloses starting from the cell lumen (brown rot). A further variation includes attack of the high lignin containing compound middle lamella regions by certain white rot fungi.

Fungal Colonization of Wood

All fungi that colonize and decay wood have a filamentous growth form and belong either to the deuteromycetes (fungi imperfecti), ascomycetes, or basidiomycetes groups. When fungi come into contact with wood by spores or through direct contact as in ground situations, colonization can take place. This may be rapid as seen for blue stain and mold fungi on freshly cut timber or debarked round wood or take time if the wood has been treated with preservatives or chemically modified. Colonization is normally strongly related to the moisture content of the wood (Zabel and Morrell, 1992). Fungal colonization of wood can take place through open cells (e.g., cut-ends of wood) but is more frequently through rays in larger wooden constructions (e.g., poles). The rays are natural pathways in wood often providing fungi with food reserves (starch) or soluble sugars to accelerate colonization and penetration into the wood structure. In addition, the rays frequently contain parenchyma/epithelial cells that are nonlignified and readily degraded by most fungi including mold fungi like *Aspergillus* and *Penicillium* spp. (Fig. 8.2a–c). Once inside the wood, fungi penetrate from one cell to another through natural openings like pits (i.e., simple, half-bordered, bordered pits) or penetrate directly through the cell walls by producing bore holes (Figs 8.2d–f). The types (original and final size) and size of bore holes produced vary between the main fungal groups (blue stain, white, brown, and soft rots) and fungal species although the terminology used reflects transverse penetration across wood cells rather than axial development along the fiber axis (Figs 8.2d–f, 8.3a–g, and 8.4a, b). In hardwoods, both the rays and vessels are the major pathways of colonization throughout the wood structure

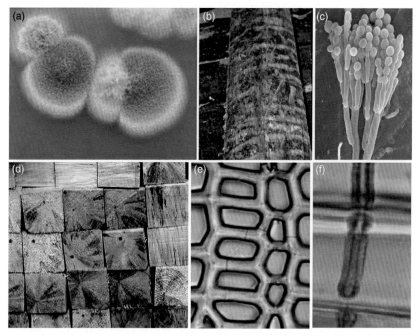

FIGURE 8.2 Examples of mold and blue stain colonization of wood. (a) Colonies of *Aspergillus niger* isolated from wood and growing on agar; (b) gray and green mold of *Aspergillus* and *Penicillium* spp. on debarked round wood; (c) SEM of *Penicillium* spp. coniophore from (b); (d) blue stain colonization of pallets blocks cut from a tropical hardwood; (e) transverse colonization through wood fiber cell walls by very fine blue stain penetration (bore) hyphae; (f) fine hyphal penetration through tracheid cell walls and full size melanized hyphae in the fiber cell lumina.

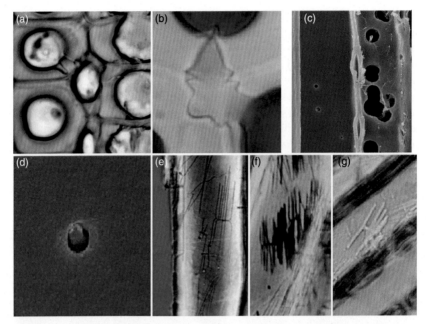

FIGURE 8.3   Different types of hyphal penetration through wood cell walls. (a–d) Bore hole traversing the middle lamella and secondary cell walls of pine as seen with light microscopy (a, b) and SEM (c, d). The effect of the cell wall microfibril angle is seen in (b), where decay is more rapid in the S1 than in the middle lamella and S2 layers producing a characteristic wedged-shaped decay pattern. (e) Multiple T-branching in the S2 wall of pine by *O. mucida*. (f, g) Multiple T-branching of an unknown white rot fungus attacking pine wood in a forest soil.

by all fungal groups. All fungi can colonize wood directly assuming conditions are suitable (e.g., this is shown with wood decay tests carried out with monocultures of fungi). While it is generally accepted that wood exposed under natural conditions (e.g., in soils) is first colonized by rapidly growing primary colonizers (e.g., from fungi imperfecti), which are then succeeded by higher fungi (true wood degrading fungi), this is not an absolute requirement but rather dependent on local conditions and fungal inoculate present. Fungi from the different groups under natural conditions will often colonize and degrade the same wood material and even be present and attack the same wood cells.

Figure 8.1 provides a schematic overview of generalized wood cells using light and scanning electron microscopy images of sections from *P. abies* and *B. verrucosa* to show typical wood cellular details so as to emphasize where fungal hyphae grow and cause decay. In brief, wood cells are composed of secondary and primary cell walls. In the secondary cell wall, the S2 layer is dominating (80–90% in softwoods; ~1.7–3.7 μm in spruce) in all wood cells with the cellulose microfibrils usually aligned (i.e., microfibril angle, MFA) in a steep helix (11–20°) along the fiber axis (Fig. 8.1a). The S2 layer is enclosed on the inside by a thin S3 layer (i.e., lumen wall) (e.g., ~0.09–0.14 μm in spruce) and to the

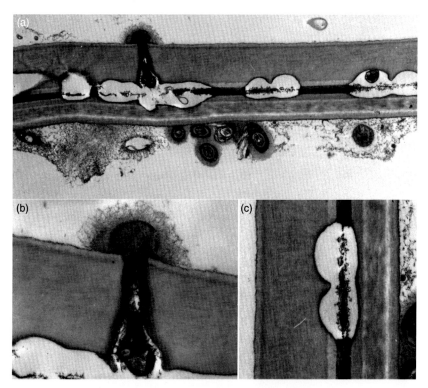

FIGURE 8.4   Fiber wall penetration and decay of middle lamella regions by an unknown white rot fungus from forest soils. (a) Cavities developing along the compound middle lamella/S1 into the adjacent S2 layer; (b) cell wall penetration into the middle lamella region; (c) attack causes almost entire disruption of the compound middle lamella region.

outside by a slightly thicker S1 layer ($\sim$0.25–0.35 μm in spruce). The S1 and S3 layers have cellulose microfibrils with a flat MFA (e.g., S1, 50–80°; S3, 60–80°) (Fig. 8.1a). During biosynthesis the individual wood cells develop together and have a common compound middle lamella ($\sim$0.05–0.16 μm in spruce) composed of the true middle lamella and thin primary cell walls of adjacent cells (Fig. 8.1). Chemically the highest lignin content in g/g is always found in the middle lamella while the S2 layer contains the greatest total volume of lignin, cellulose, and hemicelluloses. The primary cell wall is characterized by high pectin content and randomly orientated cellulose microfibrils. At high magnification the secondary cell wall layers have the cellulose, hemicelluloses, and lignin components organized into a concentric or sometimes radial macromolecular structure, which is one of the bases for determining how decay patterns in wood cells develop (Figs 8.5b, 8.6b, 8.7b, c, and 8.8b) (Daniel, 1994, 2003). During decay, fungi normally grow in the cell lumen from where they can attack the surrounding wood cell wall (e.g., by erosion) producing different patterns.

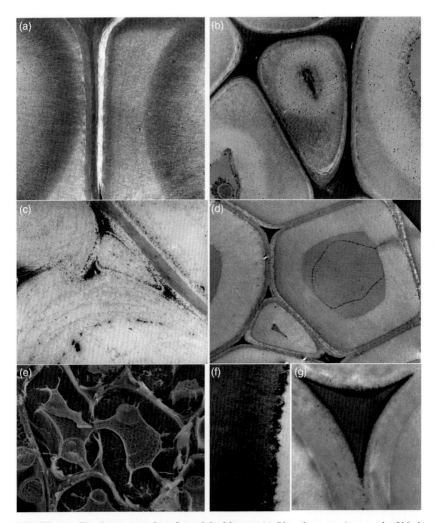

FIGURE 8.5 Simultaneous and preferential white rot. (a) *Pha. chrysosporium* attack of birch fibers with decay zones. Fiber cell walls are thinned progressively outward from the lumen. (b) Preferential decay by the white rot mutant *Phl. radiata* Cel 26. Decay zones progress across the cell wall into adjoining fibers with lignin and hemicelluloses removed. (c) *Phl. radiata* (wild) total decay of fibers with only middle lamella corners remaining; the structure held together by extracellular slime. (d) Advanced *Phl. radiata* Cel 26 attack with only lignified middle lamella cell corners remaining. (e) Similar stage as (d) observed using Cryo-SEM to emphasize slime. (f) Dark staining Mn deposits associated with lumen cell wall with decay beneath. (g) Attack by *Phl. radiata* Cel 26 of remaining middle lamella cell corners with thin decay zones present.

Depending on decay type, the hyphae may grow in the S2 cell wall or occasionally the S1 layer and more rarely in the compound middle lamella region (Fig. 8.4a, c). Fungal hyphae are not known to grow in the S3 or primary cell wall layers because of their thickness compared to that of fungal hyphae. When

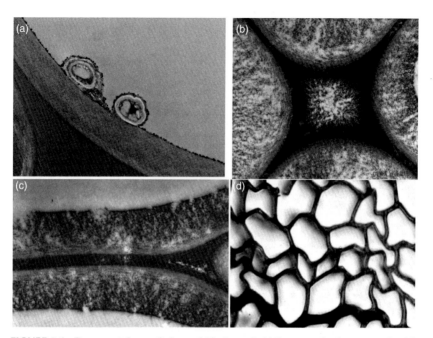

FIGURE 8.6 Brown rot decay of pine and birch wood. (a) Brown rot hyphae encapsulated in slime and attached to the lumen wall of a pine tracheid at early stage of attack. (b, c) Advanced brown rot of birch fibers with most of the cellulose and hemicellulose removed leaving a lignin skeleton; the secondary wall structure very open. (d) Advanced brown rot decay of pine earlywood tracheids with loss of wall integrity.

hyphae develop in the S2 and S1 layers they often follow the orientation of the cellulose MFA in some decay groups (fungi imperfecti, ascomycetes; Figs 8.7a, b and 8.8b) while in others (certain basidiomycetes) this may occur but is not a requisite for decay.

BLUE STAIN (SAPSTAIN) AND MOLD FUNGI ON WOOD

Blue stain and mold fungi primarily use the nutrient reserves in wood (e.g., in rays and axial parenchyma) and usually do not cause significant cell wall decay. Normally they cause only superficial discoloration of the wood surface (molds) or blue stain of the sapwood (Fig. 8.2d–f) but not in the heartwood due to inadequate moisture content. The wood can take on various colors including shades of blue, brown, green, and even black depending on the fungus involved and wood species. Discoloration is usually reflected by the color/pigment of the fungal hyphae themselves but secretion of extracellular materials also occurs. A number of blue-stain fungi (e.g., *Lasiodiplodia theobromae*) are also capable of producing soft rot of wood (e.g., 12% mass loss) (Encinas and Daniel, 1995) particularly in low-lignified hardwoods with

FIGURE 8.7   Soft rot decay of wood. (a) Utility pole with outer regions degraded by soft rot. (b, c) Soft rot Type I with cavities orientated along the cellulose microfibrils in the S2 layer. (d, e) Transverse sections of pine at early (cavities) and late stages of soft rot Type I. At late stages, only the middle lamellae and S3 layers remain in softwoods. Transverse sections of pine (f) and birch (g) showing soft rot cell wall erosion (Type II). Ultimately, only the middle lamellae remain.

tension wood (Encinas and Daniel, 1997). The reason is that low-lignified wood species and tension wood fibers have readily available cellulose and hemicelluloses for decay and any filamentous fungus with cellulolytic/hemicellulolytic ability are able to cause decay. Since the mycelia of the fungi are often strongly colored, after colonization the wood also becomes discolored. For example, the blue-stain fungus *L. theobromae* is capable of growing 1 cm/day under optimal conditions and turns the sapwood of *Pinus caribaea* (Encinas and Daniel, 1995) blue–black within a very short period of time.

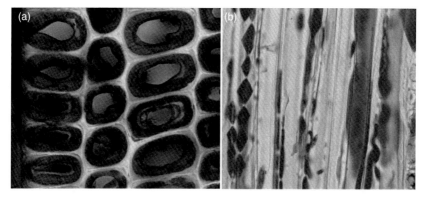

FIGURE 8.8   Diffuse soft rot Type I decay of pine and birch wood by *Phi. dimorphospora*. (a) Soft rot hyphae within the secondary cell walls (S2) with diffuse widespread decay not restricted to cavities. (b) Cavities (stained blue to right of photo) are often very large and most of the secondary wall degraded. Note: Subpart (a) is not stained, but colored black through melanized hyphae and melanin secretions.

Coloration of the wood occurs through melanization of the hyphae, a material that is highly inert and persists even after death of the fungus. Soft rot fungi also cause discoloration of the wood in the same way, but in this case the fungi can also cause strong decay of wood cells (Figs 8.7e and 8.8a). Table 8.1 compares the effects of mold and blue stain on wood in comparison to true wood-degrading fungi.

TRUE WOOD-DEGRADING FUNGI

The major feature of wood-degrading fungi is their ability to overcome the lignin barrier and cause decay and often complete mineralization of lignified wood cell walls. These fungi are recognized as the "true" wood-destroying fungi and have been historically divided into a number of artificial groups, namely, the white-, brown-, and soft rot fungi based on visual macroscopic changes induced in the wood structure. The classification is rather imprecise in view of the thousands of wood-degrading fungi known (Gilbertson, 1980; Eastwood et al., 2011) and great variation exists within each of the groups. In the present chapter only the major characteristics for the different groups are outlined with respect to attack and changes caused in wood cell walls and enzymes employed. Additional details on chemical and biochemical mechanisms of attack can be found in recent reviews (Hatakka and Hammel, 2011; Arantes and Goodell, 2014; Daniel, 2014).

Wood Cell Wall Degradation by White Rot Fungi

Research on white rot decay is more advanced than for all other major forms of wood decay. Developments have been driven by the potential of the enzymes produced by these fungi in biotechnology processes in the pulp and paper (e.g.,

TABLE 8.1 Classification of Morphological Effects of Fungal Attack of Wood Cell Walls

Decay type	Morphological changes of wood cell walls	Cell wall components attacked	Taxonomic grouping	Typical examples
White rot (simultaneous)	Hyphal bore holes enlarge Cell wall thinning from lumen Middle lamellae degraded Development of cavities in some species	Cellulose, lignin, hemicelluloses Extractives	Basidiomycetes Higher ascomycetes	T. versicolor* Heterobasidium annosus Xylaria polymorpha** Daldinia concentrica***
White rot (preferential)	Hyphal bore holes enlarge Cell wall attack from lumen Middle lamellae degraded Fiber defibration and separation	Hemicelluloses, lignin Extractives	Basidiomycetes	Ceriporiopsis subvermispora Heterobasidium annosum Phl. radiata Cel 26
Brown rot	Rapid attack of cell walls Cell wall attack from lumen	Depolymerization of cellulose, hemicelluloses; lignin modified	Basidiomycetes	C. puteana* Ol. (Postia) placenta*
Soft rot Type I	Hyphal bore holes remain small Longitudinal cavities Middle lamellae remain	Cellulose, hemicelluloses lignin modified/degraded	Ascomycetes Fungi imperfecti	Chaetomium globosum* Phialophora mutabilis
Soft rot Type I "Diffuse type"	Hyphal bore holes remain small Coalescence of longitudinal cavities Middle lamellae remain	Cellulose, hemicelluloses lignin modified/degraded	Ascomycetes Fungi imperfecti	Phi. dimorphospora Ch. globosum Bispora betulina
Soft rot Type II	Hyphal bore holes remain small Cell wall thinning from lumen Middle lamella remains	Cellulose, hemicelluloses lignin modified/degraded	Ascomycetes Fungi imperfecti	Ch. globosum Phialop. mutabilis
Blue stain fungi	Fine bore holes Small cavities/erosion troughs	Primarily nonlignified cells (ray parenchyma), extractives	Ascomycetes Fungi imperfecti	Ophiostoma piceae L. theobromae
Mold fungi	Growth in surface regions	Soluble sugars, extractives	Ascomycetes Fungi imperfecti Zygomycetes	Ch. globosum Penicillium brevicompactum Aspergillus versicolor Rhizopus spp.

*Frequently used as test fungi for assessing new types of wood protection – for example, preservatives, wood modification.

**Higher Ascomycete cause white rot but are not known to degrade middle lamella regions.

fiber modification, bleaching in kraft pulping) and related forest industries (e.g., ecofriendly wood protection using laccase from white rot fungi in laccase-catalyzed iodination of wood surfaces (Schubert et al., 2012)). White rot fungi (e.g., *Physisporinus vitreus*) have also been assessed for "bioincising" to increase the permeability of refractory wood species (e.g., Norway spruce) to improve the penetration of wood preservatives and also for improving the acoustic properties of musical instruments (Schwarze and Landmesser, 2000; Lehringer et al., 2011; Schwarze and Schubert, 2011). In native situations, white rot fungi are typically found colonizing and degrading hardwoods (Fig. 8.9a) producing a variety of decay patterns referred as "white pocket," "stringy," or "spongy rot." The fungi are generally restricted to terrestrial situations with only a few marine species known (Moss, 1986; Mouzouras et al., 1988). Generally, with only a few exceptions, white rot fungi show poor tolerance to conventional metal-containing wood preservatives although they have been reported on bis(tri-*n*-butyltin)oxide (TnBTO)–benzalkyl trimethyl ammonium chloride (AAC) formulations under test conditions (Daniel and Bergman, 1997). White rot has possibly the most economic importance in standing trees, wood fencing (Fig. 8.9b), and during storage of wood and lumber.

The majority of white rot fungi (i.e., several thousand species) belong to the higher fungi (i.e., basidiomycetes) and are unique in that they are the only taxonomic group known to efficiently biomineralize lignin from both softwoods and hardwoods in addition to the wood polysaccharides to CO_2 and H_2O under aerobic conditions (Gilbertson, 1980). This process involves a wide range of physiological and biocatalytic activities with diverse enzymes and nonenzymatic systems implicated (Eriksson et al., 1990; Goodell et al., 1997; Daniel, 2003; Hatakka and Hammel, 2011) (shown later). A number of higher ascomycetes from the Xylariaceous group (e.g., *Xylaria*, *Daldinia*, and *Hypoxylon*) can also produce decay patterns (i.e., cell wall thinning) similar to white rot in hardwoods like birch, although their inability to degrade middle lamellae or vessels (i.e., both have guaiacyl lignin) in these wood species or decay softwoods suggests their lignin-degrading abilities are more similar to that of soft rot fungi (Nilsson et al., 1989). In addition to commonly known white rot causing decay of wood, the majority of litter degrading basidiomycetes from forest soils also cause white rot.

Morphological Aspects of Wood Cell Decay

White rot fungi can produce a wide range of morphological decay patterns in wood that can vary with fungal species, wood type, and physiological status (Blanchette et al., 1990; Daniel, 1994, 2003; Schwarze, 2007). Principally, the great variability in decay patterns produced reflects the order and/or proportions during which the major wood components (cellulose, hemicelluloses, and lignin) are removed from wood cells (Table 8.1). Generally, the wood structure becomes bleached from its normal tan color at advanced stages of decay through the decay of lignin; however, before arriving at this stage there is a wide range of white–brown colorations developed dependent on wood type and fungal species, rate of decay, and environmental conditions.

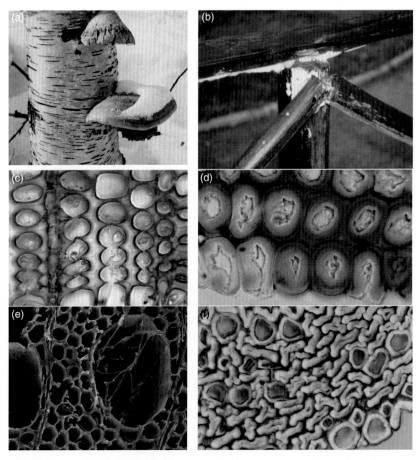

FIGURE 8.9 (a, b) Typical fruit bodies of white rot fungi on standing trees and a garden fence. (c, d) Light microscopy images of transverse sections of pine wood degraded by *T. versicolor* (c, simultaneous decay) and *Phl. radiata* Cel 26 (*preferential decay*). (e, f) SEM image of simultaneous decay of birch wood by *Heterobasidium annosum* and light microscopy image of birch fibers remaining after preferential white rot by *Phl. radiata* Cel 26.

As described earlier, the principal path for white rot invasion into wood is via the ray canals that provide rapid penetration into the radial structure of the wood and provision of nutrients stored in parenchyma cells. Hyphae then develop throughout the wood structure using natural pathways such as pits or develop specialized bore hyphae (for traversing across wood cells; Fig. 8.3a–c). The majority of hyphae cause decay of wood cell walls while located in the cell lumina (Figs 8.5b, e and 8.9c) although more specialized types of attack exist where hyphae can cause prominent decay in the compound middle lamella regions between wood cells (e.g., Fig. 8.4a, c). A typical example is *Pycnoporus sanquineous* (Daniel unpublished observation).

The speed and order in which white rot fungi degrade and remove components from wood cell walls have been classified into two principal types of decay based on microscopic and chemical analyses of remaining wood components. The two decay patterns are known as "simultaneous" and "selective" (preferential) and reflect the morphological appearance of wood fibers during decay (Figs 8.5a–g and 8.9c, d). Simultaneous white rot results in a rather uniform depletion of cellulose, hemicelluloses (xylan, mannan), and lignin resulting in a bleached appearance and sometimes fibrous appearance of the wood. With simultaneous white rot (e.g., *Trametes versicolor*, *Phlebia radiata* (wild), *Phanerochaete chrysosporium*), all the main wood components are degraded more or less simultaneously from the cell lumen outward (Figs 8.5a, c and 8.9c, e) (Blanchette et al., 1990; Eriksson et al., 1990; Daniel, 1994; Daniel and Nilsson, 1998) and depending on fiber shape and thickness of the cell walls, rupture of tangential walls occurs first (Fig. 8.9c, e) particularly in softwoods. Microscopically, decay occurs by a gradual erosion of the surrounding cell wall by lumina-based hyphae resulting in progressively thinner cell walls (Fig. 8.9c, e). As decay continues even the lignin-rich middle lamella regions between fibers are degraded with decay often progressing into adjacent fibers (Fig. 8.9c). Ultimately only the lignin-rich middle lamella cell corners remain (Fig. 8.5c) but even these can be degraded at advanced stages. One of the most characteristic features of decay is the development at the cell wall level of a progressive decay zone, which reflects lignin/hemicellulose mineralization in "time and space" (Figs 8.5a and 8.9c). The zone is readily visible at the light microscope level after staining with safranin (Fig. 8.9c) and at the transmission electron microscopy (TEM) level as an electron-dense layer followed by an electron-lucent region (Fig. 8.5a); the thicker the wood cell wall, the more apparent the zone. The zones of decay are apparent in all wood cell wall layers but are very thin in the remaining middle lamella cell corner regions because of the higher amounts and density of the lignin present.

In contrast, with selective white rot, the lignin and hemicelluloses are preferentially degraded allowing for fiber separation (i.e., defibration) in advanced stages leaving fibers with modified cellulose (Fig. 8.9f). Degradation of lignin is concomitant with loss of hemicellulose, which presumably provides the energy source for decay. Preferential white rot has received great interest over the years as a possible way to reduce the lignin content of wood chips before chemical (kraft) pulping or energy during the refining step of mechanical (thermomechanical) pulping. Both native and mutant (i.e., strains impaired or with reduced cellulase ability; e.g., *Cerioporiopsis subvermispora*, *Dichomitus squalens*, *Phl. radiata* Cel 26) white rot species have been studied (Blanchette et al., 1987; Blanchette, 1991; Daniel, 1994, 2003; Messner and Srebotnik, 1994; Daniel et al., 2004) in-depth to understand the events of attack in detail. Decay of cell walls is in principle similar to simultaneous attack but differs in that the decay zones are normally very distinct as a broad band using light and electron microscopy (Figs 8.5b and 8.9d). The outer perimeter of the zones is where decay of lignin and hemicelluloses takes place with the inner regions of the zones no longer staining with $KMnO_4$ (stains lignin) (Fig. 8.9d) and losing their electron density in the TEM (Fig. 8.5b, d). Ultimately, the zones meet from adjacent fibers by passing through the compound middle lamellae resulting in delignified fibers and fiber separation (Figs 8.5d and 8.9f). The middle lamella cell corners are the last regions to be attacked and here like with simultaneous white rot, very thin decay zones are produced in the corners (Fig. 8.5g). Manganese deposits are sometimes found associated with white rot decay (Daniel and Bergman, 1997) and areas around them often show signs of decay (Fig. 8.5f). Under laboratory conditions preferential white rot can be influenced significantly by prior impregnation of the wood with sugars.

A great deal of variability exists in decay strategies and some white rot fungi cause simultaneous and preferential attack in the same wood material indicating the complex nature of the response and importance of the fungus genome and local environmental conditions.

Besides the patterns of simultaneous and preferential white rot decay, a number of white rot fungi also develop thin cavities (i.e., along the axis of fibers) in the secondary cell walls of wood cells (Nilsson and Daniel, 1988; Daniel et al., 1992; Schwarze et al., 1995; Worrall et al., 1997; Schwarze et al., 2000; Daniel, 2003) in which the hyphae may or may not be aligned with the cellulose microfibrils (Fig. 8.3e–g). Typical examples include *Oudemansiella mucida* (Fig. 8.3e) (Daniel et al., 1992), *Auricularia auricular-judea* (Worrall et al., 1997), *Flammularia velutipes*, and *Chondrostereum purpureum* (Nilsson and Daniel, 1988). This may represent a remaining adaption to lignified cell walls where cavity formation as seen in the ascomycetes and fungi imperfecti may be a strategy to overcome the effect of lignin.

Wood Cell Wall Degradation by Brown Rot Fungi

Brown rot decay is caused exclusively by basidiomycete fungi. Decay is characterized by extensive depolymerization and fragmentation of wood cellulose and hemicelluloses leaving lignin as a modified residue (Figs 8.6b–d and 8.10a, b). The ability to selectively remove the wood polysaccharides is unique and distinguishes it from all other known types of fungal wood decay. In advanced stages of brown rot, the cellulose and hemicelluloses can be almost completely removed leaving only modified lignin, which is brown in color and hence provides the name of the decay type. Because of cellulose depolymerization, brown-rotted wood has a high lignin content and little residual strength. Brown-rotted wood on drying cracks characteristically into cubical pieces and is easily crushed between the fingers to powder. Even after advanced brown rot and loss of polysaccharides, wood cells tend to maintain their form but are more pliable (before drying) due to mostly lignin remaining (Figs 8.6d and 8.10d). The most serious aspect of brown rot is wood cellulose depolymerization, which can be rapid even at early stages of decay (i.e., at a time of little mass loss; e.g., 10%;

FIGURE 8.10   Brown rot decay of wood fibers. Preservative-treated stake (a) and pole (b) showing brown rot as dark staining blotches on the stake and brownish discoloration of the pole beneath the surface. (c) Hyphal colonization (blue) of *G. trabeum* via rays in pine. (d) Autofluorescence of pine tracheids with brown rot showing radial cracking of cell walls plus dark zones of decay. (e) Early brown rot with slight discoloration of the secondary cell wall (stained with safranin). (f) Loss of birefringence of pine tracheids under polarized light indicating cellulose depolymerization.

(Cowling, 1963)) and thus can have serious implications for construction wood in-service. In contrast with soft- and white-rot decay, lignin type and content appears to have only minor influence on the rate of decay by brown-rot fungi. Hardwoods (i.e., contain syringyl and guiaicyl lignin) and softwoods (contain only guiaicyl lignin) and even compression wood (i.e., softwoods with slightly higher lignin content than normal wood) can be degraded at the same rate (Nilsson et al., 1988). This suggests a well-developed mechanism for overcoming the lignin barrier in wood. Attack and depolymerization of the cellulose in wood is easily shown using polarized light microscopy where loss in crystallinity causes a loss in wood cell wall birefringence (Fig. 8.10f). In this way progress of decay across the wood and in individual cells can be easily visualized.

Brown rot represents the most serious hazard for wood under service situations in terrestrial environments in ground- and above-ground contact situations and can be a major concern in houses with cellars. Brown rot is not known, however, in aquatic situations.

Morphological Aspects of Cell Wall Decay by Brown Rot Fungi

Decay is caused by hyphae growing in the cell lumina of wood cells from where they cause attack of the surrounding cell wall carbohydrates (Figs 8.6a–c and 8.10c–f). Initial attack of wood cell walls is difficult to observe using light microscopy but may be detected as slight reddish staining with safranin and appearance of concentric layers in the S2 layer of softwoods (Fig. 8.10e). It is generally reported that the first real signs of brown rot attack are seen at the interfaces of the S1 and S2 layers with little effect on the S3 layer. However, this requires more study since it could simply reflect the diffusion of active agents along the longitudinal axial of fibers rather than across the transverse fiber wall, with initiation and penetration of agents taking place via the compound middle lamella/S1 regions at pit regions or from bore hyphae (Daniel, unpublished observation). Our recent ultrastructure (TEM) and immunocytochemistry studies with spruce and ash sapwood degraded by the brown rot fungus *Oligoporus* (*Postia*) *placenta* suggest that incipient decay may be initiated in the compound middle lamella (CML)/middle lamella cell corner (MLcc) regions of xylem cells remote from hyphal colonization in the cell lumen and that the S3 layer is also attacked at early stages (Kim et al., 2015).

Severe degradation of pectins (i.e., 1,5-α-arabinan, xyloglucan, and methyl-esterified homogalacturonan) and hemicelluloses have been detected in CML/MLcc regions even though secondary cell walls (except for S3 layers) only showed minor changes in the microdistribution of hemicelluloses at early stages of decay (Kim et al., 2015). Ultrastructural observations also showed significant differences in the decay of spruce and ash secondary cell walls by *Ol. placenta* after initial and similar attack of CML/MLcc regions and the S3 layer. The main secondary tracheid cell walls of spruce showed gradual degradation from the S3 to S1 layers (i.e., no initial degradation in the S2 layer). In contrast, degradation of ash fibers like that observed earlier in birch (Daniel unpublished observation) and other hardwoods (Highley et al., 1985) occurred primarily in the middle of the S2 layer while the S1 and outer/inner S2 layer regions appeared intact. In more advanced decay stages, a similar effect was noted by attack of the S1 layer while the outer regions of the S2 layer appeared intact. Evidence for the selective removal of hemicelluloses from the different regions of the secondary walls was provided by differences in loss of xylans and mannans using specific labeling of the polymers. In advanced stages of decay, the secondary cell walls of fibers in birch and ash have a very open structure and the cells are retained together by remaining lignin in middle lamella regions (Daniel, 1994;

Kim et al., 2015). Figure 8.6 b, c shows typical TEM images of birch fiber cell walls at advanced decay by *Ol. placenta* when most of the polysaccharides have been depleted from the S2 layer. The cell walls show a very open structure and in particular indications of radial orientated openings (electron-lucent regions) in the wall. While the presence of lignin may not actually influence the decay rate as seen for white- and soft rot attack, probably its presence has an indirect effect by reducing the amount of polysaccharides available. Thus, parenchyma cells from hardwoods, like sycamore, birch, oak, and robina (Daniel, 1994; Schwarze, 2007), have frequently been shown to have higher resistance to brown rot by *Fomitopsis pinicola*, *Laetiporus sulphureus*, and *Ol. placenta* in contrast to parenchyma cells in conifers. This would seem related to the presence of guaiacyl lignin in hardwood parenchyma cells although this may not be entirely correct since in a study of Norway spruce, the parenchyma cells were more resistant than tracheids (Kim et al., 2015). The very open nature of the secondary cell walls of fibers in hardwoods strongly indicates that there should not be any restriction to the diffusion of hydrolytic and oxidative enzymes involved in cellulose/hemicellulose decay (Fig. 8.6b, c).

While brown rot fungi are classified as causing decay developed from hyphae only present in the cell lumina of wood cells (Fig. 8.6a) (Table 8.1), there are reports of brown rot hyphae growing within the secondary cell walls of hard- and softwoods in much the same way as the white rot fungus *O. mucida*. Examples include *Coniophora puteana* (Kleist and Schmitt, 2001) and *Gloeophyllum trabeum* (Daniel, unpublished observation).

Wood Cell Wall Degradation by Soft Rot Fungi

Soft rot is caused by a diverse range of microfungi from the ascomycetes and fungi imperfecti and has ubiquitous distribution throughout the world. Soft rot can have severe economic consequences since several of the species involved attack wooden constructions including preservative-treated wood under terrestrial (Fig. 8.7a) and aquatic (marine and sweet water) situations (Gersonde and Kerner-Gang, 1976; Leightley, 1977; Leightley and Eaton, 1978; Zabel et al., 1985; Daniel and Nilsson, 1989). For example, soft rot has been documented as one of the major forms of attack of preservative-treated (i.e., with waterborne preservatives like copper–chrome–arsenic and creosote) utility poles (Gersonde and Kerner-Gang, 1976; Henningsson and Nilsson, 1976; Zabel et al., 1985). Visually soft-rotted wood becomes gray and even black in advanced stages of decay due to colonization by fungi hyphae in the wood. The surface of soft-rotted wood also tends to crack in a similar way as brown rot. The term "soft rot" was originally used to describe the decay from waterlogged wood, which was soft to touch (Savory, 1954). However, this definition is not correct and wood from terrestrial environments can be very heavily degraded but still remain hard as for example seen with utility poles. When such poles are strained, typical brash fractures are produced in contrast to that of brown rot, which produces brittle fractures due to depolymerization of the cellulose. Thus, soft rot can give rise to significant reductions in the strength of wood at comparable small weight losses although not to the level of brown rot.

Soft rot decay of wood is strongly affected by both lignin type and concentration in contrast to brown rot fungi and more significantly than white rot decay. Thus, softwoods (e.g., spruce and pine) show greater resistance to attack than hardwoods especially low lignin hardwoods (e.g., aspen and birch beech). In addition, at the cellular level in hardwoods, the variation in microdistribution of syringyl- and guaiacyl lignin that occurs between cell types (vessels/fibers/parenchyma) very often creates variations in decay morphology (Daniel and Nilsson, 1998).

Morphological Aspects of Cell Wall Decay by Soft Rot Fungi

Soft rot produces two distinct forms of attack of wood decay known as Type I in which characteristic cavities are produced in cell walls, and Type II where hyphae localized in the cell lumina cause cell wall erosion (i.e., cell wall thinning) (Figs 8.7 and 8.11) (Table 8.1). In particular, cavity formation is very diagnostic for soft rot decay although many of the soft rot fungi cause both decay types simultaneously, sometimes in the same cells. A few species only cause Type II attack (Nilsson, 1973). Type I decay results in the formation of characteristic cavities produced by hyphae within secondary cell walls that align along the cellulose microfibrils (Figs 8.7b–d, 8.8b, and 8.11c, d), their presence best observed using longitudinal sections under polarized light (Fig. 8.7c). Type I decay is initiated by specialized microhyphae ($\sim$0.5 μm) or fine penetration hyphae that grow from the cell lumen into the S2 layer of wood cell walls (Leightley, 1977; Crossley, 1979; Hale and Eaton, 1985a, 1985b; Daniel and Nilsson, 1989). Such hyphae can penetrate directly through adjacent cells without forming a cavity and expand again to the native size in the adjacent lumen (i.e., rather like blue-stain fungi). It is only when the hypha reorientates itself in the S2 layer by forming either a T-branch or L-bend that cavity formation is initiated. When orientated along the cellulose microfibrils in the cell wall, the hypha stops growing and produces a cavity (Hale and Eaton, 1985a, 1985b; Daniel and Nilsson, 1988). When the cavity has enlarged, growth can be resumed by the hypha either from one end (L-bending) or from both ends (T-branching) simultaneously. After the cavities have enlarged, the process repeats itself and chains of cavities are produced (Figs 8.7b, c and 8.8b). The process of cavity formation has been very well documented and the size and shape of the cavities are known to vary with fungal species and wood type (Daniel and Nilsson, 1998). T-branching is not limited to single branches and soft rot fungi are frequently known to develop multi-T-branching in cell walls (Fig. 8.11b) as observed with some white rot basidiomycetes. The fact that cavity formation always follows the orientation of the cellulose microfibrils is shown by the circular orientation of hyphae that invade pit chamber walls and between pits (Khalili et al., 2000). Cavities are most often observed first in the thick latewood cells (Fig. 8.7d, e) of

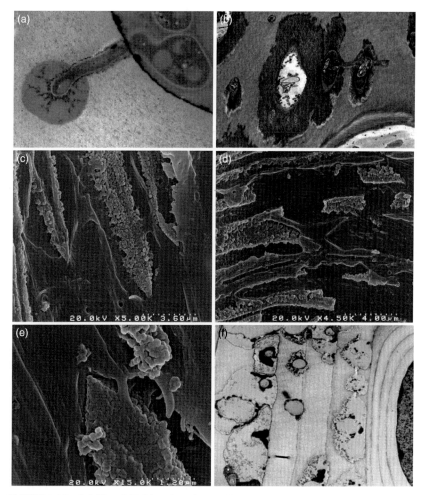

FIGURE 8.11   Soft Type I. TEM sections through birch (a), pine (b), and *H. foetium* (f) wood degraded by soft rot Type I. Sites of T-branching, multiple T-branching, and cavity formation in the S2 layer (a, b). (c–e) SEM images of soft rot cavities in S2 (c, e) and S1 layers (d). Hyphae are covered in melanin deposits and lignin breakdown products. Multilayered fiber of *H. foetium* showing effect of lignin on soft rot attack with half-moon shaped cavities produced against thin concentric layers containing high lignin in the S2 layer (f).

wood spreading thereafter to earlywood. The relationship here is thought related to the size of the hyphae as cavities are rarely observed in the S1 and never in the S3 layer (i.e., it is too thin and has high lignin in softwoods), although in advanced stages of decay the S1 and S3 are degraded through hyphal attack from the adjoining S2 layer.

The effect of lignin content and lignin type is shown at the cellular level in advanced stages of attack where only the high lignin middle lamellae remain in hard- and softwoods (Fig. 8.7e). In addition, the hyphae and cavities often have darkly staining materials closely associated consisting of lignin remnants and melanin, the latter directly attached to hyphae or extracellular within the cavities (Figs 8.7e, 8.8a, and 8.11c–f). A similar effect can also be noted in fibers showing very thin concentric layers containing higher lignin levels than the surrounding cell wall (e.g., *Homalium foetium*; Fig. 8.11f). Here, "half-moon" cavities are produced with flat axis orientated along the more lignified layer. That soft rot fungi need to orientate along the cellulose to cause cavity formation suggests it is an adaptation to overcome the effect of lignin coating the cellulose at the macromolecular level. Since the effect is shown in soft- and hardwoods with different lignin types and content suggests that it is the cellulose orientation that is most important. This is shown quite easily by slight delignification of wood, which enhances not only its susceptibility to soft rot but changes the nature of the decay process by modification or loss of cavity formation.

Type II soft rot is very similar to simultaneous white rot attack and can result in a complete removal of the secondary cell walls but in contrast the middle lamellae remain (Fig. 8.7f, g). This decay type is very frequently observed in low-lignin hardwood (e.g., aspen and birch) especially under high moisture situations.

A third decay form known as "diffuse cavity formation" has been described in which attack is similar to Type I during the initial stages but after cavity formation solubilization of the polysaccharides is more widespread and diffuse in the S2 layer and more like brown rot attack (Fig. 8.8a, b). The decay type is best observed in low-lignin containing hardwoods (e.g., birch) although it does occur in softwoods. Typical soft rot species producing this decay pattern include *Phialocephala dimorphospora* (Anagnost et al., 1994).

In Table 8.1, the effects on wood cell walls are compared with true rot fungi.

FUNGAL ENZYMATIC SYSTEMS INVOLVED IN WOOD DECAY

Numerous studies have been conducted to understand the biochemical/chemical mechanisms involved in fungal decay of lignocellulose, the majority aimed at biotechnological goals rather than using the understanding to produce better protective measures of wood in-service. In principle the majority of these studies have been carried out using liquid cultures of fungal monocultures together with lignocellulose in various forms (e.g., as particles, sawdust, or as flakes), purified cellulose/hemicelluloses, or lignin monomers (e.g., synthetic/natural). While these studies give information on the types of enzymes that may be produced under various physiological conditions (e.g., temperature, pH, and shaking/static), it may not reflect the situation that occurs in wood substrates under native conditions, but rather the potential ability of the fungi involved. For example, few studies have actually been involved in measuring enzyme activities in wood undergoing decay because of the difficulties in extracting sufficient amounts of "active proteins" in order to carryout enzymatic assays. Notable exceptions include the studies by Daniel et al. (1994) on the extraction of lignin degrading enzyme (LiP, MnP, laccase) produced by *Pha. chrysosporium*, *T. versicolor*, and *O. mucida* in birch wood. A further problem concerns the sensitivity of the enzyme assays, which may not be sufficiently sensitive to detect the minor amount of proteins that can be extracted even in highly degraded wood materials. For example, extraction of proteins from white-rotted wood in which profuse hyphae growth is often seen is easier than from either brown- or soft-rotted wood. In recent years, a variety of genomic, transcriptome, and secretome analytical approaches have been developed allowing for profile overviews of the enzyme systems available or employed by various fungi when grown on purified wood components (e.g., cellulose) (Martinez et al., 2009). This approach gives a more in-depth view of the enzymes or enzyme systems potentially involved and the changes in enzyme profiles over time during wood decay can be followed. Evidence for the upregulation of a gene does not however mean it is actually involved in decay but rather that it has a potential. Possibly one of the most important approaches although indirect for proving the involvement of enzymes at sites of wood decay is by using antibodies produced against the purified proteins in microscopy assays. The easiest approach is by indirect labeling and the use of secondary antibodies whereby the sites of primary antibody labeling of enzymes *in situ* are recognized by the secondary antibodies carrying a fluorescence or gold label, which can then be visualized using fluorescence/confocal microscopy or electron microscopy. Examples of such approaches can be found in later chapters in this book. These studies were very important in proving the extracellular secretion of enzymes involved in decay and their remote distribution from hyphae during the wood decay process (Daniel et al., 1989, 1990, 1994, 2007; Ruel et al., 1990; Srinivasan et al., 1995). A wide variety of enzymes may be directly involved in hydrolytic activities (e.g., cellulases, hemicellulases, and pectinases) or act as oxidases in the production of oxidants that indirectly affect wood components (e.g., ·OH radicals from H_2O_2; pyranose 2-oxidase, and glucose 1-oxidase). A further important characteristic concerns the nature of the enzymes and whether they are associated with extracellular matrix materials such as slime and other polysaccharide-based materials (Daniel, 2014).

To provide a complete overview of the enzymes/enzyme systems involved in the various kinds of wood decay is out of the scope of this chapter and only some brief aspects are covered.

White Rot Fungi: Enzymes Involved and Some Biochemical Aspects of Decay

White rot fungi can produce a wide variety of polysaccharide- (cellulose/hemicellulose/pectins) and lignin-degrading enzymes and are thus capable of complete degradation of wood and lignocellulose materials.

White rot fungi and fungi from several of the other taxonomic groups (shown later) can produce a number of endoglucanases (EC 3.2.14), exoglucanases (*syn*: cellobiohydrolases; EC 3.2.1.91), as well as β-glucosidases (EC 3.2.1.21) for breaking down cellulose. The endoglucanases cause indiscriminant cleavage of the backbone of cellulose chains, exoglucanases attack cellulose chains from the reducing or nonreducing ends, while the β-glucosidases attack cellobiose or cello-oligosaccharides producing glucose. The cellulases are thought to function in close cooperation at sites of wood cell wall decay. A number of white rot fungi (e.g., *Pha. chrysosporium*, *T. versicolor*, and *Pycnoporus cinnabarinus*) are also known to produce cellobiose dehydrogenase (CDH) for oxidizing products produced through hydrolytic activities of cellulases and other hydrolytic enzymes. In comparison to lignin-degrading enzymes, considerably less work has been done on the cellulase and hemicellulase systems of white rot fungi with the possible exception of *Pha. chrysosporium*. For more details on the biodegradation of cellulose by white rot fungi, readers should consult a review by Baldrian and Valášková (2008).

Unlike polymeric cellulose and hemicelluloses that can be degraded by nearly all fungi from all decay groups, complete degradation of lignin is only known for the white rot fungi. In wood, lignin is not linearly orientated but rather randomly orientated with variable linkages. Lignin biomineralization is carried out by random oxidative reactions by a limited number of extracellular oxidative peroxidase and laccase enzymes in conjunction with associated enzymes and cofactors (Hofrichter, 2002; Hammel and Cullen, 2008; Hatakka and Hammel, 2011). Large variations exist between different white rot species not only in the ability to produce the different enzymes but also they can vary over time. This presumably relates to the types of decay patterns they can produce in wood. Three main peroxidase systems have been characterized: (1) lignin peroxidase (*syn* = ligninase) (LiP; EC1.11.1.14) (Tien and Kirk, 1984); (2) manganese peroxidase (MnP; EC1.11.1.13) (Glenn and Gold, 1985; Glenn et al., 1986); and (3) the versatile peroxidases (VPs EC 1.11.1.16) (Ruiz-Dueñas et al., 2001; Pérez-Boada et al., 2005). The enzymes are heme-containing glycoproteins with variable molecular weight all requiring H_2O_2 as the oxidant (Martínez, 2002). Numerous screening studies conducted on white rot fungi have shown MnP as the most frequently detected peroxidase and therefore often regarded as the most important (Orth et al., 1993; Hatakka, 1994; Hofrichter, 2002). In contrast, the VPs have to date only been reported from edible white rot fungi including *Bjerkandera* (*adusta*) and *Pleurotus* (*eryngii*). MnPs and LiPs have been shown using electron microscopy and labeling approaches associated with wood cell wall decay for a variety of simultaneous and preferential white- rot fungi (Daniel, 2014). It is, however, still unknown whether peroxidases are able to penetrate decay zones to the outermost sites of lignin biomineralization as shown for example in Figure 8.5a–c, or whether this is performed by small oxidants, because of restrictions in porosity of wood cell walls or the known molecular size of the enzymes involved (~35–60,000 daltons). Laccases are also known to be involved in fungal physiological processes (Käärik, 1965; Bollag and Leonowicz, 1984; Baldrian, 2006), although their true effect on macromolecular lignin and their ability to cause delignification of wood remains controversial. Despite this, laccases are frequently localized extracellularly at sites of decay (Daniel, 1994; Daniel et al., 2004) and are often produced together with MnP (Hatakka and Hammel, 2011).

Morphological, white and brown rot show similarities in decay patterns as shown in Figs 8.5c and 8.10f. In preferential white rot fungi (Figs 8.9d and 8.5b–d), the lignin and the hemicelluloses are degraded leaving modified cellulose fibers, while with simultaneous white rot a thin zone of decay in which all the lignocellulose components are degraded progresses across cell walls. With brown rot, the cellulose and hemicelluloses are degraded and a modified lignin (i.e., demethylated) skeleton left over. The restricting factor in all three types of decay is the wood cell wall porosity, which limits the penetration and diffusion of hydrolytic and oxidoreductase enzymes. To explain this phenomenon of "decay at a distance" a variety of oxidants have been implicated (Hammel et al., 2002). Basically these extracellular fungal systems include the following: (1) MnP system where Mn(III) chelates are produced that cleave lignin phenolic structures in lignin (Wariishi et al., 1988, 1991); (2) MnP and Mn(III) oxidize organic acids and unsaturated lipids to produce reactive oxygen species like peroxyl radicals (.OOR) (Kapich et al., 1999); (3) the LiP system with H_2O_2 for removing electrons from nonphenolic lignin structures; (4) the CDH system, which reduces Fe(III) chelates in lignocelluloses to Fe(II) oxalates that can react with Fenton reactions with H_2O_2 to produce hydroxyl radicals (H_2O_2 + Fe (II) + H+ --- H_2O + Fe(III) + .OH); and (5) the vetratryl alcohol system where extracellular LiP oxidizes fungal-secreted VA to produce cation radicals that diffuse into the wood cell wall to open lignin structures via electron transfer (Harvey et al., 1985; Candeias and Harvey, 1995; Bietti et al., 1998).

There is little doubt that our understanding of the true biochemical events of white rot are still evolving but that the systems utilized vary with fungal species and over time. Enzymes (hydrolytic/oxidative) presumably play major roles for biomineralization of the main polymers in wood as shown by their localization at sites of decay using microscopy techniques. While it is expected that nonenzymatic systems also play major roles in white rot, whether it is only for opening the wood cell wall for enzymatic penetration or for other processes has yet to be fully confirmed. It is unlikely that the involvement of such nonenzymatic systems will be confirmed using liquid cultures of the fungi but rather studies should be done on wood substrates.

Brown Rot Fungi: Enzymes Involved and Some Biochemical Aspects of Decay

Studies on the enzymatic/nonenzymatic systems employed by brown rot fungi have been driven for many years to understand how this group of fungi can cause rapid loss in wood strength at minimal mass loss. Since brown rot fungi preferentially attack coniferous woods and most construction wood is made of this material, brown rot is the primary cause of decay of wood products in service. In comparison to white-rot fungi, brown-rot fungi have been shown using functional genomics to possess a more limited range of enzymes despite being evolved earlier from white rot saprophytes (Hibbett and Donoghue, 2001; Martinez et al., 2009; Eastwood et al., 2011). This reduction in enzyme diversity is consistent with the more limited effect of brown-rot fungi on lignin even though they can cause extensive modification via demethoxylation. The fact that brown-rot fungi can cause rapid depolymerization of wood despite a reduced enzymatic capability has however led to speculation on whether they are "more" or "less" evolved than white-rot fungi.

Enzymatic and nonenzymatic systems have been implicated in the decay patterns illustrated for brown rot in Figs 8.6b, c and 8.10d, f. In principle, decay is thought to reflect two phases occurring simultaneously or in sequence in which the wood cell wall is "opened up" and during which cellulose depolymerization (i.e., nonenzymatic system) occurs followed by a second phase involving enzymatic decay of the polysaccharides. Evidence for the involvement of a low molecular system comes entirely from the limited ability of "known" polysaccharide degrading enzymes to penetrate lignified wood cell walls and because of the rapidity of depolymerization, which can lead to a rapid reduction in strength as assessed using modulus of rupture (MOR) and elasticity (MOE) testing procedures (Wilcox et al., 1974; Curling et al., 2001). The principal nonenzymatic system postulated as playing a major role in brown rot depolymerization is based on Fenton's chemistry whereby hydroxyl radicals produced via reaction of H_2O_2 with ferrous iron (Fe^{2+}) cause oxidative degradation of cellulose (Koenigs, 1972, 1974). For this mechanism to function *in situ*, the fungi must have a source of Fe^{2+} and H_2O_2 and many of the studies conducted have been directed to proving their origin and involvement. For Fenton's chemistry to function *in situ*, it needs to be activated at a distance from the fungal hyphae thereby not only protecting the fungus from potentially damaging radicals but also to explain the attack of wood components remote from the hyphae, which may be several micrometers depending on the thickness of the cell wall. Also because of the short lifetime of the radicals they need to be produced in very close proximity to the wood components. The hypotheses put forward to date to explain this phenomenon involves the existence of a pH gradient from hyphae in the cell lumen to the outer wood cell wall. Stabile iron (Fe^{3+})-oxalate complexes prevent production of Fe^{2+} and ·OH radicals in the acid environment (i.e., low pH < 3.0) near to hyphae but as the oxalate:Fe complexes diffuses across the wood cell wall, the oxalate has less affinity for Fe^{3+} allowing transfer to the cellulose and hemicelluloses (Goodell et al., 2008; Arantes et al., 2009).

Hydrogen peroxide systems have also been implicated in brown rot decay of wood, with evidence of the oxidant being produced extracellularly by either enzymatic oxidation of methanol (Daniel et al., 2007; Wymelenberg et al., 2010) during lignin demethylation or by reduction of O_2 (Enoki et al., 1997; Hyde and Wood, 1997; Kerem et al., 1999; Daniel et al., 2007). Enzymatically, alcohol (methanol) oxidase (EC:1.1.3.13) are induced by the presence of cellulose in the brown rot fungus *Ol. placenta* (Martinez et al., 2009) and were found extracellularly associated with the hyphae and extracellular slime of *G. trabeum* in liquid cultures and on wood (Daniel et al., 2007).

A key mechanism for driving Fenton's chemistry during brown rot decay is the availability of Fe^{2+} brought about by the reduction of Fe^{3+}. Numerous Fe^{3+} oxidants have been postulated involved including low molecular weight aromatic compounds produced by fungi and through products of decay (Arantes et al., 2012; Arantes and Goodell, 2014). Such Fe^{3+} reductants are advantageous as they are not restricted like enzymes to penetrate wood cell wall matrices and have been isolated from several brown-rot fungi (Shimokawa et al., 2004; Wei et al., 2010). For detailed information, readers should consult a recent review on the topic by Arantes and Goodell (2014). The presence of large quantities of reductants in wood cell walls degraded by brown-rot fungi can be easily demonstrated by treating the wood with osmium tetroxide, which causes production of osmium blacks (i.e., precipitates) throughout degraded regions (Messner et al., 1985). Enzymatic recycling of Fe^{3+} reductants has been postulated to occur via a quinone oxidoreductase (EC: 1.12.5.1) system (Hammel et al., 2002) although greater evidence has been given for chemical mineralization of Fe^{3+} to CO_2 and producing Fe^{2+} (Pracht et al., 2001). Additional evidence for the involvement of intracellular NADH:quinone oxidoreductases has also been presented although at this time it is not known how this enzyme could function *in situ* as Fe^{3+} would need to diffuse to hyphae in the wood cell lumen to be regenerated. Very often with brown rot attack (e.g., *G. trabeum*), decay is widespread throughout the wood structure although the amount of total hyphal development is rather limited so it is not obvious how such a system could function without at least one hypha per cell lumen. A completely different system for Fe^{3+} reduction has been proposed for the brown rot fungus *C. puteana* involving CDH (EC 1.1.99.18) where oxidation of cellodextrins is coupled to the conversion of Fe^{3+} to Fe^{2+} (Hyde and Wood, 1997) by hyphae in the cell lumen. The system is also thought to function by operation of a pH gradient from the hyphae across wood cell walls with diffusion of Fe^{2+}-oxalate and auto-oxidation forming H_2O_2 distant from the hyphae. To date however, the CDH system has only been postulated for *C. puteana* and no other brown rot fungus. Interestingly, *C. puteana* is the only brown rot fungus known to date to possess a complete cellulase system comparable with white rot fungi with endoglucanases, cellobiohydrolase and β-glucosidases (Schmidhalter and Canevascini, 1993a, 1993b).

In comparison to white rot decay, studies on polysaccharide-degrading enzymatic systems of the brown rot fungi are more limited. Figures 8.6a–c and 8.10d–f show phases in the "opening" up of wood cells walls from a solid biopolymeric matrix to a very open cell wall structure. It is apparent from the images that at late stages of decay, the wood cell wall is open for diffusion of polysaccharide-degrading enzymes into the wood cell wall. These observations are consistent with the rapid loss in DP (Cowling, 1963; Highley and

Dashek, 1998) and increase in overall crystallinity due to preferential removal of the hemicelluloses (Howell et al., 2009; Schilling et al., 2009; Fackler et al., 2010; Fackler and Schwanninger, 2012) and amorphous regions of the cellulose microfibrils. Figures 8.6b, c show birch cell walls (after ~45% weight loss) with a very open structure reflecting the original high concentration of hemicelluloses (primarily xylan) and cellulose in the fiber cell walls and low syringyl lignin content. With softwoods the wall structure does not show the same openness because of the comparable higher lignin content and more condense nature of guaiacyl lignin. Brown rot fungi are known to produce a wide variety of hemicellulases (Ritschkoff et al., 1994), endoglucanases, and pectinases (Green et al., 1995) but lack cellobiohydrolases apart from *C. puteana* (Schmidhalter and Canevascini, 1993a, 1993b; Martinez et al., 2009; Hori et al., 2013). Progressive endoglucanases have also been reported produced to compensate for the lack of cellobiohydrolase by Chen et al. (2005). Interestingly, degradation of crystalline cellulose under liquid culture conditions with monocultures or by using purified enzymes has not yet been proven although, crystalline cellulose (e.g., cotton) can be degraded when bound to wood in decay tests such as EN113 (Daniel, unpublished observation) suggesting that the wood substrate may induce special conditions for the fungi.

In summary, evidence suggests the brown rot decay system involves a two-phase attack in which the wood cell wall is initially opened up by a rapid effecting oxidative low molecular weight system, which then provides conditions for an enzymatic system involving hydrolytic and/or oxidative (*C. puteana*) enzyme systems. The large openings visible by TEM in the wood cell wall indicate that diffusion of known cellulase enzymes (i.e., in the range 45–65,000 daltons) into the wall should not be inhibited.

Soft Rot Fungi: Enzymes Involved and Some Biochemical Aspects of Decay

Compared with white and brown rot fungi, much less is known about the degradative enzyme systems produced by ascomycetes and fungi imperfecti during soft rot attack of wood in which cavity and erosion decay occur. Cavity formation in hardwoods is easier as shown by the larger number of cavities formed in a fixed period of time. Currently, there are no indications that enzymatic systems used in cavity formation differ from those used in cell wall erosion. The most important requirement for soft rot cavity formation is the alignment of hyphae with the cellulose microfibrils; the alignment probably inducing the secretion of cell wall degrading cellulases. This indicates that the enzyme systems involved are likely to be closely associated with the fungal hyphae and initially at least present on the hyphal surfaces (Fig. 8.11a–f). The fact that the bioconical cavities are formed by hyphae aligned with the cellulose microfibrils in wood cell walls and not by the thin penetration (i.e., bore hyphae) traversing cell walls also suggests involvement of different or the absence of enzyme systems. Studies indicate that most cellulolytic microfungi cause some erosion of hardwood cell walls even those essentially regarded as typical mold fungi (e.g., *Aspergillus*, *Penicillium*, *Trichoderma* spp.); however, their ability to cause the same attack on softwoods is greatly limited. Frequently, *Trichoderma* spp. are reported as a typical soft rot fungi of wood and thus used as an example for comparison with white and brown rot fungi. However, as indicated earlier, *Trichoderma* cannot be really classified as a true soft rot fungus as it has not been unequivocally confirmed to produce cavities in the secondary cell walls of softwoods.

Few studies have been conducted on the chemical changes in wood following soft rot attack. It is clear however, that since soft rot fungi have an inability to degrade the middle lamella regions in wood, their ability to degrade lignin is more limited than that of white rot fungi and thus an increase in the relative lignin content in wood due to preferential removal of polysaccharides is expected. Lignin degradation has been reported, but it seems related to the fungal species involved. Chemical analyses of lignin from beech wood degraded by soft rot fungi have shown a lower methoxyl content and greater acid-solubility than in undegraded wood (Levi and Preston, 1965). Recent studies on birch wood degraded by a range of fungi imperfecti and ascomycetes have shown decay characterized by lower lignin (Klason) loss compared to white rot and much lower alkali solubility compared with brown rot (Worrall et al., 1997). Lignin peroxidase has been isolated and purified from the ascomycete *Chrysonilia sitophilia* (Durán et al., 1987; Rodríguez et al., 1997), although the fungus has not been shown to degrade wood. Similarly, a range of phenolic and lignin-related compounds have been shown degraded by soft rot fungi (Haider and Trojanowski, 1975; Bugos et al., 1988), but this does not confirm they degrade lignin in wood or cause wood decay. A number of thermophilic ascomycetes (Machuca et al., 1998) and some heat-tolerant soft rot fungi like *Talaromycetes termophilus* and *Thielavia terrestris* are also weakly ligninolytic but their true effect of wood is limited (Dix and Webster, 1995). Observations show extensive removal of wood cell wall materials surrounding cavities produced in the secondary walls (S2, S1 layers) of both hard- and softwoods in advanced stages (Figs 8.7d, e and 8.11a, b, f). However, in both wood species and depending on soft rot fungus, there are generally always large amounts of residual electron materials surrounding and attached to the hyphae (e.g., Fig. 8.11f) (Daniel and Nilsson, 1989). These materials represent partially degraded lignin remaining after preferential cellulose/hemicellulose removal as well as melanin.

The ability of soft fungi to preferentially remove carbohydrates and leave lignin in cavities and in the cell wall during erosion decay indicates the involvement of a diffusive cellulase/hemicellulase system. This is consistent with several large screening studies carried out on soft rot fungi from terrestrial, marine, and freshwater situations showing enzymatic "clearing" (i.e., decay) of ball mill cellulose/cellulose agar and hemicellulose (xylan) agar during growth of these microfungi (Nilsson, 1973; Bucher et al., 2004; Duncan et al., 2006, 2008; Simonis et al., 2008). These enzyme-agar studies showed that clearing zones could occur a considerable distance from the fungal hyphae indicating a highly diffusible enzymatic system that vary with fungal species. The fact that cavity formation and erosion decay is a frequent form of decay in aquatic situations emphasizes not only the success of this form of decay compared to higher fungi, but also that the hydrated cell wall may be advantageous for the enzymatic systems involved. Very little research has been conducted on the cellulolytic and hemicellulolytic systems on true soft rot fungi and few studies on the types of enzymes involved. Screening studies indicate, however, the effective endoglucanase activities, which is consistent with the alignment of hyphae with the cellulose microfibrils in wood cell walls with cavity formation. Soft rot fungi producing cavities and erosion in preservative-treated wood cell walls also have an effective system for immobilization of heavy metals in addition to causing decay.

For an overview of carbohydrate-degrading enzymes produced by wood decay fungi, readers should consult the CAZy database (www.cazy.org) and for fungal oxidoreductases (i.e., lignin-degrading enzymes) the FOLy (Fungal Oxidative Lignin Enzymes) database (https://foly-db.esil.univ-mrs.fr/) (Levasseur et al., 2008), and the more recently updated CAZy database that includes auxiliary activities and covers redox enzymes (http://www.cazy.org/Auxiliary-Activities.html). The aim of the databases is to provide an overview of cellulose- and lignin-degrading enzymes for biotechnical applications.

ACKNOWLEDGMENT

The author gratefully acknowledges funding provided by FORMAS projects 2008-1399, 2009-582 & 418 2011-416.

REFERENCES

Anagnost, S.E., Worrall, J.J., Wang, C.K., 1994. Diffuse cavity formation in soft rot of pine. Wood Sci. Technol. 28, 199–208.

Arantes, V., Goodell, B., 2014. Current understanding of brown-rot fungal biodegradation mechanisms: a review. In: Schultz, T.P., Goodell, B., Nicholas, D.D. (Eds.), Deterioration and Protection of Sustainable Biomaterials. American Chemical Society, Washington DC, pp. 3–21.

Arantes, V., Qian, Y., Milagres, A.M.F., Jellison, J., Goodell, B., 2009. Effect of pH and oxalic acid on the reduction of Fe^{3+} by a biomimetic chelator and on Fe^{3+} desorption/adsorption onto wood: implications for brown-rot decay. Int. Biodeterior. Biodegrad. 63, 478–483.

Arantes, V., Jellison, J., Goodell, B., 2012. Peculiarities of brown-rot fungi and biochemical Fenton reaction with regard to their potential as a model for bioprocessing biomass. Appl. Microbiol. Biotechnol. 94, 323–338.

Baldrian, P., 2006. Fungal laccases – occurrence and properties. FEMS Microbiol. Rev. 30, 215–242.

Baldrian, P., Valášková, V., 2008. Degradation of cellulose by basidiomycetous fungi. FEMS Microbiol. Rev. 32, 501–521.

Bietti, M., Baciocchi, E., Steenken, S., 1998. Lifetime, reduction potential and base-induced fragmentation of the veratryl alcohol radical cation in aqueous solution. Pulse radiolysis studies on a ligninase "mediator". J. Phys. Chem. A 102, 7337–7342.

Blanchette, R.A., 1991. Delignification by wood-decay fungi. Annu. Rev. Phytopathol. 29, 381–398.

Blanchette, R.A., Otjen, L., Carlson, M.C., 1987. Lignin distribution in cell walls of birch wood decayed by white rot basidiomycetes. Phytopathology 77, 684–690.

Blanchette, R.A., Nilsson, T., Daniel, G., Abad, A., 1990. Biological degradation of wood. In: Rowell, R.M., Barbour, R.J. (Eds.), Archaeological Wood - Properties, Chemistry, and Preservation. American Chemical Society, Washington DC, pp. 141–174.

Blanchette, R.A., Obst, J.R., Timell, T.E., 1994. Biodegradation of compression wood and tension wood by white and brown rot fungi. Holzforschung 48, 34–42.

Bollag, J.-M., Leonowicz, A., 1984. Comparative studies of extracellular fungal laccases. Appl. Environ. Microbiol. 48, 849–854.

Bucher, V.V.C., Hyde, K.D., Pointing, S.B., Reddy, C.A., 2004. Production of wood decay enzymes, mass loss and lignin solubilization in wood by marine ascomycetes and their anamorphs. Fungal Divers. 15, 1–14.

Bugos, R.C., Sutherland, J.B., Adler, J.H., 1988. Phenolic compound utilization by the soft rot fungus *Lecythophora hoffmannii*. Appl. Environ. Microbiol. 54, 1882–1885.

Butcher JA, Nilsson T., 1982. Influence of variable lignin content amongst hardwoods of soft-rot susceptibility and performance of CCA preservatives. The International Research Group on Wood Preservation (IRG/WP No. 1151).

Candeias, L.P., Harvey, P.J., 1995. Lifetime and reactivity of the veratryl alcohol radical cation. Implications for lignin peroxidase catalysis. J. Biol. Chem. 270, 16745–16748.

Carlquist, S., 2001. Comparative Wood Anatomy: Systematic, Ecological, and Evolutionary Aspects of Dicotyledon Wood. Springer-Verlag, Berlin.

Cohen, R., Suzuki, M.R., Hammel, K.E., 2005. Processive endoglucanase active in crystalline cellulose hydrolysis by the brown rot basidiomycete *Gloeophyllum trabeum*. Appl. Environ. Microbiol. 71, 2412–2417.

Côté, W.A., 1977. Wood ultrastructure in relation to chemical composition. In: Lorwus, F.A., Runeckles, V.C. (Eds.), The Structure, Biosynthesis, and Degradation of Wood. Plenum Press, New York, pp. 1–44.

Cowling, E.B., 1963. Structural features of cellulose that influence its susceptibility to enzymatic hydrolysis. In: Reese, E.T. (Ed.), Advances in Enzymic Hydrolysis of Cellulose and Related Materials: including a Bibliography for the Years 1950-1961. Pergamon Press, London, pp. 1–32.

Crossley, A., 1979. The use of electron microscopy to compare wood decay mechanisms. PhD thesis, Imperial College of Science and Technology, UK.

Curling, S., Clausen, C.A., Winandy, J.E., 2001. The effect of hemicellulose degradation on the mechanical properties of wood during brown rot decay. The International Research Group on Wood Preservation (IRG/WP No. 01-20219).

Daniel, G., 1994. Use of electron microscopy for aiding our understanding of wood biodegradation. FEMS Microbiol. Rev. 13, 199–233.

Daniel, G., 2003. Microview of wood under degradation by bacteria and fungi. In: Goodell, B., Nicholas, D.D., Schultz, T.P. (Eds.), Wood Deterioration and Preservation: Advances in Our Changing World. American Chemical Society, Washington DC, pp. 34–72.

Daniel, G., 2009. Wood and fibre morphology. In: Monica, E.K., Gellerstedt, G., Henriksson, G. (Eds.), Pulp and Paper Chemistry and Technology Vol. 1: Wood Chemistry and Wood Biotechnology. Walter de Gruyter, Göttingen, pp. 45–75.

Daniel, G., 2014. Fungal and bacterial biodeterioration: white rots, brown rots, soft rots and bacteria. In: Schultz, T.P., Goodell, B., Nicholas, D.D. (Eds.), Deterioration and Protection of Sustainable Biomaterials. American Chemical Society, Washington DC, pp. 23–58.

Daniel, G., Bergman, Ö., 1997. White rot and manganese deposition in TnBTO-AAC preservative treated pine stakes from field tests. Holz. Roh-Werkst. 55, 197–201.

Daniel, G., Nilsson, T., 1988. Studies on preservative tolerant *Phialophora* species. Int. Biodeterior. 24, 327–335.

Daniel, G., Nilsson, T., 1989. Interactions between soft rot fungi and CCA preservatives in *Betula verrucosa*. J. Inst. Wood Sci. 11, 162–171.

Daniel, G., Nilsson, T., 1998. Developments in the study of soft rot and bacterial decay. In: Bruce, A., Palfreyman, J.W. (Eds.), Forest Products Biotechnology. CRC Press, London, pp. 37–62.

Daniel, G., Nilsson, T., Pettersson, B., 1989. Intra- and extracellular localization of lignin peroxidase during the degradation of solid wood and wood fragments by *Phanerochaete chrysosporium* by using transmission electron microscopy and immuno-gold labeling. Appl. Environ. Microbiol. 55, 871–881.

Daniel, G., Pettersson, B., Nilsson, T., Volc, J., 1990. Use of immunogold cytochemistry to detect Mn (II)-dependent and lignin peroxidases in wood degraded by the white rot fungi *Phanerochaete chrysosporium* and *Lentinula edodes*. Can. J. Bot. 68, 920–933.

Daniel, G., Volc, J., Nilsson, T., 1992. Soft rot and multiple T-branching by the basidiomycete *Oudemansiella mucida*. Mycol. Res. 96, 49–54.

Daniel, G., Volc, J., Kubatova, E., 1994. Pyranose oxidase, a major source of H_2O_2 during wood degradation by *Phanerochaete chrysosporium*, *Trametes versicolor*, and *Oudemansiella mucida*. Appl. Environ. Microbiol. 60, 2524–2532.

Daniel, G., Volc, J., Niku-Paavola, M.-L., 2004. Cryo-FE-SEM & TEM immuno-techniques reveal new details for understanding white-rot decay of lignocellulose. C.R. Biologies 327, 861–871.

Daniel, G., Volc, J., Filonova, L., Plíhal, O., Kubátová, E., Halada, P., 2007. Characteristics of *Gloeophyllum trabeum* alcohol oxidase, an extracellular source of H_2O_2 in brown rot decay of wood. Appl. Environ. Microbiol. 73, 6241–6253.

Dix, N.J., Webster, J., 1995. Colonization and decay of wood. In: Dix, N.J., Webster, J. (Eds.), Fungal Ecology. Springer, Wallington, pp. 145–171.

Duncan, S.M., Farrell, R.L., Thwaites, J.M., Held, B.W., Arenz, B.E., Jurgens, J.A., Blanchette, R.A., 2006. Endoglucanase producing fungi isolated from Cape Evans historic expedition hut on Ross Island. Antarctica. Environ. Microbiol. 8, 1212–1219.

Duncan, S.M., Minasaki, R., Farrell, R.L., Thwaites, J.M., Held, B.W., Arenz, B.E., Jurgens, J.A., Blanchette, R.A., 2008. Screening fungi isolated from historic Discovery Hut on Ross Island, Antarctica for cellulose degradation. Antarct. Sci. 20, 463–470.

Durán, N., Ferrer, I., Rodríguez, J., 1987. Ligninases from *Chrysonilia sitophila* (TFB-27441 strain). Appl. Biochem. Biotechol. 16, 157–167.

Eastwood, D.C., Floudas, D., Binder, M., Majcherczyk, A., Schneider, P., Aerts, A., Asiegbu, F.O., Baker, S.E., Barry, K., Bendiksby, M., et al.,2011. The plant cell wall – Decomposing machinery underlies the functional diversity of forest fungi. Science 333, 762–765.

Eaton, R.A., Hale, M.D., 1993. Wood: Decay, Pests and Protection. Chapman and Hall Ltd, London, New York.

Encinas, O., Daniel, G., 1995. Wood cell wall biodegradation by the blue stain fungus *Botryodiplodia theobromae*. Pat. Mater. Org. 29, 255–272.

Encinas, O., Daniel, G., 1997. Degradation of the gelatinous layer in aspen and rubberwood by the blue stain fungus *Lasiodiplodia theobromae*. IAWA J. 18, 107–115.

Enoki, A., Itakura, S., Tanaka, H., 1997. The involvement of extracelluar substances for reducing molecular oxygen to hydroxyl radical and ferric iron to ferrous iron in wood degradation by wood decay fungi. J. Biotechnol. 53, 265–272.

Eriksson, K-E.L., Blanchette, R.A., Ander, P., 1990. Microbial and Enzymatic Degradation of Wood and Wood Components. Springer-Verlag, Berlin.

Fackler, K., Schwanninger, M., 2012. How spectroscopy and microspectroscopy of degraded wood contribute to understand fungal wood decay. Appl. Microbiol. Biotechnol. 96, 587–599.

Fackler, K., Stevanic, J.S., Ters, T., Hinterstoisser, B., Schwanninger, M., Salmén, L., 2010. Localisation and characterisation of incipient brown-rot decay within spruce wood cell walls using FT-IR imaging microscopy. Enzyme Microb. Technol. 47, 257–267.

Gersonde, M., Kerner-Gang, W., 1976. Review of information available for development of method for testing wood preservatives with soft rot fungi. Int. Biodeterior. Bull. 12, 5–13.

Gilbertson, R.L., 1980. Wood-rotting fungi of North America. Mycologia 72, 1–49.

Glenn, J.K., Gold, M.H., 1985. Purification and characterization of an extracellular Mn (II)-dependent peroxidase from the lignin-degrading basidiomycete, *Phanerochaete chrysosporium*. Arch. Biochem. Biophys. 242, 329–341.

Glenn, J.K., Akileswaran, L., Gold, M.H., 1986. Mn (II) oxidation is the principal function of the extracellular Mn-peroxidase from *Phanerochaete chrysosporium*. Arch. Biochem. Biophys. 251, 688–696.

Goodell, B., Jellison, J., Liu, J., Daniel, G., Paszczynski, A., Fekete, F., Krishnamurthy, S., Jun, L., Xu, G., 1997. Low molecular weight chelators and phenolic compounds isolated from wood decay fungi and their role in the fungal biodegradation of wood. J. Biotechnol. 53, 133–162.

Goodell, B., Nicholas, D.D., Schultz, T.P., 2003. Wood Deterioration and Preservation: Advances in our Changing World. American Chemical Society, Washington DC.

Goodell, B., Qian, Y., Jellison, J., 2008. Fungal decay of wood: soft rot-brown rot-white rot. In: Schultz, T.P., Nicholas, D.D., Militz, H., Freeman, M.H., Goodell, B. (Eds.), Development of Commercial Wood Preservatives: Efficacy, Environment, and Health Issues. American Chemical Society, Washington DC, pp. 9–31.

Green, III, F., Clausen, C.A., Kuster, T.A., Highley, T.L., 1995. Induction of polygalacturonase and the formation of oxalic acid by pectin in brown-rot fungi. World J. Microbiol. Biotechnol. 11, 519–524.

Haider, K., Trojanowski, J., 1975. Decomposition of specifically ^{14}C-labelled phenols and dehydropolymers of coniferyl alcohol as models for lignin degradation by soft and white rot fungi. Arch. Microbiol. 105, 33–41.

Hale, M.D., Eaton, R.A., 1985a. The ultrastructure of soft rot fungi. II. Cavity-forming hyphae in wood cell walls. Mycologia 77, 594–605.

Hale, M.D., Eaton, R.A., 1985b. The ultrastructure of soft rot fungi. I. Fine hyphae in wood cell walls. Mycologia 77, 447–463.

Hammel, K.E., Cullen, D., 2008. Role of fungal peroxidases in biological ligninolysis. Curr. Opin. Plant Biol. 11, 349–355.

Hammel, K.E., Kapich, A.N., Jensen, K.A., Ryan, Z.C., 2002. Reactive oxygen species as agents of wood decay by fungi. Enzyme Microb. Technol. 30, 445–453.

Hartig, R., 1874. Wichtige Krankheiten der Waldbäume. Beiträge zur Mykologie und Phytopathologie für Botaniker und Forstmänner. Springer, Berlin Heidelberg, New York.

Harvey, P.J., Schoemaker, H.E., Bowen, R.M., Palmer, J.M., 1985. Single-electron transfer processes and the reaction mechanism of enzymic degradation of lignin. FEBS Lett. 183, 13–16.

Hatakka, A., 1994. Lignin-modifying enzymes from selected white-rot fungi: production and role from in lignin degradation. FEMS Microbiol. Rev. 13, 125–135.

Hatakka, A., Hammel, K.E., 2011. Fungal biodegradation of lignocelluloses. Hofrichter, M. (Ed.), The Mycota – Industrial Applications, 10, 2nd ed. Springer-Verlag, Berlin Heidelberg, pp. 319–340.

Henningsson, B., Nilsson, T., 1976. Some aspects on microflora and the decomposition on preservative-treated wood in ground contact. Mater. Org. 3, 307–318.

Hibbett, D.S., Donoghue, M.J., 2001. Analysis of character correlations among wood decay mechanisms, mating systems, and substrate ranges in homobasidiomycetes. Syst. Biol. 50, 215–242.

Highley, T.L., Dashek, W.V., 1998. Biotechnology in the study of brown- and white-rot decay. In: Bruce, A., Palfreyman, J.W. (Eds.), Forest Products Biotechnology. CRC Press, London, pp. 15–36.

Highley, T.L., Murmanis, L., Palmer, J., 1985. Micromorphology of degradation in western hemlock and sweetgum by the brown-rot fungus *Poria placenta*. Holzforschung 39, 73–78.

Hofrichter, M., 2002. Review: lignin conversion by manganese peroxidase (MnP). Enzyme Microb. Technol. 30, 454–466.

Hori, C., Gaskell, J., Igarashi, K., Samejima, M., Hibbett, D., Henrissat, B., Cullen, D., 2013. Genome wide analysis of polysaccharides degrading enzymes in 11 white- and brown-rot Polyporales provides insight into mechanisms of wood decay. Mycologia 105, 1412–1427.

Howell, C., Hastrup, A.C.S., Goodell, B., Jellison, J., 2009. Temporal changes in wood crystalline cellulose during degradation by brown rot fungi. Int. Biodeterior. Biodegrad. 63, 414–419.

Hyde, S.M., Wood, P.M., 1997. A mechanism for production of hydroxyl radicals by the brown-rot fungus *Coniophora puteana*: Fe(III) reduction by cellobiose dehydrogenase and Fe(II) oxidation at a distance from the hyphae. Microbiology 143, 259–266.

Jane, F.W., 1956. The Structure of Wood. Adam and Charles Black, London.

Käärik, A., 1965. The identification of the mycelia of wood-decay fungi by their oxidation reactions with phenolic compounds. Stud. For. Suec. 31, 1–80.

Kapich, A., Hofrichter, M., Vares, T., Hatakka, A., 1999. Coupling of manganese peroxidase-mediated lipid peroxidation with destruction of non-phenolic lignin model compounds and ^{14}C-labeled lignins. Biochem. Biophy. Res. Comm. 259, 212–219.

Kerem, Z., Jensen, K.A., Hammel, K.E., 1999. Biodegradative mechanism of the brown rot basidiomycete *Gloeophyllum trabeum*: evidence for an extracellular hydroquinone-driven Fenton reaction. FEBS Lett. 446, 49–54.

Khalili, S., Daniel, G., Nilsson, T., 2000. Use of soft rot fungi for studies on the microstructure of kapok (*Ceiba pentandra* (L.) Gaertn.) fibre cell walls. Holzforschung 54, 229–233.

Kim, J.S., Gao, J., Daniel, G., 2015. Ultrastructure and immunocytochemistry of degradation in spruce and ash sapwood by the brown rot fungus *Postia placenta*: characterization of incipient stages of decay and variation in decay process. Int. Biodeterior. Biodegrad. 103, 161–178.

Kleist, G., Schmitt, U., 2001. Characterisation of a soft rot-like decay pattern caused by *Coniophora puteana* (Schum.) Karst. in Sapelli wood (*Entandrophragma cylindricum* Sprague). Holzforschung 55, 573–578.

Koenigs, J.W., 1972. Effects of hydrogen peroxide on cellulose and on its susceptibility to cellulase. Mater. Org. 7, 133–147.

Koenigs, J.W., 1974. Hydrogen peroxide and iron: a proposed system for decomposition of wood by brown-rot basidiomycetes. Wood Fiber Sci. 6, 66–80.

Kollmann, F.F., Cote, Jr., W.A., 1968. Principles of Wood Science and Technology. Vol I. Solid Wood. Springer-Verlag, New York.

Lehringer, C., Koch, G., Adusumalli, R.-B., Mook, W.M., Richter, K., Militz, H., 2011. Effect of *Physisporinus vitreus* on wood properties of Norway spruce. Part 1: aspects of delignification and surface hardness. Holzforschung 65, 711–719.

Leightley, L.E., 1977. Mechanisms of timber degradation by aquatic fungi. PhD thesis, Portmouth University, Portsmouth, UK.

Leightley, L., Eaton, R., 1978. Mechanisms of Decay of Timber by Aquatic Micro-organisms. Proceedings of the Annual Convention of British Wood Preserving Association. British Wood Preserving Association, Cambridge, pp. 221–246.

Levasseur, A., Piumi, F., Coutinho, P.M., Rancurel, C., Asther, M., Delattre, M., Henrissat, B., Pontarotti, P., Asther, M., Record, E., 2008. FOLy: an integrated database for the classification and functional annotation of fungal oxidoreductases potentially involved in the degradation of lignin and related aromatic compounds. Fungal Genet. Biol. 45, 638–645.

Levi, M.P., Preston, R.D., 1965. A chemical and microscopic examination of the action of the soft-rot fungus *Chaetomium globosum* on beechwood (*Fagus sylvaticus*). Holzforschung 19, 183–190.

Lhate, I., Cuvilas, C., Terziev, N., Jirjis, R., 2010. Chemical composition of traditionally and lesser used wood species from Mozambique. Wood Mater. Sci. Eng. 5, 143–150.

Machuca, A., Aoyama, H., Durán, N., 1998. Production and characterization of thermostable phenol oxidases of the ascomycete *Thermoascus aurantiacus*. Biotechnol. Appl. Biochem. 27, 217–223.

Martínez, A.T., 2002. Molecular biology and structure-function of lignin-degrading heme peroxidases. Enzyme Microb. Technol. 30, 425–444.

Martinez, D., Challacombe, J., Morgenstern, I., Hibbett, D., Schmoll, M., Kubicek, C.P., Ferreira, P., Ruiz-Duenas, F.J., Martinez, A.T., Kersten, P., et al.,2009. Genome, transcriptome, and secretome analysis of wood decay fungus *Postia placenta* supports unique mechanisms of lignocellulose conversion. Proc. Natl. Acad. Sci. USA 106, 1954–1959.

Messner, K., Srebotnik, E., 1994. Biopulping: an overview of developments in an environmentally safe paper-making technology. FEMS Microbiol. Rev. 13, 351–364.

Messner, K., Foisner, R., Stachelberger, H., Röhr, M., 1985. Osmiophilic particles as a typical aspect of brown and white rot systems in transmission electron microscope studies. Trans. Br. Mycol. Soc. 84, 457–466.

Moss, S.T., 1986. The Biology of Marine Fungi. Cambridge University Press, Cambridge.

Mouzouras, R., Jones, E.B.G., Venkatasamy, R., Holt, D.M., 1988. Microbial decay of lignocellulose in the marine environment. In: Thompson, M.F., Sarojini, R., Nagabhushanam, R. (Eds.), Marine Biodeterioration, Advanced Techniques Applicable to the Indian Ocean. Oxford & IBH publishing, New Delhi, pp. 329–354.

Nilsson, T., 1973. Studies on wood degradation and cellulolytic activity of microfungi. Stud. For. Suec. 104, 1–40.

Nilsson, T., Daniel, G., 1988. Micromorphology of the decay caused by *Chodrostereum purpureum* (Pers. Fr) Pouzar and *Flammulina velutipes* (Curst. Fr.) Singer. The International Research Group on Wood Preservation (IRG/WP No. 1358).

Nilsson, T., Obst, J.R., Daniel, G., 1988. The possible significance of the lignin content and lignin type on the performance of CCA-treated timber in ground contact. The International Research Group on Wood Preservation (IRG/WP No. 1357).

Nilsson, T., Daniel, G., Kirk, T.K., Obst, J.R., 1989. Chemistry and microscopy of wood decay by some higher ascomycetes. Holzforschung 43, 11–18.

Nilsson, T., Daniel, G., Rowell, R., Barbour, R., 1990. Structure and the aging process of dry archaeological wood. In: Rowell, R.M., Barbour, R.J. (Eds.), Archaeological Wood: Properties, Chemistry, and Preservation. American Chemical Society, Washington DC, pp. 67–86.

Orth, A.B., Royse, D.J., Tien, M., 1993. Ubiquity of lignin-degrading peroxidases among various wood-degrading fungi. Appl. Environ. Microbiol. 59, 4017–4023.

Panshin, A.J., Zeeuw, Cd., 1970. Textbook of Wood Technology, Vol. I: Structure, Identification, Properties and Uses of the Commercial Woods of the United States and Canada. McGraw-Hill College, New York.

Pérez-Boada, M., Ruiz-Dueñas, F.J., Pogni, R., Basosi, R., Choinowski, T., Martínez, M.J., Piontek, K., Martínez, A.T., 2005. Versatile peroxidase oxidation of high redox potential aromatic compounds: site-directed mutagenesis, spectroscopic and crystallographic investigation of three long-range electron transfer pathways. J. Mol. Biol. 354, 385–402.

Pracht, J., Boenigk, J., Isenbeck-Schröter, M., Keppler, F., Schöler, H., 2001. Abiotic Fe(III) induced mineralization of phenolic substances. Chemosphere 44, 613–619.

Ritschkoff, A.-C., Buchert, J., Viikari, L., 1994. Purification and characterization of a thermophilic xylanase from the brown-rot fungus *Gloeophyllum trabeum*. J. Biotechnol. 32, 67–74.

Rodríguez, J., Ferraz, A., Nogueira, R.F., Ferrer, I., Esposito, E., Durán, N., 1997. Lignin biodegradation by the ascomycete *Chrysonilia sitophila*. Appl. Biochem. Biotechnol. 62, 233–242.

Ruel, K., Odier, E., Joseleau, J-p., 1990. Immunocytochemical observation of fungal enzymes during degradation of wood cell walls by *Phanerochaete chrysosporium*. In: Kirk, T.K., Chang, H.-M. (Eds.), Biotechnology in Pulp and Paper Manufacture: Applications and Fundamental Investigations. Butterworth-Heinemann, Boston, pp. 83–97.

Ruiz-Dueñas, F., Camarero, S., Pérez-Boada, M., Martinez, M.J., Martinez, A.T., 2001. A new versatile peroxidase from *Pleurotus*. Biochem. Soc. Trans. 29, 116–122.

Saka, S., Whiting, P., Fukazawa, K., Goring, D.A.I., 1982. Comparative studies on lignin distribution by UV microscopy and bromination combined with EDXA. Wood Sci. Technol. FEBS Lett. 16, 269–277.

Savory, J.G., 1954. Breakdown of timber by ascomycetes and fungi imperfecti. Ann. Appl. Biol. 41, 336–347.

Schilling, J.S., Tewalt, J.P., Duncan, S.M., 2009. Synergy between pretreatment lignocellulose modifications and saccharification efficiency in two brown rot fungal systems. Appl. Microbiol. Biotechnol. 84, 465–475.

Schmidhalter, D.R., Canevascini, G., 1993a. Purification and characterization of two exocellobiohydrolases from the brown-rot fungus *Coniophora puteana* (Schum ex Fr) Karst. Arch. Biochem. Biophys. 300, 551–558.

Schmidhalter, D.R., Canevascini, G., 1993b. Isolation and characterization of the cellobiose dehydrogenase from the brown-rot fungus *Coniophora puteana* (Schum ex Fr.) Karst. Arch. Biochem. Biophys. 300, 559–563.

Schubert, M., Engel, J., Thöny-Meyer, L., Schwarze, F., Ihssen, J., 2012. Protection of wood from microorganisms by laccase-catalyzed iodination. Appl. Environ. Microbiol. 78, 7267–7275.

Schwarze, F.W., 2007. Wood decay under the microscope. Fungal Biol. Rev. 21, 133–170.

Schwarze, F.W., Landmesser, H., 2000. Preferential degradation of pit membranes within tracheids by the basidiomycete *Physisporinus vitreus*. Holzforschung 54, 461–462.

Schwarze, F.W., Schubert, M., 2011. *Physisporinus vitreus*: a versatile white rot fungus for engineering value-added wood products. Appl. Microbiol. Biotechnol. 92, 431–440.

Schwarze, F., Lonsdale, D., Fink, S., 1995. Soft rot and multiple T-branching by the basidiomycete *Inonotus hispidus* in ash and London plane. Mycol. Res. 99, 813–820.

Schwarze, F.W., Engels, J., Mattheck, C., 2000. Fungal Strategies of Wood Decay in Trees. Springer-Verlag, Berlin Heidelberg.

Shimokawa, T., Nakamura, M., Hayashi, N., Ishihara, M., 2004. Production of 2,5-dimethoxyhydroquinone by the brown-rot fungus *Serpula lacrymans* to drive extracellular Fenton reaction. Holzforschung 58, 305–310.

Simonis, J.L., Raja, H.A., Shearer, C.A., 2008. Extracellular enzymes and soft rot decay: are ascomycetes important degraders in fresh water. Fungal Divers. 31, 135–146.

Srinivasan, C., Dsouza, T.M., Boominathan, K., Reddy, C.A., 1995. Demonstration of laccase in the white rot basidiomycete *Phanerochaete chrysosporium* BKM-F1767. Appl. Environ. Microbiol. 61, 4274–4277.

Tien, M., Kirk, T.K., 1984. Lignin-degrading enzyme from *Phanerochaete chrysosporium*: purification, characterization, and catalytic properties of a unique H_2O_2-requiring oxygenase. Proc. Natl. Acad. Sci. USA 81, 2280–2284.

Wariishi, H., Akileswaran, L., Gold, M.H., 1988. Manganese peroxidase from the basidiomycete *Phanerochaete chrysosporium*: spectral characterization of the oxidized states and the catalytic cycle. Biochemistry 27, 5365–5370.

Wariishi, H., Valli, K., Gold, M.H., 1991. *In vitro* depolymerization of lignin by manganese peroxidase of *Phanerochaete chrysosporium*. Biochem. Biophys. Res. Commun. 176, 269–275.

Wei, D., Houtman, C.J., Kapich, A.N., Hunt, C.G., Cullen, D., Hammel, K.E., 2010. Laccase and its role in production of extracellular reactive oxygen species during wood decay by the brown rot basidiomycete *Postia placenta*. Appl. Environ. Microbiol. 76, 2091–2097.

Wilcox, W.W., Parameswaran, N., Liese, W., 1974. Ultrastructure of brown rot in wood treated with pentachlorophenol. Holzforschung 28, 211–217.

Worrall, J.J., Anagnost, S.E., Zabel, R.A., 1997. Comparison of wood decay among diverse lignicolous fungi. Mycologia 89, 199–219.

Wymelenberg, A.V., Gaskell, J., Mozuch, M., Sabat, G., Ralph, J., Skyba, O., Mansfield, S.D., Blanchette, R.A., Martinez, D., Grigoriev, I., 2010. Comparative transcriptome and secretome analysis of wood decay fungi *Postia placenta* and *Phanerochaete chrysosporium*. Appl. Environ. Microbiol. 76, 3599–3610.

Zabel, R.A., Morrell, J.J., 1992. Wood Microbiology: Decay and its Prevention. Academic Press, San Diego, CA.

Zabel, R.A., Lombard, F.F., Wang, C.J.K., Terracina, F.C., 1985. Fungi associated with decay in treated southern pine utility poles in the eastern United States. Wood Fiber Sci. 17, 75–91.

Chapter 9

Bacterial Degradation of Wood

Adya P. Singh*, Yoon Soo Kim**, Tripti Singh*
*Manufacturing and Bioproducts Group, Scion (New Zealand Forest Research Institute), Rotorua, New Zealand; **Department of Wood Science and Engineering, Chonnam National University, Gwangju, South Korea

Chapter Outline

INTRODUCTION

Bacteria are ubiquitous in nature and can utilize a wide range of substrates for their nutrition, from simple sugars and complex molecules to highly toxic substances. Therefore, it is not surprising that certain types of bacteria can inhabit standing trees and also colonize and attack harvested wood and wood products placed in service. Whereas, fungi play a dominant role in wood decay, certain groups of bacteria can also degrade lignin-containing cell walls (reviewed in Blanchette et al., 1990; Singh and Butcher, 1991; Daniel and Nilsson, 1998;

Kim and Singh, 2000). The degradation patterns produced by bacteria are strikingly different from the decay patterns produced by fungi, which makes it possible to distinguish bacterial degradation from fungal degradation, and causal bacteria have been named tunneling and erosion bacteria (reviewed in Singh and Butcher, 1991; Daniel and Nilsson, 1998).

Compared to white and brown rot fungi, wood-degrading bacteria (WDB) are regarded to possess greater tolerance to toxic chemicals (Clausen, 1996), for example, heartwood extractives (Nilsson et al., 1992) and wood preservatives (copper-chrome-arsenate, pentachlorophenol), as they can degrade preservative-treated wood exposed in service under a range of conditions, such as in contact with soil (Singh et al., 1994) and as part of cooling towers (Singh et al., 1992). Tunneling bacteria can also degrade a range of lignocellulosic substrates with high lignin content, such as the tropical hardwood species *Alstonia scholaris* and *Homalium foetidum*, stems of the fern *Pteridium*, highly lignified middle lamella in conifer wood (Singh and Butcher, 1991; Daniel and Nilsson, 1998), and lignin-rich outer S2 wall in compression wood of *Pinus radiata* (Singh, 1997a).

WDB are widely present in nature in terrestrial and aquatic environments, and bacterial degradation of cell wall has been reported in wood samples obtained from different parts of the world. While tunneling bacteria (TB) seem to require the presence of oxygen for their activity, erosion bacteria (EB) can tolerate conditions with extremely low levels of oxygen (Singh et al., 1990). Wooden objects placed in situations where they become saturated with water resulting in anoxic conditions or in environments supporting limited oxygen availability are predominantly attacked by erosion bacteria (Donaldson and Singh, 1990; Blanchette et al., 1991a, 1991b; Kim and Singh 1994; Blanchette 1995; Kim et al. 1996; Björdal et al., 1999; Björdal, 2000; Kim and Singh, 2000; Singh et al., 2003; Hoffmann et al., 2004; Schmitt et al., 2005; Klaassen, 2008; Singh, 2012; Rehbein et al., 2013; Cha et al., 2014). In situations with adequate moisture and oxygen, often mixed soft rot and bacterial attacks represent. For example, in preservative-treated timbers exposed in horticultural soils, combined attacks by EB and TB have been observed, and in timbers placed in cooling towers mixed attacks by TB and soft rot fungi have been reported (Singh et al., 1992).

Understanding bacterial degradation of wood is important not only for scientific but also economic reasons. For example, evaluating conditions that support bacterial attacks on wood and degradation of cell walls can be helpful in developing protection strategies for enhancing the service life of wooden products. Bacteria are the primary degraders of wood and wooden structures of historical and cultural importance, particularly those recovered from oxygen-deficient environments. Detailed information on the state of cell walls at the micro- and ultrastructural levels (Singh, 2012) can be useful in developing and fine-tuning restoration methods. Although the main focus of this review is on wood-degrading bacteria, a brief account of certain cellulolytic bacteria is included.

WOOD-DEGRADING BACTERIA AND DEGRADATION PATTERNS

Based primarily on ultrastructural descriptions of the micromorphology of the patterns produced during cell wall degradation, two distinctly different bacterial decay types have been recognized and named tunneling and erosion and correspondingly bacteria causing the decay tunneling and erosion. The classification is based on an examination of a large number of wood samples from natural environments as well as laboratory studies using mixed bacterial cultures to reproduce bacterial decay patterns (reviewed in Blanchette et al., 1990; Daniel and Nilsson, 1998).

Since the application of EM to unravel bacterial decay patterns, wood scientists and technologists have been able to distinguish bacterial degradation from fungal decay patterns using light microscopy if the degradation is not too advanced. However, in advanced stages of degradation, where cell wall integrity may be essentially lost and bacteria may no longer be present in extensively degraded wood tissue regions, recognition as to a particular type of bacterial degradation can greatly benefit from combining LM with EM, and particularly transmission electron microscopy (TEM) even in samples containing mixed bacterial or bacterial and fungal attacks, as is the case in wood exposed to natural environments or in service (Singh et al., 1992).

SLIME

In terrestrial and aquatic environments, bacteria attach to a solid substrate with the help of a slimy material they produce extracellularly in generous quantities. The slime is specific to bacterial species, which also differ in the amount of slime they can produce. Chemical characterization suggests that bacterial slime is composed of polyanionic polysaccharides. Support also comes from electron microscopic studies employing ruthenium red in the fixative to stain the slime. The slime stains intensely with this reagent, suggesting the presence of acidic polysaccharides. However, the slimes produced by WDB have yet to be chemically characterized. Extracellular slime production in response to contact with a solid substrate is not only a feature of bacteria but also of fungi. In addition to adhesion, slimes are considered to have a variety of other roles. For example, the proposed functions for the slime produced by fungi and bacteria are the protection from desiccation and other environmental stresses, concentration of substrates, efficient delivery of enzymes, and immobilization of heavy metals such as copper (reviewed in Singh and Butcher, 1991). Using TEM-EDAX of copper-chrome-arsenate (CCA) treated wood, it has been demonstrated that the extracellular slime of TB bind CCA components, a possible mechanism to prevent toxic CCA components from coming in contact with vital bacterial cell components.

COLONIZATION

The pattern of colonization of wood is similar for TB and EB (reviewed in Blanchette et al., 1990; Singh and Butcher, 1991; Daniel and Nilsson, 1998). Following attachment to wood surfaces, WDB enter wood tissues generally via rays. Cell walls of ray parenchyma in many wood species are not lignified, and thus can be more easily degraded compared to lignin-containing cell walls. Ray parenchyma are also a source of abundant readily hydrolyzable food for bacteria. Bacterial entry into ray contact cells (tracheids, vessels) is likely to be via pits. Pit membranes consist of cellulose, hemicelluloses, and pectins (West et al., 2012), and in sapwood is not lignified, and thus can be readily degraded by WDB. Because TB have the capacity to degrade highly lignified middle lamella, direct penetration through the common cell wall is also likely, although aspects of the movement of WDB into contact cells from rays have not been closely examined. EM images demonstrating penetration of TB across the common middle lamella suggest that movement of TB between wood elements can also occur through direct passage across the double cell wall. EB, on the other hand, can degrade the secondary cell wall but not the middle lamella. Thus, cell-to-cell movement by EB can only occur via pits.

CELL WALL DEGRADATION

Micromorphological features associated with cell wall degradation by TB differ markedly from those produced during cell wall degradation by EB, which will be described separately.

TUNNELING BACTERIA AND TUNNELING TYPE DEGRADATION

TB attack the cell wall from the cell lumen where they initially multiply. The detailed description of the initial bacterial attachment to the exposed face of the cell wall and the stages of entry into and movement within the cell wall during its degradation have been presented in several reports, and reviewed in Singh and Butcher (1991). TB attachment to the cell wall is facilitated by an extracellular slime, which helps retain the bacterium in contact with the cell wall. TB undergo remarkable shape changes from the initial point penetration through progressive stages of entry into the cell wall (Fig. 9.1). The point penetration is typically associated with a pear-shaped form, with partial cell wall entry achieved by the pointed front of the bacterium, where cell wall degrading elements (enzymes, radicals) may become more concentrated. The S3 layer is the first point of contact and bacterial attachment. In many wood species this layer is more highly lignified than the S2 layer, and this may have a relation to the initial pear shape, as this is the first cell wall structure to be breached prior to complete bacterial entry into the secondary cell wall. It would be interesting to investigate if this is a consistent morphological feature of TB in relation to the initial attack on the cell wall, or whether the shapes vary in relation to the substrates, differing particularly in the

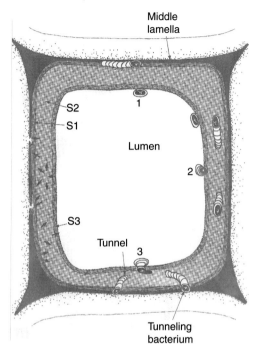

FIGURE 9.1   **Stages (1, 2, 3) of TB entry into cell wall from the lumen.** Arrows indicate directions of TB movement within the cell wall.

type and concentration of lignin, a highly recalcitrant component of wood cell walls. Interestingly, with progressively deeper bacterial penetration into the cell wall, the form of TB changes from the initial pear shape (Fig. 9.1). The contour of the cavity formed into the cell wall due to degradation closely corresponds to the advancing front of the bacterium. This feature and the close spatial relationship between the bacterial cell and the cell wall region undergoing degradation suggests that the enzymes produced may be tightly associated with the bacterial surface, an adaptation to minimize loss of enzymes through diffusion in the moist or water-saturated environments, where TB attacks are common. EM has shown the presence of tiny vesicles associated with degrading cell wall, and it has been suggested that vesicles may carry enzymes, as is known for some cellulolytic bacteria. This could be another mechanism to prevent/minimize loss of enzymes, but confirmation will first require characterization of enzymes.

TB Movement and Tunneling of Cell Wall

Bacteria belonging to this group have been aptly named tunneling bacteria, because they produce tunnels as they degrade the cell wall (Nilsson and Singh, 2014). TB are Gram-negative short rods. However, they have the capacity for shape changes (pleomorphism) and EM has revealed a range of shapes

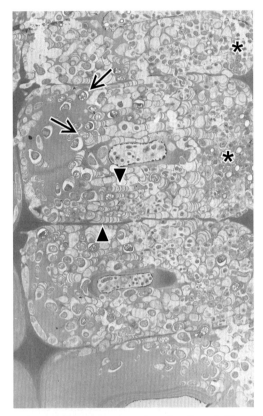

FIGURE 9.2   TEM micrograph from a transverse section through *Homalium foetidum* fibers degraded by TB. Scavenging bacteria (asterisk); TB (arrows); tunnels (arrowheads).

during cell wall degradation and bacterial movement within and between cell wall layers varying in lignin concentration. Particularly distinctive is the "dumbbell shape" as TB traverse highly lignified middle lamella during movement between neighboring cells (Kim and Singh, 2000). The impression gained is that such shape changes enable TB to maximize degradation while conserving their energy and cell wall degrading elements (enzymes, radicals). EM images of intact tunnels reveal an intriguing micromorphology. Tunnels contain periodic bands along their length, as observed in sectional views under TEM (Figs 9.2 and 9.3). These may be convex discs in three dimensions, as SEM images would suggest. SEM images also suggest that tunnels may be slime tubes, which develop from the slime released by TB, and closely correspond to the form and diameter of the encased bacteria. The fibrillar slime is sparsely distributed within the tube except for banded regions, where the slime is rather dense (Singh, 1989; Fig. 9.3). The impression gained is that TB degrade cell wall not by continuous movement but by "moving–stopping" action; however, the slime

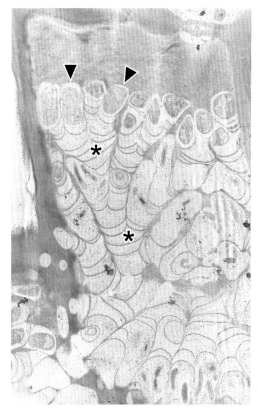

FIGURE 9.3   TEM micrograph showing extensive branching of tunnels, resulting from repeated bacterial division and tunneling within the cell wall. TB (arrowheads); tunnels (asterisks).

is likely to be continually released. The slime is sparsely distributed when TB are moving, and becomes more heavily deposited in places where TB stop to "charge" their battery. The compact appearance of the bands (Fig. 9.3) is probably because of the heavy concentration of slime in banded regions. Dehydration during sample preparation for TEM may also contribute to slime compaction. The slime, in addition to assisting TB glide during their movement, may have other important functions. It is well known that microbial slime can bind heavy metals, and this knowledge has served to make use of certain groups of bacteria and fungi in removing toxic metals from contaminated soils and aquatic environments (reviewed in Singh and Singh, 2014). TEM-EDAX has shown that the sequestered components of CCA (copper-chrome-arsenate) during degradation of cell walls in CCA-treated timbers bind to the tunnel slime. This may be a mechanism that allows TB to keep CCA from penetrating and damaging their vital cell components.

Light microscopy has shown that TB not only multiply and increase the size of their colonies in cell lumens but also within the wood cell wall. The form of colonies may vary, but the colonies typically have a "rosette" formation, with branches radiating from the center of the colony. Favorable TEM images also clearly show this pattern (Fig. 9.3). Although the initial entry into the cell wall facing the lumen is usually by point-penetration, all cell wall regions, including the highly lignified middle lamella, can be penetrated and degraded by TB (Fig. 9.2). Once in contact with the S2 layer, TB can move in all directions and degrade the cell wall without regard to the microfibrillar angle (Figs 9.1 and 9.2), as strikingly observable in microscopic images showing the pathways of TB movement in situations where mixed TB and soft-rot attacks are present within a particular region of the cell wall (Santhakumaran and Singh, 1992; Singh et al., 1992). TEM images show the presence of tunnels even among closely placed soft-rot cavities, and the impression gained is that the ability of TB to move in any direction within the cell wall allows these bacteria to maximally utilize the available cell wall resource for their nutrition. However, there appears to be a preference for cell wall regions with lower lignin concentration, namely, the S1 and S2 layers, at least in the initial stages of degradation. In later stages, when the supply of the S1 and S2 wall is exhausted either through extensive degradation by TB alone or in competition with other wood-degrading microorganisms, such as erosion bacteria and soft rot fungi, tunneling of middle lamella is observable (Singh et al., 1992). These observations suggest that TB may have a capacity to extensively degrade not only cell wall polysaccharides but also lignin.

Enzymes, Vesicles, Radicals

It has not been possible to isolate TB in pure culture and to characterize their enzymes. However, micromorphological features associated with cell wall degradation provide some clues. High magnification TEM images reveal a close spatial relationship between TB and the sound wood cell wall, with only a thin layer of slime present in between bacterium and the cell wall. This suggests that enzymes upon release have only to traverse the thin slime layer to reach the cell wall substrate. However, there are also TEM images showing the presence of small membrane-bound vesicles along the bacterial surface. Vesicles may be an artifact of chemical fixation during specimen preparation for TEM or may carry enzymes, an adaptation to prevent enzyme losses via diffusion in the moist or water-saturated environments where TB attacks are prevalent. The presence of enzymes in vesicles produced by certain cellulolytic bacteria has been reported. However, confirmation of any of these mechanisms has to wait until it becomes possible to characterize bacterial enzymes.

Microbial enzymes are rather large molecules and cannot penetrate sound wood cell walls, which contain nanoscale pores (about 2 nm in kiln-dried wood). Wood-degrading basidiomycete fungi have developed a mechanism for circumventing this barrier. The tiny radicals generated (Hammel et al., 2002) loosen the cell wall, thus facilitating enzyme penetration. It is not known whether TB also deploy a similar mechanism, but is a distinct possibility.

Occurrence in Nature and Wood Substrates

TB are widely present in aquatic and terrestrial environments. The substrates examined include straw, cotton fibers, pulped fibers, fern stems, and wood. These substrates widely vary in their cell wall composition, from cellulosic and slightly and moderately lignified to highly lignified cell walls. Furthermore, some wood substrates examined were chemically modified, preservative-treated, or contained abundant extractives. The fact that TB can degrade lignified and unlignified cell walls suggests that the presence of lignin in cell walls is not a prerequisite for TB attack. The ability of TB to degrade wood substrates resistant to white and brown rot fungi, such as wood treated to high CCA retentions and extractive-rich heartwood of belian, is an indication that TB have evolved strategies to degrade recalcitrant substrates, thus avoiding competition with faster-degrading basidiomycete fungi. However, microscopic images showing cohabitation of TB with soft rot fungi and erosion bacteria (Nagashima et al., 1990; Santhakumaran and Singh, 1992; Singh et al., 1992) in wood samples examined from terrestrial and aquatic environments indicate that these organisms do not pose a threat to TB, and TB are able to tunnel through and degrade the cell wall despite their presence within the same wood tissues and even in the same wood cell wall. The physiological and nutritional relationships among these coexisting microorganisms remain unknown.

Tunneling is a Unique Adaptation

TB have evolved an adaptation that enables them to obtain their food supply from lignocellulosic cell walls without being exposed to harsh external environments and the threat of predation by other organisms, such as amoebae, which also inhabit decaying wood. The competitive advantage of tunneling over the mechanisms other microorganisms have developed to degrade wood cell walls may be that this mode of attack enables TB to be in close vicinity to cell wall polymers, separated only by a thin layer of slime sheath they produce to glide on during cell wall degradation. Microbial slime is likely to be a highly porous structure, permitting rather unobstructed diffusion of wood-degrading agents (enzymes, radicals) produced by TB. Thus, these agents can readily diffuse through the slime lining tunnels and the close contact of the slime with cell wall polymers prevents their loss, which can otherwise occur in the moist and aquatic environments where TB actively degrade cell walls. The tunneling strategy also enables TB to explore all available cell wall areas in situations involving cohabitation and cell wall attack by other wood-degrading microorganisms, such as EB and soft rot fungi, aided by their ability to change shape and move in all directions within the cell wall. This adaptation enables TB to efficiently utilize their nutrition without entering in direct competition with cohabiting microorganisms. The ability to attack wood substrates recalcitrant to wood-degrading basidiomycete fungi, such as acetylated wood, wood treated with CCA, creosote and other preservatives to high retentions, and extractive-rich heartwoods of tropical timbers, has given TB a distinct advantage for obtaining nutrition despite their slower rate of cell wall degradation.

Environmental and Ecological Significance

In nature TB are likely to play an important role in carbon recycling, particularly in environments and under conditions not conducive to the growth and activity of wood-degrading basidiomycete fungi. Discarded preservative-treated wood products placed in land-fill situations, which cannot be readily attacked by wood-degrading basidiomycete fungi, are not immune to degradation by TB. The toxic and heavy metal binding ability of TB has the potential to be exploited for the removal of such metals from contaminated land and aquatic environments; enormous benefits to nature can be expected especially if it is proved that TB can also breakdown such substances, reducing or eliminating their toxicity. TB-degraded wood can serve as a source of shelter and nutrients for a wide variety of organisms, such as millipedes in terrestrial environments and *Limnoria* in aquatic environments.

EROSION BACTERIA AND EROSION TYPE DEGRADATION

Like TB, EB have a world-wide distribution. EB are present in terrestrial and aquatic environments and can colonize and degrade wood exposed in oxygenated environments as well as under conditions with extremely low levels of oxygen (Singh et al., 1990; Björdal, 2000; Kim and Singh, 2000). In nature, particularly where oxygen availability is not restrictive such as in agricultural and horticultural soils, EB often cohabit wood with other wood-degrading microorganisms such as TB and soft rot fungi. Support for this comes from TEM observations demonstrating the presence of EB attack together with TB attack and also combined EB, TB, and soft rot attacks in the same areas of wood tissue and even associated with the same wood cell wall (Singh et al., 1992). During EB degradation, erosion of cell wall occurs, and in the process erosion channels parallel with cell wall microfibrils are produced (reviewed in Singh and Butcher, 1991; Daniel and Nilsson, 1998).

Comparison of EB and TB

EB, like TB, are nonflagellate Gram-negative rods. TEM has confirmed the presence of a thin membranous wall (Singh and Butcher, 1991), a typical feature of Gram-negative bacteria. Although TB and EB appear to be similar in size (1.5–2.00 μm in length and 0.5 μm in diameter) and both types can attack a variety of substrates, distinct differences in their form and mode of attack are observable. Comparison of the forms of TB and EB attacking various substrates as well as their coattack on the same cells within a wood substrate has revealed some useful distinguishing features. While TB tend to be rounded at their ends, EB possess flat to angular or conical ends. TB can assume variable shapes in response to substrate composition and during entry into and movement within wood cell wall, changes in the form of EB appear to be much less drastic. TB primarily degrade cell walls by way of tunneling, and thus are surrounded by the substrate. EB degrade by eroding the exposed faces of the cell wall, and erosion channels are produced as a consequence of cell wall degradation. TB can move in all directions within the cell wall regardless of microfibrillar orientation. EB appear to strictly follow the long direction of microfibrils, as indicated by the parallel orientation of erosion channels with microfibrils. TB have a capacity to degrade all cell wall regions regardless of lignin concentration. EB are unable to degrade the lignin-rich middle lamella in normal wood and the highly lignified outer S2 regions of compression wood cell walls. Judging by their absence in deeply buried and waterlogged wood structures, such as in sunken ships, it is considered that TB activity requires adequate supply of oxygen. EB are active in environments supporting the presence of varying levels of oxygen, from situations with readily available oxygen to those with extremely low levels of oxygen. It is not known, however, whether EB can degrade lignocellulosic substrates under complete anoxic conditions. TB can tolerate extremely high levels of extractives, such as in highly durable tropical heartwoods, and high retentions of CCA preservative in treated wood. EB have been observed to coexist with TB and degrade CCA-treated *Pinus radiata* wood, indicating that EB can also tolerate the highly toxic CCA preservative, at least the levels used to treat timbers for application in ground contact.

Cell Wall Degradation and Erosion Channel Formation

Although bacterial tunneling type degradation received considerable attention in the 1980s because of the unique, intricate nature of degradation via tunneling within the cell wall and the practical and economic relevance of TB attack on preservative-treated wood, in later years the focus shifted toward studying EB type degradation, mainly because of the discovery that EB can degrade wood under near-anaerobic conditions (Singh et al., 1990). This finding led to a flurry of investigations to understand the cause of deterioration of wooden structures of historical and cultural importance that had been waterlogged due to burial in the ocean, river, and lake sediments, or had been a part of foundation piles supporting various structures in aquatic or water-saturated environments (Blanchette et al., 1991a, 1991b; Kim and Singh, 1994, 1999; Kim et al., 1996; Björdal et al., 1999; Björdal, 2000; Singh et al., 1991, 2003; Hoffmann et al., 2004; Schmitt et al., 2005; Klaassen, 2008; Rehbein et al., 2013; Cha et al., 2014).

EB were the primary degraders in all wood samples examined in these studies. The information has been presented in several reviews (Björdal, 2000; Blanchette, 2000; Kim and Singh, 2000; Singh, 2012).

The mode of colonization of wood by EB is similar to that for TB. The main pathway of initial entry into wood is the ray, as described in an earlier section. Light microscopy of hand-cut or sliding microtome-cut sections from EB-attacked wood pieces show a pattern distinctive to bacterial erosion. The sections contain parallel striations or stripes (stripy erosion), which, while aligned parallel to cell wall microfibrils as with soft-rot cavities, being much thinner can be readily distinguished from soft-rot cavities. The information on cell wall degradation presented here is based on observations of EB-attacked wood obtained from various environments and conditions, including archaeological sites supporting active cell wall degradation, as well as laboratory experiments. In all cases, the pattern of degradation was the same, that is, EB degraded cell walls in a way that produced channels (troughs) in the faces of cell walls being eroded – a feature considered very characteristic of EB attack.

As is the case for TB, EB colonize cell lumens where they multiply and from where they attack the cell wall. This is accomplished with the help of an extracellular slime these bacteria produce, which helps retain EB in close proximity to the cell wall, and ensures that bacteria are not dislodged in the moist or wet environments that support EB attack. Microscopic images suggest that the slime must be produced abundantly, because even after dehydration in alcohol or acetone during sample preparation for TEM, some slime is retained and is visible as a fibrillar material around EB (Singh et al., 1992). Cryoelectron microscopy, where dehydration of samples is not required, would be an ideal way of examining the nature and extent of slime and its spatial relationship to bacteria and the substrate cell wall. Subsequent to adhesion, EB align themselves along the long direction of cellulose microfibrils and degrade the cell wall from lumen outward, that is, toward the middle lamella (Fig. 9.4). In the process, erosion troughs are produced visible as channels in longitudinal views and as crescent-shaped erosion zones in transversely cut sections (Singh et al., 1990, 1992; Singh and Butcher, 1991; Daniel and Nilsson, 1998; Björdal, 2000; Schmitt et al., 2005; Figs 9.4 and 9.5).

During the initial attack, the S3 layer is the first cell wall structure that EB encounter. The impression gained from TEM images of EB attack on conifer woods is that the degradation of this layer may be confined to places (Fig. 9.4), judging by the presence of long stretches of the S3 layer even when the degradation of the S2 layer is relatively advanced (Singh et al., 1990; Björdal, 2000). The main reason for the relative resistance of the S3 layer to EB is likely to be its high lignin content in conifer woods. Once the S3 layer has been breached, EB degrade the S2 layer, which appears to be the preferred substrate, judging by the presence of a large number of EB along this wall (Björdal, 2000; Fig. 9.5), giving an impression of a coordinated attack. As the degradation progresses, erosion channels deepen

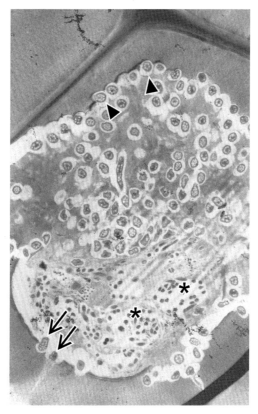

FIGURE 9.4 TEM micrograph showing localized (arrows) and wide erosion of a tracheid cell wall by EB (arrowheads). Scavenging bacteria (asterisks).

with EB sinking in them. It is not clear from the static electron microscopic images whether EB continue to erode the cell wall at the same location they initially settle on or can also move back and forth within deepening channels. Live imaging using confocal laser scanning microscopy can shed light on this aspect of the degradation process. The types of enzymes produced and whether or not they are cell wall-bound are the aspects of the mechanism that need clarification. The presence of a slime coat around EB indicates that bacterial cell wall is not in direct contact with wood cell wall undergoing degradation, and the cell wall degrading agents (enzymes, radicals) released at the cell surface must diffuse through the slime sheath to reach the degradation site. Vesicles are often present in the region intervening between the bacterium and wood cell wall, and may carry packaged enzymes to degradation sites, as discussed in an earlier section. Enzymes and radicals produced have yet to be characterized.

    A close relationship between EB and the direction of cell wall microfibrils is maintained throughout the degradation process. Right from the initial stages of

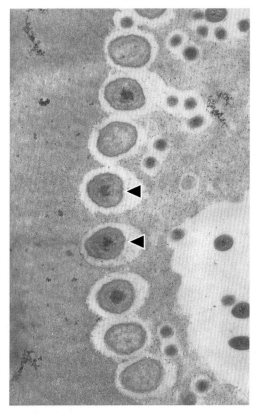

FIGURE 9.5   TEM micrograph from a transverse section through several erosion channels along the cell wall. EB (arrowheads).

degradation and erosion channel formation EB align themselves in parallel with microfibrils, and thus channels produced follow the same course. This relationship is kept regardless of the composition of the substrate, such as cell wall lignin concentration, and the environmental conditions of EB attack, from oxygenated to near-anoxic conditions. This feature of EB degradation is clearly different from TB degradation, where there appears to be no strict relationship between TB-produced tunnels and microfibrillar orientation. This suggests possible differences between EB and TB in the cellulose degrading enzymes they produce.

Whereas EB can degrade secondary cell wall regions, the middle lamella appears to be resistant, as the presence of middle lamella even after the degradation of the entire secondary cell wall would indicate. This feature is most striking in heavily degraded buried and waterlogged archaeological wood samples where often the middle lamella is the only part of the cell wall that remains within the large masses of residues (Blanchette et al., 1991a, 1991b; Kim and Singh, 1994; Kim et al., 1996; Cha et al., 2014). The middle lamella is a highly lignified region of wood cell wall, particularly in conifers, and the impression gained is that the resistance of this cell wall structure comes from its high lignin concentration. Other highly lignified wood components, such as the cell wall of ray tracheids in *Pinus sylvestris* (Singh et al., 1996), the outer S2 wall of compression wood (Singh, 1997a), and initial pit borders in *Pinus radiata* (Singh, 1997b), also show resistance to EB attack.

While, except for the presence of periodic slime bands, TB tunnels appear empty or contain scant residues, cell wall regions eroded by EB contain copious amounts of granular material (Fig. 9.4), which has been described as residual material and is considered to represent cell wall degradation products (Singh et al., 1990). In TEM preparations, the residual material stains intensely with potassium permanganate (Singh et al., 1990; Björdal, 2000; Schmitt et al., 2005; Rehbein et al., 2013), a reagent used widely as a stain to contrast lignin in lignified cell walls (Maurer and Fengel, 1990), which suggests that the residual material is either lignin or contains lignin degradation products. Indirect support for this also comes from polarization light microscopic images. The residual material shows no birefringence, a feature that has been useful in distinguishing bacterial erosion at the light microscope level, particularly in heavily degraded archaeological woods. Birefringence is a property of cellulose and not lignin. A recent UV-microspectrophotometric study (Rehbein et al., 2013) has confirmed the lignin nature of the residual material.

Enzymes, Vesicles, Radicals

Much of the description presented for TB also applies to EB. Because of the inability to obtain EB in pure culture, it has not been possible to characterize EB-produced enzymes. One of the striking differences between TB and EB is that while TB tunnels generally contain little or no degraded cell wall residues, substantial amounts of lignin residues are left in EB-degraded regions of the cell wall, suggesting differences between TB and EB in their lignin degrading capabilities. The presence of membranous vesicles also during EB degradation points to similarities between TB and EB in the roles vesicles may play in facilitating delivery of wood-degrading agents to cell wall degradation sites. EB are also likely to deploy radicals to facilitate cell wall degradation, as proposed for TB.

Occurrence in Nature and Wood Substrates

EB have a wide distribution in nature, and can degrade wood cell walls in terrestrial and aquatic environments. The presence of mixed TB and EB attacks in CCA-treated wood exposed in fertile soils suggests that EB, like TB, can tolerate highly toxic CCA preservative. However, it is not clear whether EB slime can also bind CCA components, as has been demonstrated for TB. A major difference between TB and EB with regard to habitat conditions is that while TB activity requires aerobic conditions, EB can degrade wood cell walls in oxygenated environments as well as those with extremely low levels of oxygen. Reported degradation of chemically pulped fibers (~3% residual lignin) suggests that the presence of lignin is not a requirement for EB degradation. In contrast to TB, EB are unable to degrade highly lignified cell wall regions, such as the middle lamella and the outer S2 wall in compression wood tracheids, and lignin-rich cell walls of ray tracheids in conifers.

Environmental and Ecological Significance

EB are present in a range of terrestrial and aquatic environments. Whereas TB require oxygenated conditions for their activity, EB can function in habitats with adequate oxygen availability, as their cohabitation with TB in wooden structures exposed in fertile horticultural soils would indicate, as well as in those containing extremely low levels of oxygen, such as in ocean bottom sediments, where bacterial erosion is the most dominant form of wood degradation (reviewed in Björdal, 2000; Kim and Singh, 2000; Singh, 2012). Thus, EB are important players in the recycling of carbon from wood sources even in extreme environments from which other wood decay microorganisms may be excluded. EB degradation results in the accumulation of large quantities of residues containing lignin degradation products, mixed with bacterial slime and possibly also some polysaccharide degradation products (Singh and Butcher, 1991), which can support the nutrition of other bacterial forms (scavenging bacteria) that are not primary wood degraders but are often present in cells undergoing bacterial erosion (Fig. 9.4), where by utilizing readily metabolizable left-over residues scavenging bacteria may promote cell wall erosion activity of EB.

Chemical Characterization

Much of the information on the chemical characteristics of bacterial degraded wood comes from studies undertaken on buried and waterlogged wooden objects, mainly because of the relevance to conservation of archaeological and historically important wooden structures. Under such conditions wood is primarily attacked by EB, and cell wall degradation is very slow; wooden structures can survive for hundreds and even thousands of years. Chemical information can serve as a useful measure of the extent of degradation (Gelbrich et al., 2012) and can thus aid in developing strategies for conserving the historically important wooden treasures appropriately, particularly in combination with information from microscopic (Singh, 2012) and physical characterization.

Chemical studies on buried and waterlogged woods, using conventional characterization methods, have been carried out for quite some time (reviewed in Kim and Singh, 2000), and in later years UV, FTIR, and NMR techniques have also been applied. The consensus is that generally hemicelluloses are degraded first, followed by cellulose, hardwood lignin, and then softwood lignin. Prolonged exposure of wood to such conditions can cause extensive losses in polysaccharides, and sometimes the wood may be completely depleted of polysaccharides (Kim and Singh, 2000). Although lignin appears to largely resist degradation in such environments, modifications in the native lignin structure may occur. TEM examination of potassium permanganate ($KMnO_4$) stained sections has revealed intense staining of the cell wall residues in the samples of wood degraded by EB (Singh et al., 1990; Kim and Singh, 1994; Björdal, 2000; Hoffmann et al., 2004; Schmitt et al., 2005; Rehbein et al., 2013). As mentioned in an earlier section, $KMnO_4$ shows strong affinity for lignin in lignified tissues, and has therefore been widely used to contrast lignin in cell walls in TEM studies of plant and wood tissues (Maurer and Fengel, 1990; Singh et al., 2002). In fact, staining intensity in EB-degraded cell wall regions is greater than that of native cell wall, which could be due to greater accessibility of $KMnO_4$ stain in the modified, more porous EB-degraded cell wall regions. In a UV microspectrophotometric study of EB-degraded wood samples from harbor foundation piles increased UV absorption correlated with those cell wall regions that were degraded (Rehbein et al., 2013), thus confirming results from earlier studies that the residual cell wall material represents a mix of lignin and lignin degradation products.

CONCLUSIONS

Unprotected wood in natural environments is rapidly degraded by basidiomycete fungi, such as white and brown rot fungi. Certain groups of bacteria have evolved mechanisms and adaptations to also degrade wood, and produce degradation patterns (tunneling and erosion) that serve as diagnostic features for the identification of bacteria-caused cell wall degradation. Although bacterial degradation of wood is markedly slower compared to basidiomycete fungi, the unique adaptations of TB and EB in nature enables them to utilize lignocellulosic substrates without entering in direct competition with these fungi. For example, TB can tolerate wood preservatives and heartwood extractives highly toxic to basidiomycete fungi, and their ability to tunnel within the cell wall during degradation is a remarkable strategy for avoiding direct competition with other wood-degrading microorganisms and escaping predation in natural environments where a multitude of organisms can inhabit decaying wood. While TB require adequate oxygen for their growth and activity, EB can function in aerobic environments as well as under conditions with extremely low levels of oxygen, such as in deep sediments at ocean floors from where historically and culturally important wooden artefacts have been recovered in relatively good conditions even after hundreds of years of exposure. Thus, TB and EB have evolved mechanisms and adaptations to obtain their nutrition from lignocellulosic substrates under conditions considered extreme for basidiomycete white and brown rot fungi, and while such strategies are well suited to their well-being and survival, there are also environmental benefits through carbon recycling under extreme conditions. When it becomes possible to taxonomically identify these bacteria and learn more about their physiology and the biochemical and molecular mechanisms of their wood-degrading activity and tolerance to extreme conditions, wood-degrading bacteria can potentially be exploited in biotechnological applications, from toxic waste removal from contaminated environments to more direct health related issues.

POTENTIAL BIOTECHNOLOGICAL APPLICATIONS OF WOOD-DEGRADING BACTERIA

Bacteria and their enzymes hold great promise in biotechnological applications. Some of the areas of successful exploration are degradation and removal of environmental pollutants (Clausen, 2000; Al-Wasify and Hamed, 2014), microbial fuel cells (He and Angenent, 2006), pitch control in paper production processes (Burnes et al., 2000), food and beverage production (Raspor and Goranovic, 2008), plant nutrition, biofuel, and medicine. It is envisaged that WDB can serve to strengthen existing biotechnological applications or promote the prospects of new applications. Based on the information on the characteristics of WDB available to date, some predictions for possible applications can be made. One possible area of biotechnological application for WDB is removal of toxic CCA components from CCA-treated wood recovered after service. Using TEM-EDAX, the presence of CCA components in the bacterial slime has been demonstrated during TB attack on CCA-treated wood. Apparently, TB are capable of sequestering CCA components from CCA-containing cell walls, and the dislodged CCA is held within the slime sheath upon contact. It is not known whether TB can also degrade CCA components. WDB can tolerate conditions considered extreme for wood-degrading basidiomycete fungi. For example, TB are tolerant to high concentrations of toxic chemicals, such as CCA and heartwood extractives, and EB can degrade wood cell walls under near-anaerobic conditions. Such adaptations may be because of the unique set of proteins these bacteria produce, which may find applications in environmental, pharmaceutical, and health areas. However, understanding the taxonomy, physiology, and molecular biology of WDB will be a critical step in assessing their biotechnological potential.

CELLULOLYTIC BACTERIA

Certain types of bacteria can degrade cellulosic components of plant and wood tissues but are unable to degrade lignocellulosic cell walls. A brief description of two particular groups of cellulolytic bacteria, pit membrane degrading and rumen bacteria, is included here because of their practical relevance.

PIT MEMBRANE DEGRADING BACTERIA

The surfaces of ponded and water-sprinkled wood is rapidly colonized by natural bacterial populations, which in time can penetrate wood tissues via pits because they can preferentially colonize and degrade pit membranes (Blanchette et al., 1990; Singh et al., 2000; Burnes et al., 2000; Nijdam et al., 2004 and references therein). Bacterial degradation of pit membranes enhances wood permeability, particularly in conifer species where timber drying causes bordered pit membranes to aspirate, resulting in the blockage of the pit aperture by the compact central part of the pit membrane called torus. Pit aspiration adversely affects permeability and treatability of wood. Bacterial ability to degrade pit membranes has prompted exploration of the use of natural bacterial microflora as well as enzymes and enzyme mixtures in practical applications to enhance the treatability of timbers with preservatives and other wood property enhancing substances (references in Singh et al., 2000 and Nijdam et al., 2004). Pit membrane degradation occurs due to the combined action of cellulose and pectin-degrading enzymes, which pit membrane degrading bacteria produce. Certain pit membrane degrading bacteria can also degrade wood extractives, and thus can be employed for the removal of extractives (pitch) from wood chips prior to paper making (Burnes et al., 2000).

RUMEN BACTERIA

A complex mix of diverse microorganisms present in the rumen of animals feeding on plants probably work together to achieve efficient digestion of masticated plant tissues. Among rumen microorganisms, bacterial populations have been studied extensively because of their importance in the degradation of plant biomass (Li et al., 2013). The breakdown of cell walls by bacteria has been examined in the rumen ecosystem as well as in experimental systems by incubating plant tissues with known rumen bacterial strains. Studies on molecular diversity among rumen bacterial populations using gene technology have enabled the identification of most dominant bacterial communities for specific rumen systems (Li et al., 2013).

In forage plants, some cell walls may be entirely composed of polysaccharides or may also contain lignin. Rumen bacteria possess enzymatic machinery to degrade polysaccharides but not lignin (Lee et al., 2002 and references therein). Therefore, the presence of lignin in forage tissues has an effect on ruminal forage digestion. However, the amount of lignin relative to carbohydrates in forage plants is small, although the lignin level can vary with plant maturity and species. The mechanism of bacterial adhesion to the substrate and enzyme release at degradation sites may involve packaging of multienzyme complexes within bacterial surface-associated vesicular structures described as cellulosomes and cellulosome-like structures (Kim et al., 2001; Miron et al., 2001). A detailed understanding of the taxonomy and molecular biology of rumen bacteria is important for manipulating ruminal feed to improve its digestion.

REFERENCES

Al-Wasify, R.S., Hamed, S.R., 2014. Bacterial biodegradation of crude oil using local isolates. Int. J. Bacteriol. 2014, 1–8.

Björdal, C.G., 2000. Waterlogged archaeological wood: biodegradation and its implications for conservation. Doctoral Thesis. Swedish University, Agricultural Sciences, Uppsala, Sweden.

Björdal, G.C., Nilsson, T., Daniel, G., 1999. Microbial decay of waterlogged archaeological wood found in Sweden: applicable to archaeology and conservation. Int. Biodeterior. Biodegrad. 43, 63–71.

Blanchette, R.A., 1995. Biodeterioration of archaeological wood. Biodeterior. Abstr. 9, 113–127.

Blanchette, R.A., 2000. A review of microbial deterioration found in archaeological wood from different environments. Int. Biodeterior. Biodegrad. 46, 189–204.

Blanchette, R.A., Nilsson, T., Daniel, G., 1990. Biological degradation of wood. Rowell, R.M., Barbour, J. (Eds.), Archaeological Wood: Properties, Chemistry and Preservation, Advances in Chemistry, vol. 225, American Chemical Society, Washington, DC.

Blanchette, R.A., Cease, K.R., Abad, A.R., Koastler, R.J., Simpson, E., Sams, G.K., 1991a. An evaluation of different forms of deterioration found in archaeological wood. Int. Biodeterior. Biodegrad. 28, 3–22.

Blanchette, R.A., Iiyama, K., Abad, A.R., Cease, K.R., 1991b. Ultrastructure of ancient buried wood from Japan. Holzforschung 45, 161–168.

Burnes, T.A., Blanchette, R.A., Farrell, R.L., 2000. Bacterial biodegradation of extractives and patterns of bordered pit membrane attack in pine wood. Appl. Environ. Microbiol. 66, 5201–5205.

Cha, M.Y., Lee, K.H., Kim, Y.S., 2014. Micromorphological and chemical aspects of archaeological bamboos under long-term waterlogged condition. Int. Biodeterior. Biodegrad. 86, 115–121.

Clausen, C.A., 1996. Bacterial association with decaying wood: a review. Int. Biodeterior. Biodegrad. 37, 101–107.

Clausen, C.A., 2000. Isolating metal-tolerant bacteria capable of removing copper, chromium, and arsenic from treated wood. Waste Manage. Res. 18, 264–268.

Daniel, G., Nilsson, T., 1998. Developments in the study of soft rot and bacterial decay. In: Bruce, A., Palfreymann, J.W. (Eds.), Forest Products Biotechnology. Taylor & Francis Ltd, London, pp. 37–62.

Donaldson, L.A., Singh, A.P., 1990. Ultrastructure of Terminalia wood from an ancient Polynesian canoe. IAWA Bull. 11, 195–202.

Gelbrich, J., Mai, C., Militz, H., 2012. Evaluation of bacterial wood degradation by Fourier Transform Infrared (FTIR) measurements. J. Cult. Herit. 13S, S135–S138.

Hammel, K.E., Kapich, A.N., Jensen, K.A., Ryan, Z.C., 2002. Reactive oxygen species as agents of wood decay by fungi. Enzyme Microb. Technol. 30, 445–453.

He, Z., Angenent, L.T., 2006. Application of bacterial biocathodes in microbial fuel cells. Electroanalysis 18, 2009–2015.

Hoffmann, P., Singh, A.P., Kim, Y.S., Wi, S.G., Kim, I.-J., Schmitt, U., 2004. The Bremen Cog of 1380 – an electron microscopic study of its degraded wood before and after stabilization with PEG. Holzforschung 58, 211–218.

Kim, Y.S., Singh, A.P., 1994. Ultrastructural aspects of bacterial attacks of a submerged ancient wood. Mokuzai Gakkaishi 40, 554–562.

Kim, Y.S., Singh, A.P., 1999. Micromorphological characteristics of compression wood degradation in waterlogged archaeological pine wood. Holzforschung 53, 381–385.

Kim, Y.S., Singh, A.P., 2000. Micromorphological characteristics of wood biodegradation in wet environments: a review. IAWA J. 21, 135–155.

Kim, Y.S., Singh, A.P., Nilsson, T., 1996. Bacteria as important degraders of waterlogged archaeological woods. Holzforschung 50, 389–392.

Kim, Y.S., Singh, A.P., Wi, S.G., Myung, K.H., Karita, S., Ohmiya, K., 2001. Cellulosome-like structures in ruminal cellulolytic bacterium *Ruminococcus albus* F-40 as revealed by electron microscopy. Asian-Aust. J. Anim. Sci. 14, 1429–1433.

Klaassen, R.K., 2008. Bacterial decay in wooden foundation piles – patterns and causes: a study of historical pile foundations in the Netherlands. Int. Biodeterior. Biodegrad. 61, 45–60.

Lee, S.S., Kim, C.-H., Ha, J.K., Moon, Y.H., Choi, N.J., Cheng, K.-J., 2002. Distribution and activities of hydrolytic enzymes in the rumen compartments of Hereford bulls fed alfalfa based diet. Asian-Aust. J. Anim. Sci. 15, 1725–1731.

Li, J.P., Liu, H.L., Li, G.Y., Bao, K., Wang, K.Y., Xu, C., Yang, F.H., Wright, A.-D., 2013. Molecular diversity of rumen bacterial communities from tannin-rich and fiber-rich forage fed domestic Sika deer (Ceruus nippon) in China. BMC Microbiol. 13, 151.

Maurer, A., Fengel, D., 1990. A process for improving the quality and lignin staining of ultrathin sections from wood tissues. Holzforschung 44, 453–460.

Miron, J., Ben-Ghedalia, D., Morrison, M., 2001. Invited review: adhesion mechanisms of rumen cellulolytic bacteria. J. Dairy Sci. 84, 1294–1309.

Nagashima, Y., Fukuda, K., Sato, S., Moroboshi, N., Haraguchi, T., 1990. Coexistence of micro-fungal and bacterial degradation in a single wood cell wall. Mokuzai Gakkaishi 36, 480–486.

Nijdam, J.J., Lehmann, E., Keey, R.B., 2004. Application of neutron radiography to investigate changes in permeability in bacteria treated *Pinus radiata* timber. Maderas Cienc. Technol. 6, 19–31.

Nilsson, T., Singh, A.P., 2014. Tunnelling bacteria and tunnelling of wood cell walls. McGraw-Hill 2014 Yearbook of Science and TechnologyMcGraw-Hill, New York, pp. 395–399.

Nilsson, T., Singh, A.P., Daniel, G., 1992. Ultrastructure of the attack of *Eusideroxylon zwageri* wood by tunnelling bacteria. Holzforschung 46, 361–367.

Raspor, P., Goranovic, D., 2008. Biotechnological applications of acetic acid bacteria. Crit. Rev. Biotechnol. 28, 101–124.

Rehbein, M., Koch, G., Schmitt, U., Huckfeldt, T., 2013. Topochemical and transmission electron microscopic studies of bacterial decay in pine (*Pinus sylvestris* L.) harbour foundation piles. Micron 44, 150–158.

Santhakumaran, L.N., Singh, A.P., 1992. Destruction of two tropical timbers by marine borers and microorganisms in Goa waters (India). Inter. Research Group on Wood Preservation. Document No. 4176-92.

Schmitt, U., Singh, A.P., Thieme, H., Friedrich, P., Hoffmann, P., 2005. Electron microscopic characterization of cell wall degradation of the 400,000-year-old wooden Schöningen spears. Holz Roh-Werkst. 63, 118–122.

Singh, A.P., 1989. Certain aspects of bacterial degradation of Pinus radiata wood. IAWA Bull. 10, 405–415.

Singh, A.P., 1997a. The ultrastructure of the attack of *Pinus radiata* mild compression wood by erosion and tunnelling bacteria. Can. J. Bot. 75, 1095–1102.

Singh, A.P., 1997b. Initial pit borders in *Pinus radiata* are resistant to degradation by soft rot fungi and erosion bacteria but not tunnelling bacteria. Holzforschung 51, 15–18.

Singh, A.P., 2012. A review of microbial decay types found in wooden objects of cultural heritage recovered from buried and waterlogged environments. J. Cult. Herit. 13S, 520–576.

Singh, A.P., Butcher, J.A., 1991. Bacterial degradation of wood cell walls: a review of degradation patterns. J. Inst. Wood Sci. 12, 143–157.

Singh, A.P., Singh, T., 2014. Biotechnological applications of wood rotting fungi: a review. Biomass Bioenergy 62, 198–206.

Singh, A.P., Nilsson, T., Daniel, G., 1990. Bacterial attack of *Pinus sylvestris* wood under near-anaerobic conditions. J. Inst. Wood Sci. 11, 237–249.

Singh, A.P., Nilsson, T., Daniel, G., 1991. Ultrastructure of Premnopitys ferruginea wood from a buried forest in New Zealand. International Research Group on Wood Preservation. Document No. 1489.

Singh, A.P., Hedley, M.E., Page, D.R., Han, C.S., Atisongkroh, K., 1992. Microbial decay of CCA-treated cooling tower timbers. IAWA Bull. 13, 215–231.

Singh, A.P., Wakeling, R.N., Drysdale, J.A., 1994. Microbial attack of CCA-treated *Pinus radiata* timber from a retaining wall. Holzforschung 48, 458–462.

Singh, A.P., Nilsson, T., Daniel, G., 1996. Variable resistance of *Pinus sylvestris* wood components to attack by wood degrading bacteria. In: Donaldson, L.A., Singh, A.P., Butterfield, B.G., Whitehouse, L. (Eds.), Recent Advances in Wood Anatomy. New Zealand Forest Research Institute, Rotorua, New Zealand, pp. 408–416.

Singh, A.P., Schmitt, U., Kim, Y.S., Dawson, B., 2000. The knowledge of wood structure critical to understanding process performance. In: Kim, Y.S. (Ed.), New Horizons in Wood Anatomy. Chonnam National University Press, Gwangju, South Korea, pp. 306–314.

Singh, A.P., Daniel, G., Nilsson, T., 2002. High variability in the thickness of the S3 layer in *Pinus radiata* tracheids. Holzforschung 56, 111–116.

Singh, A.P., Kim, Y.S., Wi, S.G., Lee, K.H., Kim, I.-J., 2003. Evidence of the degradation of middle lamella in a waterlogged archaeological wood. Holzforschung 57, 115–119.

West, M.A., Vaidya, A., Singh, A.P., 2012. Correlative light and electron microscopy of the same sections gives new insights into the effects of pectin lyase on bordered pit membranes in Pinus radiata wood. Micron 43, 916–919.

Part III

Economic Application of Secondary Xylem

Chapter 10

Genetic Engineering for Secondary Xylem Modification: Unraveling the Genetic Regulation of Wood Formation

Jae-Heung Ko*, Won-Chan Kim**, Daniel E. Keathley[†], Kyung-Hwan Han[†,‡]

*Department of Plant and New Resources, Kyung Hee University, Yongin, Korea; **School of Applied Biosciences, College of Agriculture and Life Sciences, Kyungpook National University, Daegu, South Korea; [†]Department of Horticulture, Michigan State University, East Lansing, MI, USA; [‡]Department of Forestry, Michigan State University, East Lansing, MI, USA

Chapter Outline

INTRODUCTION

Tree growth resulting from cell division in the vascular cambium, which is called 'secondary growth,' leads to the production of secondary vascular tissues, secondary phloem, and xylem (i.e., wood) (Mauseth, 1998). As a result of this unique ability to produce a massive woody body through secondary growth, trees comprise over 90% of the terrestrial biomass of the earth and serve as a primary feedstock for biofuel, fiber, solid wood products, and various natural compounds. The demand for these wood products and for wood energy is expected to increase continuously (2009 UN Report on World Forests), and meeting the construction and fiber needs of an expanding population,

Secondary Xylem Biology. http://dx.doi.org/10.1016/B978-0-12-802185-9.00010-3
193

in consideration of diminishing forestland, requires that future forestry must achieve higher biological productivity of trees and more efficient utilization and recycling of the biomass they produce as forest products. Marginal agricultural land and degraded forestlands have the potential to be replanted to sustainably produce crops of woody biomaterials, and thus could play a significant role in meeting the escalating demand. With genetically designed planting stock, such plantation-grown woody biomaterials could not only support existing industries, but also lead to the development of new industries based on cellulose nanocrystals, wood-derived composites, and biofuels. Maximizing yield and modifying the chemical composition of the biomaterials being produced are essential in making biomaterials plantations an economically viable alternative to other land uses and cost competitive with nonrenewable resources, such as fossil fuels. However, development of such tailored planting stocks has been a largely intractable question because the cellular and molecular regulation of wood formation remains so poorly understood. Several reasons account for the lack of knowledge on this very important process, including the following: (1) difficulty of experimentally observing the processes of wood formation; (2) no 'easy-to-work-with' tree model system is available for study; (3) the large physical size and long generation cycle of trees; and (4) the lack of true breeding, inbred lines, or readily available mutant populations for functional genomics studies. The recent advances in molecular biological technology and genome sequencing of tree species, such as *Eucalyptus*, *Populus*, and *Prunus*, provide a unique opportunity to overcome these inherent problems associated with tree studies (Tuskan et al., 2006; Grattapaglia and Kirst, 2008; Rengel et al., 2009; Zhang et al., 2012; Verde et al., 2013; Myburg et al., 2014). This chapter describes our current understanding of the genetic control of wood formation and its implication for biotechnological improvement of tree crops.

SECONDARY GROWTH AND WOOD FORMATION

During secondary growth, cell division in the vascular cambium and subsequent cell differentiation result in the production of secondary xylem and phloem elements. The vascular cambium normally consists of 5 to 15 cambium initial cells occurring as a continuous ring of cells between the xylem and the phloem throughout the length of fully expanded shoots and roots (the so-called cambial zone) (Larson, 1994; Mauseth, 1998) (Fig. 10.1). Two types of initials are present in the cambium: (1) the fusiform initials leading to the axial system and (2) the ray initials, which produce the cells that differentiate into the system of rays throughout the wood of the stem (Lev-Yadun and Aloni, 1995). These initials serve as a conduit for radial (across the cambium) and longitudinal (along the cambium) transfer of developmental signals and nutrients. Adjusting to the demands of water transport required by the leaf biomass and of the mechanical strength necessary to support the crown and to withstand wind forces (Zimmermann and Brown, 1971), cambial growth promotes an increase in stem enlargement by the production of functional vascular elements through radial

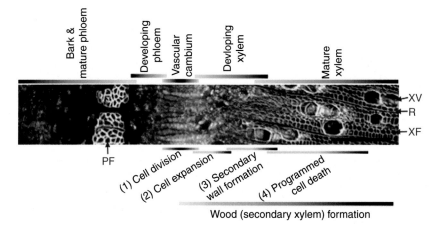

FIGURE 10.1   Cross-section of a poplar stem showing the organization of the cambial region and wood formation progress. The bars above the stem section describe approximate regions of indicated developmental tissues. Vascular cambial zone has meristematic cells (i.e., fusiform initials and ray initials), which produce phloem mother cells outside and xylem mother cell inside. Sequential wood formation stages are shown. PF, phloem fiber; XV, xylem vessel; XF, xylary fiber; R, ray cell. Poplar stem (hybrid aspen clone 717 INRA) cross-sections stained with Calcofluor, auramine O, and propidium iodide were observed using confocal laser microscopy. Scale bars represent 200 mm.

(or anticlinal) and tangential (or periclinal) divisions (Catesson et al., 1994). Diameter growth is also coordinated with changes in crown architecture and plant height (Larson, 1963), indicating a signaling system that integrates these growth responses. The exact molecular mechanisms underlying the regulation of cambial growth have not been elucidated.

Wood is produced by the successive addition of secondary xylem, which differentiates from the vascular cambium (Plomion et al., 2001). For wood formation, the cells on the xylem side of the cambium pass through four sequential developmental stages: (1) division of the xylem mother cells, (2) expansion of the derivative cells to their final size, (3) lignification and secondary cell wall formation (i.e., cell maturation), and (4) programmed cell death (Uggla et al., 1996, 1998; Chaffey, 1999) (Fig. 10.1). The resulting mature secondary xylem includes xylem parenchyma, fibers, vessels, and tracheary elements. This development of secondary xylem (i.e., xylogenesis) appears to be regulated by positional information that controls the cambial growth rate by defining the width of the cambial zone and, therefore, the radial number of dividing cells. Growth regulators, such as auxin, may be the source of this positional information (Wolpert, 1996; Bhalerao and Fischer, 2014), given IAA's polar basipital transport and the reported correlation of the IAA concentration gradient with cambial growth rate (Uggla et al., 1998). Gibberellin and the activation of its signaling pathway have also been shown to directly stimulate xylogenesis in *Arabidopsis* (Ragni et al., 2011).

Simultaneous increases in the radial number of dividing cells and the rate of cambial cell division result in increased productivity. Cambial growth and the subsequent differentiation of its derivatives appear to be under strict spatial and temporal control (Larson, 1994). Therefore, the quantity and quality of the final wood product is determined by a patterned control of numbers, places, and planes of cambial cell division, and a subsequent coordinated differentiation of the cambial derivatives into xylem tissues (Mauseth, 1998). This patterned growth requires that every cell must express the appropriate genes in a tightly coordinated manner upon receipt of positional information. As this regulation is under strong genetic control (Zobel and Jett, 1995), it should then be possible to genetically manipulate the quality and quantity of wood that is produced. Environmental factors, such as temperature, early season drought, and photoperiod, also affect wood formation, cell enlargement, and secondary wall thickening (Antonova and Stasova, 1997; Arend and Fromm, 2007).

While several plant hormones have been implicated in the regulation of wood formation, auxin appears to serve as a positional signal for the production of xylem and phloem by the vascular cambium (Little and Sundberg, 1991; Uggla et al., 1996, 1998; Sachs, 2000; Leyser, 2006; Bhalerao and Fischer, 2014). While gibberellins (GAs) are required for longitudinal growth (Wang et al., 1995). Uggla et al. (1996) observed a steep radial gradient of auxin across the cambial region in *Pinus sylvestris*, indicating that auxin acts as a positional signal that informs cambial derivatives of their radial position and regulates cambial growth rate by determining the radial population of dividing cambial-zone cells. In the presence of cytokinin, auxin induces xylem tracheary element differentiation in suspension culture cells of *Zinnia* (Fukuda, 1997). Klee et al. (1987) observed that auxin-overproducing transgenic petunia plants doubled in the amount of xylem and phloem production. Locally applied auxin can induce the formation of new vascular strands from parenchymatic cells (Sachs, 1981). Downregulation of auxin efflux carriers reduced auxin polar flow and consequently vascular cambium activity in the basal portions of the inflorescence stems (Zhong and Ye, 2001). Several *Arabidopsis* mutants with auxin transport or signaling defects show apparent interference with various aspects of vascular development (Hardtke and Berleth, 1998; Berleth and Sachs, 2001; Ko et al., 2004). The notion of auxin serving as a positional signal for wood formation, given its basipital movement, is consistent with the observation that stem-diameter growth is often greatest within the young crown and decreases gradually down the stem in forest trees.

REACTION WOOD

When the natural position of the stem is disrupted by gravitational or mechanical stimuli, reaction wood is formed through stimulation of cambial growth either at the lower side of the leaning stem in gymnosperm trees (called compression wood) or at the upper side in angiosperm trees (called tension wood). In tension wood the altered growth results in cells differentiating an additional layer at the inner face of the secondary cell wall of tension-stressed tissues (Timell, 1969; Mellerowicz et al., 2001; Hellgren et al., 2004; Pilate et al., 2004). This cellulose-rich additional layer, called gelatinous or G-layer, is thought to be primarily constituted of crystalline cellulose, though it also contains rhamnogalacturonan I and other pectins and polysaccharides (Bowling and Vaughn, 2008), and has no, or decreased levels of, lignin and hemicellulose. In tension wood, cellulose becomes the major carbon sink and lignin biosynthesis is reduced due to the shift in carbon flux toward cellulose biosynthesis (Hu et al., 1999; Wu et al., 2000). On the other hand, in compression wood, lignin becomes the major carbon sink and cellulose content is reduced (Timell, 1986).

Tension wood offers a good model system for the study of the genetic regulation of some components of cellulose wall biosynthesis and wood formation due to the production of this cellulose-rich G-layer inside the secondary wall. Wood cell walls are largely composed of a complex mixture of cellulose, hemicellulose, and lignin. The proportion of these three major components varies depending on the plant species, growing site, climate, age, and the part of the plant harvested. On average, wood cell walls in *Populus* spp. contain approximately 45% cellulose, 25% hemicellulose, and 20% lignin (Timell, 1969; McDougall et al., 1993). The G-layer formed in the walls of fiber cells in tension wood is mostly composed of cellulose and generates high longitudinal tensile stress during cell maturation (Yamamoto et al., 2005). During tension wood development, the *de novo* synthesized cellulose G-layer microfibrils appear to be fixed by xyloglucan cross-bridges, in which xyloglucan endotransglucosylase/hyrdrolase (XTH) plays a role in linking whole xyloglucan into the preexisting wall (Nishikubo et al., 2007; Mellerowicz et al., 2008; Baba et al., 2009). This xyloglucan-mediated "wall tightening" has been proposed to reinforce the wall to withstand the tensile stress created within the cellulose G-layer microfibrils (Baba et al., 2009). Transcriptional analyses showed that various genes are differentially regulated in tension wood (Déjardin et al., 2004; Paux et al., 2005; Andersson-Gunneras et al., 2006; Bhandari et al., 2006; Lu et al., 2008; Qiu et al., 2008; Jin et al., 2011). In birch, expression of genes involved in cellulose synthesis increased while genes associated with lignin synthesis were downregulated (Wang et al., 2014). In poplar tension wood, more than 119 genes representing 37 CAZyme (carbohydrate-active enzyme) families were expressed, including glycosyl transferases (GTs), glycoside hydrolazes (GHs), polysaccharide lyases (PLs), and carbohydrate esterases (CEs) (Andersson-Gunneras et al., 2006). Two secondary wall-associated cellulose synthases, *PttCesA3-2* and *PttCesA8-2*, were upregulated in poplar tension wood, while the other cellulose synthase *PttCesA1* was downregulated compared to normal wood (Déjardin et al., 2004; Joshi et al., 2004; Andersson-Gunneras et al., 2006). Three *Eucalyptus* secondary cell wall-related cellulose synthase genes were differentially expressed in response to tension stress (Lu et al., 2008). Transcripts of lignin biosynthesis genes, such as *PttPAL1*,

PttCCoAOMT1 and 2, were decreased in tension wood, while a transcriptional repressor of lignin biosynthesis, *PttMYB21a*, was upregulated (Karpinska et al., 2004). These observations are consistent with the chemical composition of the tension wood cell walls. Auxin, a key signal in xylogenesis, has been implicated in signal transduction in response to gravistimulation and induction of the G-layer in the developing tension wood fibers (Mellerowicz et al., 2001; Andersson-Gunneras et al., 2006), although evidence against redistribution of endogenous IAA during tension wood formation points to a more complicated model (Hellgren et al., 2004). Signal perception is clearly complex and likely triggers transcriptional regulators redirecting the carbon flux from lignin and hemicellulose to cellulose.

SEASONAL REGULATION OF CAMBIAL GROWTH

In order to survive freezing temperatures and limited water availability during winter, the vascular cambium of temperate tree species shifts between active growth and vegetative dormancy in accord with seasonal climate changes (Baier et al., 1994). To achieve proper coordination of their growth and development with seasonal changes in climate, which impacts the survival and productivity, temperate perennials use cyclically changing environmental signals, such as day length and temperature (Horvath et al., 2003; Park et al., 2008; Begum et al., 2013). For instance, the ray cambial cells of white ash (*Fraxinus americana*) growing in Canada resumed cell division activity in the spring (between April 23 and May 1) and entered dormancy around September 10, while fusiform cambial cells began activity between March 30 and April 23 (Zhong et al., 1995). Proteolytic activity in the cambial zone and developing xylem of *Pinus banksiana* was the highest during the most active period of radial expansion of cambial derivatives in early spring (Iliev and Savidge, 1999). While the overarching question of the molecular mechanisms that govern the dormancy and growth changes in perennial plants remains largely unanswered, the growth-dormancy cycle appears to be tightly controlled by complex genetic regulatory networks in response to endogenous and environmental signals (Baba et al., 2011; Bhalerao et al., 2003; Druart et al., 2007; Frewen et al., 2000; Hoffman et al., 2010; Horvath et al., 2008; Park et al., 2008; Ko et al., 2011). The regulatory program seems to involve at least three sequential components. First, short-day (SD) signal perception and processing by phytochromes (PHY) and PHY-interacting transcription factors (PIFs) induce cambial dormancy (Mellerowicz et al., 1992a, 1992b; Howe et al., 1995; Olsen et al., 1997; Eklund et al., 1998; Rohde and Bhalerao, 2007; Welling et al., 2002). Photoperiod changes are monitored by means of light-sensitive pigments, including PHY, which play central roles in modulating key physiological and developmental processes (Chen and Chory, 2011; Franklin and Quail, 2010). The light-activated PHY then triggers signaling events that alter the expression of target genes, resulting in key physiological and developmental processes of plants

(Jiao et al., 2007; Quail, 2002; Rockwell et al., 2006). The PHY-interacting factor 4 (PIF4) acts as a key regulator of plant response to external signals as well as internal signals, including gibberellin and the circadian clock (Castillon et al., 2007; Duek and Fankhauser, 2005; Khanna et al., 2004; Leivar and Quail, 2011). Second, dormancy is triggered by dormancy cycle stage-specific alterations in the auxin responsiveness of the transcriptome, possibly through the action of auxin receptor F-box protein TIR1 (Baba et al., 2011). Since auxin regulates virtually every aspect of plant growth and development, it is not surprising that differential perception and interpretation of the auxin signal plays important roles in active growth-dormancy regulation. Proteins involved in auxin distribution and signaling include auxin efflux machinery components (e.g., PIN proteins), influx carriers (e.g., LAX proteins), and Aux/IAA proteins that are unstable transcriptional repressors of auxin responses (Chapman and Estelle, 2009). Auxin response factors (ARFs) are a family of transcription factors that bind auxin-responsive promoters and regulate transcription (Guilfoyle and Hagen, 2007). When interacted, Aux/IAA proteins keep ARFs from forming functional dimers, resulting in negative regulation of auxin-responsive gene expression (Benjamins and Scheres, 2008). Auxin appears to stimulate binding of an F-box protein subunit of the SCF complex SCF^{TIR1} to the AUX/IAA repressor leading to the activation of auxin-responsive genes. The expression of a poplar homolog of TIR1 was reduced by 50% by exposure to short days, suggesting its role in SD-induced growth cessation and dormancy onset in poplar (Baba et al., 2011). Third, MADS-box transcription factors are associated with endodormancy induction. These dormancy-associated MADS-box (DAM) genes were discovered from the analysis of the *evergrowing* (*evg*) mutant of peach (*Prunus persica* (L.) Batsch), which fails to cease growth and enter dormancy under dormancy-inducing conditions (Rodriguez-A et al., 1994; Wang et al., 2002). Observations in the literature suggest that DAMs may function as an internal regulator of growth cessation.

GENETIC CONTROL OF SECONDARY XYLEM (I.E., WOOD) FORMATION

Development of secondary xylem requires molecular signals that vary in accord with the relative spatial distribution of the cells within the vascular cambium and zone of differentiation (Uggla et al., 1996, 1998). This positional information coordinates the expression of the appropriate genes in the vascular cambial cells, resulting in the radial pattern of the developmental zones of division, expansion, and wall formation (Uggla et al., 1996, 1998). Forward genetic and functional genomics approaches have identified many genes affecting secondary xylem formation (Hertzberg et al., 2001; Ye, 2002; Oh et al., 2003; Fukuda, 2004; Ko et al., 2004, 2006a; Ko and Han, 2004; Demura and Fukuda, 2007; Turner et al., 2007; Du and Groover, 2010; Ohtani et al., 2011; Liu et al., 2013). As was expected, these include several pattern formation homeobox genes, such as the class III *HD-ZIP* family genes (Ohashi-Ito and Fukuda, 2003), the *KANADI* gene family (Emery et al., 2003), and *ALTERED PHLOEM DEVELOPMENT* (*APL*), which is involved in xylem–phloem switching (Bonke et al., 2003). The class III *HD-ZIP* family of transcription factors is known to regulate vascular development. An *HD-ZIP* protein INTERFASCICULAR FIBERLESS1 (IFL1), which is expressed in the expected interfascicular regions and in the vascular bundles, regulates fiber and vascular differentiation in *Arabidopsis* (Zhong and Ye, 1999). Overexpression of ATHB-8, another member of the *HD-ZIP* III family, increased the production of xylem tissue and promoted vascular cell differentiation in *Arabidopsis* (Baima et al., 2001). However, the formation of the vascular system was not affected in *athb8* mutants (Baima et al., 2001), suggesting that there might be functional redundancy provided by other similar *HD-ZIP* III gene(s) (Fig. 10.2).

NAC (no apical meristem (NAM), *Arabidopsis thaliana* transcription activation factor (ATAF1/2) and Cup-shaped cotyledon (CUC2)) proteins are plant-specific transcription factors (Olsen et al., 2005). They have been linked to

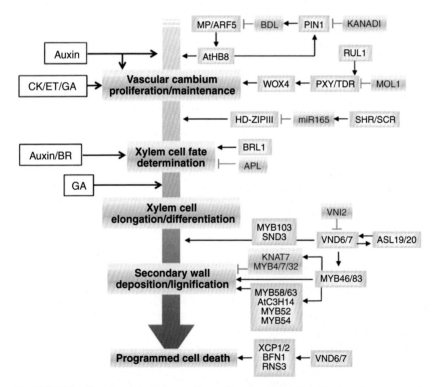

FIGURE 10.2 Genetic control of secondary xylem formation. Simplified regulatory network leading to secondary xylem formation. Hormonal regulation is shown in left and genetic control in right by incorporating representative transcription factors. Negative regulators are shown in dark gray (red in the web version). CK, cytokinin; ET, ethylene; GA, gibberellin; BR, brassinosteroids.

responses to biotic and environmental stimuli and have been shown to transcriptionally regulate various plant developmental processes, such as formation of shoot apical meristems, floral organs, and lateral roots (Aida et al., 1997; Olsen et al., 2005; Souer et al., 1996; Xie et al., 2000) and xylem differentiation (Kubo et al., 2005). Global comparative transcriptome analysis identified 52 candidate genes as regulators of wood formation and/or secondary wall biosynthesis in *Arabidopsis* (Ko et al., 2006a), which include four NAC transcription factors (*ANAC012, ANAC073, ANAC043,* and *ANAC066*) that were highly upregulated in the secondary xylem cells in *Arabidopsis*. Kubo et al. (2005) showed that two NAC transcription factors, *VASCULAR-RELATED NAC-DOMAIN6* (*VND6*) and *VND7*, function as a transcriptional switch for transdifferentiation of various cells into metaxylem- and protoxylem-like vessel elements in *Arabidopsis* and poplar (Fig. 10.2). Despite the fact that a large number of candidate genes for the regulation wood formation have been identified, a comprehensive understanding of how these genes interact to control woody growth is still lacking.

SECONDARY WALL BIOSYNTHESIS DURING WOOD FORMATION

Since the first reports that a herbaceous model species *A. thaliana* could be induced to undergo secondary growth and produce secondary xylem (i.e., wood) (Lev-Yadun, 1994), significant advances toward understanding how cells from the cambial meristem differentiate into secondary xylem have been made using this model system (Chaffey et al., 2002; Oh et al., 2003; Ko et al., 2006; Demura and Fukuda, 2007; Du and Groover, 2010). Although, as stated earlier, wood formation involves four sequential developmental stages (cell division and expansion, formation of secondary cell walls, and the autolysis of cell contents), secondary wall formation is the critical process biologically and economically. Biologically, proper formation of secondary wall is essential for the plant's survival. A defining feature of vascular land plants is the presence of xylem fibers, which provide mechanical support for their growing body, and vessels that serve as a conduit for long-distance transport of water and solutes. Secondary walls in these cells allow them to resist the gravitational forces and/or the forces of tension associated with the transpirational pull on a column of water. Economically, secondary walls represent the vast majority of plant biomass that serves as feedstock for biofuel, fiber, pulp and paper, solid wood products, and various natural compounds.

Secondary walls are formed in a highly coordinated manner by successive deposition of cellulose fibrils parallel to one another, strengthened by cementing substances such as lignin, hemicellulose, pectin, and proteins (Fukuda, 1996; Lerouxel et al., 2006; Somerville, 2006; Zhong and Ye, 2007). Cellulose, a polymer of β-1-4-linked glucose residues, is the most abundant biopolymer on Earth. It constitutes the major polysaccharide of plant cell walls (up to 50%) and is present mostly as a microfibril crystalline structure in the wall (Bhandari et al., 2006; Gardiner et al., 2003; Saxena and Brown, 2005; Suzuki et al., 2006; Tanaka et al., 2003; Taylor et al., 2003). Accounting for about 40–50% of wood, cellulose is a vital component of the load-bearing structure (e.g., xylem fiber). Its biosynthesis is catalyzed by multimeric cellulose synthase (CESA) complexes at the plasma membrane (Somerville, 2006). Since the first plant CESA gene was identified in developing cotton fibers undergoing rapid deposition of cellulose (Pear et al., 1996), numerous plant CESA genes have been identified. The *Arabidopsis* genome contains 10 cellulose synthases (CESAs), 3 of which are involved in the cellulose biosynthesis in secondary walls (CESA4, CESA7, and CESA8) (Taylor et al., 1999; Turner et al., 2001; Kim et al., 2013a, 2013b). Several CESA genes have been cloned from the wood-forming tissues of tree species, such as pine (Allona et al., 1998) and *Populus* (Wu et al., 2000; Joshi et al., 2004; Bhandari et al., 2006; Suzuki et al., 2006; Kumar et al., 2009).

The cellulose microfibrils are fastened by lignins and hemicelluloses into a honeycomb structure in the secondary wall, which is further strengthened by highly glycosylated cell wall proteins that are entwined within the microfibril arrangement. Associating with the cellulose microfibrils, hemicelluloses provide a cross-linked matrix. Several hemicellulose biosynthesis genes, such as glucomannan and xylan biosynthases, have been identified (Brown et al., 2009; Dhugga et al., 2004; Lee et al., 2009; Liepman et al., 2005; Wu et al., 2009; Zhong et al., 2005). Transcription factors regulating hemicellulose biosynthesis have been described (Kim et al., 2014a). Xylans are the dominant component of the hemicelluloses in angiosperm species. Lignin, the second most abundant biopolymer on the planet, reinforces and provides mechanical strength to the cell wall by filling the spaces between cellulose and hemicellulose. While the content and composition of lignin vary among taxa, tissues, cell types, and cell wall layers, and depend on the developmental stage of the plant as well as on the environmental conditions (Sederoff et al., 1999), it constitutes about 25–35% of wood biomass in angiosperm species (Campbell and Sederoff, 1996). Many lignin biosynthesis genes and their transcriptional regulators have been characterized and used to genetically alter the content and composition of lignin in the cell walls (Boerjan et al., 2003; Vanholme et al., 2008; Kim et al., 2014b).

GENETIC REGULATION OF SECONDARY WALL BIOSYNTHESIS

Secondary wall formation requires coordinated transcriptional activation of the biosynthetic pathways of all of the secondary wall components (i.e., cellulose, hemicellulose, and lignin). Considering the significance of secondary walls in their growth and survival, it is requisite that land vascular plants have "foolproof" mechanisms to ensure correct formation of secondary walls in the cells of vascular tissues. Indeed, multifaceted and multilayered transcriptional networks appear to be involved in the coordinated regulation of the biosynthesis of this complex structure (Demura and Ye, 2010; Ko et al., 2011, 2012, 2014; Wang and Dixon, 2012), indicating the necessity of some level of central regulation for the overall process. Several transcription factors have been identified as central regulators of secondary wall biosynthesis and appear to fill that role (for recent reviews, see Yamaguchi and Demura, 2010; Zhang et al., 2010; Zhong et al., 2010a; Wang and Dixon, 2012; Zhao and Dixon, 2011; Ko et al., 2012, 2014; Pimrote et al., 2012; Hussey et al., 2013; Schuetz et al., 2013). Particularly, *Arabidopsis* transcription factor MYB46 and its paralogue, MYB83, have been shown to function as a master switch for the secondary wall biosynthetic program in *A. thaliana* (Ko et al., 2009, 2012, 2014) (Fig. 10.2). Despite these advances in understanding transcriptional regulation of secondary wall biosynthesis, the exact nature of this multifaceted regulatory network remains to be elucidated.

GENETIC MODIFICATION OF WOOD PROPERTY

With the discovery of key genes involved in wood formation and secondary wall biosynthesis, it is possible to engineer trees to produce wood with specific mechanical characteristics or chemical composition. Several key properties, such as wood density, microfibril angle, modulus of elasticity, and carbohydrate content and composition, define wood quality and the uses for which it is well suited. Wood density varies depending on species, growth condition, and tree developmental stage. For example, eastern cottonwood wood has a density of about ~ 320 kg/m^3 while oak species often exceed ~ 550 kg/m^3 (Isenberg, 1980). Earlywood, which is formed in the spring when the cambium is very active, has lower wood density than latewood, which is produced later in the growing season. Likewise, juvenile wood has different characteristics, stemming from differences in wood density and cell wall composition from those of mature wood. It may be possible to regulate wood density and these other wood characteristics through genetic manipulation of appropriate regulator genes that direct the biochemical processes of secondary wall biosynthesis. Compared to other cell wall components, manipulation of lignin biosynthesis and its content/composition have been relatively well studied (Raes et al., 2003; Roger and Campbell, 2004; Li et al., 2006). It has been demonstrated that manipulation of key genes involved in monolignol biosynthesis could change total lignin content and composition (Huntley et al., 2003; Li et al., 2003; Stewart et al., 2004). Manipulation of key regulatory genes in cellulose and hemicellulose biosynthesis can result in significant changes in cell wall carbohydrate content (Kim et al., 2013a, 2014b), and key wood quality traits, such as density, microfibril angle, and modulus of elasticity, have been shown to have negative genetic correlations with diameter growth in juvenile wood of *Pinus radiata* (Baltunis et al., 2007). Understanding the master regulators of key elements in this total biosynthesis process would provide a means for overcoming this difficulty, although clearly any attempt to regulate cell wall composition (e.g., lower lignin content) must be carried out in the context of whole plant physiology given the fundamental role secondary wall formation plays in the growth and development of vascular plants.

REFERENCES

Aida, M., Ishida, T., Fukaki, H., Fujisawa, H., Tasaka, M., 1997. Genes involved in organ separation in *Arabidopsis*: an analysis of the cup-shaped cotyledon mutant. Plant Cell 9, 841–857.

Allona, I., Quinn, M., Shoop, E., Swope, K., Cyr, S.S., Carlis, J., Riedl, J., Retzel, E., Campbell, M.M., Sederoff, R., Whetten, R.W., 1998. Analysis of xylem formation in pine by cDNA sequencing. Proc. Natl. Acad. Sci. USA 95, 9693–9698.

Andersson-Gunneras, S., Mellerowicz, E.J., Love, J., Segerman, B., Ohmiya, Y., Coutinho, P., Nilsson, P., Henrissat, B., Moritz, T., Sundberg, B., 2006. Biosynthesis of cellulose-enriched tension wood in *Populus*: global analysis of transcripts and metabolites identifies biochemical and developmental regulators in secondary wall biosynthesis. Plant J. 45, 144–165.

Antonova, G.F., Stasova, V.V., 1997. Effects of environmental factors on wood formation in larch (*Larix sibirica* Ldb.) stems. Trees 11, 462–468.

Arend, M., Fromm, J., 2007. Seasonal change in drought response of wood cell development in poplar. Tree Physiol. 27, 985–992.

Baba, K., Park, Y.W., Kaku, T., Kaida, R., Takeuchi, M., Yoshida, M., Hosoo, Y., Ojio, Y., Okuyama, T., Taniguchi, T., Ohmiya, Y., Kondo, T., Shani, Z., Shoseyov, O., Awano, T., Serada, S., Norioka, N., Norioka, S., Hayashi, T., 2009. Xyloglucan for generating tensile stress to bend tree stem. Mol. Plant 2 (5), 893–903.

Baba, K., Karlberg, A., Schmidt, J., Schrader, J., Hvidsten, T.R., Bakó, L., Bhalerao, R.P., 2011. Activity-dormancy transition in the cambial meristem involves stage-specific modulation of auxin response in hybrid aspen. Proc. Natl. Acad. Sci. USA 108 (8), 3418–3423.

Baier, M., Goldberg, R., Catesson, A.M., Liberman, M., Bouchemal, N., Michon, V., Dupenhoat, C.H., 1994. Pectin changes in samples containing poplar cambium and inner bark in relation to the seasonal cycle. Planta 193, 446–454.

Baima, S., Possenti, M., Matteucci, A., Wisman, E., Altamura, M.M., Ruberti, I., Morelli, G., 2001. The *Arabidopsis* ATHB-8 HD-zip protein acts as a differentiation-promoting transcription factor of the vascular meristems. Plant Physiol. 126, 643–655.

Baltunis, B., Wu, H.X., Powell, M.B., 2007. Inheritance of density, microfibril angle, and modulus of elasticity in juvenile wood of *Pinus radiata* at two locations in Australia. Can. J. Forest Res. 37 (11), 2164–2174.

Begum, S., Nakaba, S., Yamagishi, Y., Oribe, Y., Funada, R., 2013. Regulation of cambial activity in relation to environmental conditions: understanding the role of temperature in wood formation of trees. Physiol. Plant 147 (1), 46–54.

Benjamins, R., Scheres, B., 2008. Auxin: the looping star in plant development. Annu. Rev. Plant Biol. 59, 443–465.

Berleth, T., Sachs, T., 2001. Plant morphogenesis: long-distance coordination and local patterning. Curr. Opin. Plant Sci. 4, 57–63.

Bhalerao, R.P., Fischer, U., 2014. Auxin gradients across wood – instructive or incidental? Physiol. Plant 151 (1), 43–51.

Bhalerao, R., Keskitalo, J., Sterky, F., Erlandsson, R., Björkbacka, H., Birve, S.J., Karlson, J., Gardestrom, P., Gustafsson, P., Lundeberg, J., Jansson, S., 2003. Gene expression in autumn leaves. Plant Physiol. 131 (2), 430–442.

Bhandari, S., Fujino, T., Thammanagowda, S., Zhang, D., Xu, F., Joshi, C.P., 2006. Xylem-specific and tension stress-responsive coexpression of KORRIGAN endoglucanase and three secondary wall-associated cellulose synthase genes in aspen trees. Planta 224 (4), 828–837.

Boerjan, W., Ralph, J., Baucher, M., 2003. Lignin biosynthesis. Annu. Rev. Plant Biol. 54, 519–546.

Bonke, M., Thitamadee, S., Mahonen, A.P., Hauser, M.T., Helariutta, Y., 2003. APL regulates vascular tissue identity in *Arabidopsis*. Nature 426, 181–186.

Bowling, A.J., Vaughn, K.C., 2008. Immunocytochemical characterization of tension wood: gelatinous fibers contain more than just cellulose. Am. J. Bot. 95, 655–663.

Brown, D.M., Zhang, Z., Stephens, E., Dupree, P., Turner, S.R., 2009. Characterization of IRX10 and IRX10-like reveals an essential role in glucuronoxylan biosynthesis in Arabidopsis. Plant J. 57, 732–746.

Campbell, M.M., Sederoff, R.R., 1996. Variation in lignin content and composition – Mechanism of control and implications for the genetic improvement of plants. Plant Physiol. 110, 3–13.

Castillon, A., Shen, H., Huq, E., 2007. Phytochrome interacting factors: central players in phytochrome-mediated light signaling networks. Trends Plant Sci. 12 (11), 514–521.

Catesson, A.M., Funada, R., Robertbaby, D., Quinetszely, M., Chuba, J., Goldberg, R., 1994. Biochemical and cytochemical cell-wall changes across the cambial zone. IAWA J. 15, 91–101.

Chaffey, N., 1999. Cambium: old challenges – New opportunities. Trees 13, 138–151.

Chaffey, N., Cholewa, E., Regan, S., Sundberg, B., 2002. Secondary xylem development in Arabidopsis: a model for wood formation. Physiol. Plant 114 (4), 594–600.

Chapman, E., Estelle, M., 2009. Mechanism of auxin-regulated gene expression in plants. Annu. Rev. Genet. 43, 265–285.

Chen, M., Chory, J., 2011. Phytochrome signaling mechanisms and the control of plant development. Trends Cell Biol. 21 (11), 664–671.

Déjardin, A., Leplé, J.C., Lesage-Descauses, M.C., Costa, G., Pilate, G., 2004. Expressed sequence tags from poplar wood tissues – a comparative analysis from multiple libraries. Plant Biol. 7, 55–64.

Demura, T., Fukuda, H., 2007. Transcriptional regulation in wood formation. Trends Plant Sci. 12 (2), 64–70.

Demura, T., Ye, Z.H., 2010. Regulation of plant biomass production. Curr. Opin. Plant Biol. 13, 299–304.

Dhugga, K.S., Barreiro, R., Whitten, B., Stecca, K., Hazebroek, J., Randhawa, G.S., Dolan, M., Kinney, A.J., Tomes, D., Nichols, S., Anderson, P., 2004. Guar seed β-mannan synthase is a member of the cellulose synthase super gene family. Science 303, 363–366.

Druart, N., Johansson, A., Baba, K., Schrader, J., Sjödin, A., Bhalerao, R.R., Resman, L., Trygg, J., Moritz, T., Bhalerao, R., 2007. Environmental and hormonal regulation of the activity–dormancy cycle in the cambial meristem involves stage-specific modulation of transcriptional and metabolic networks. Plant J. 50 (4), 557–573.

Du, J., Groover, A., 2010. Transcriptional regulation of secondary growth and wood formation. J. Integr. Plant Biol. 52 (1), 17–27.

Duek, P.D., Fankhauser, C., 2005. bHLH class transcription factors take centre stage in phytochrome signalling. Trends Plant Sci. 10 (2), 51–54.

Eklund, L., Little, C.H.A., Riding, R.T., 1998. Concentrations of oxygen and indole-3-acetic acid in the cambial region during latewood formation and dormancy development in Picea abies stems. J. Exp. Bot. 49, 205–211.

Emery, J.F., Floyd, S.K., Alvarez, J., Eshed, Y., Hawker, N.P., Izhaki, A., Baum, S.F., Bowman, J.L., 2003. Radial patterning of Arabidopsis shoots by class III HD-ZIP and KANADI genes. Curr. Biol. 13, 1768–1774.

Franklin, K.A., Quail, P.H., 2010. Phytochrome functions in Arabidopsis development. J. Exp. Bot. 61 (1), 11–24.

Frewen, B.E., Chen, T.H., Howe, G.T., Davis, J., Rohde, A., Boerjan, W., Bradshaw, H.D., 2000. Quantitative trait loci and candidate gene mapping of bud set and bud flush in Populus. Genetics 154 (2), 837–845.

Fukuda, H., 1996. Xylogenesis: initiation, progression, and cell death. Ann. Rev. Plant Physiol. Plant Mol. Biol. 47, 299–325.

Fukuda, H., 1997. Tracheary element differentiation. Plant Cell 9, 1147–1156.

Fukuda, H., 2004. Signals that control plant vascular cell differentiation. Nat. Rev. Mol. Cell Biol. 5, 379–391.

Gardiner, J.C., Taylor, N.G., Turner, S.R., 2003. Control of cellulose synthase complex localization in developing xylem. Plant Cell 15, 1740–1748.

Grattapaglia, D., Kirst, M., 2008. *Eucalyptus* applied genomics: from gene sequences to breeding tools. New Phytol. 179 (4), 911–929.

Guilfoyle, T.J., Hagen, G., 2007. Auxin response factors. Curr. Opin. Plant Biol. 10 (5), 453–460.

Hardtke, C.S., Berleth, T., 1998. The *Arabidopsis* gene MONOPTEROS encodes a transcription factor mediating embryo axis formation and vascular development. EMBO J. 2, 1405–1411.

Hellgren, J.M., Olofsson, K., Sundberg, B., 2004. Patterns of auxin distribution during gravitational induction of reaction wood in poplar and pine. Plant Physiol. 135, 212–220.

Hertzberg, M., Aspeborg, H., Schrader, J., Andersson, A., Erlandsson, R., Blomqvist, K., et al., 2001. A transcriptional roadmap to wood formation. Proc. Natl. Acad. Sci. USA 98 (25), 14732–14737.

Hoffman, D.E., Jonsson, P., Bylesjö, M., Trygg, J., Antti, H., Eriksson, M.E., Moritz, T., 2010. Changes in diurnal patterns within the *Populus* transcriptome and metabolome in response to photoperiod variation. Plant Cell Environ. 33 (8), 1298–1313.

Horvath, D.P., Anderson, J.V., Chao, W.S., Foley, M.E., 2003. Knowing when to grow: signals regulating bud dormancy. Trends Plant Sci. 8 (11), 534–540.

Horvath, D.P., Chao, W.S., Suttle, J.C., Thimmapuram, J., Anderson, J.V., 2008. Transcriptome analysis identifies novel responses and potential regulatory genes involved in seasonal dormancy transitions of leafy spurge (*Euphorbia esula* L.). BMC Genom. 9 (1), 536–1536.

Howe, G., Hackett, W., Furnier, G., Klevorn, E., 1995. Photoperiodic responses of a northern and southern ecotype of black cottonwood. Physiol. Plant 93, 695–708.

Hu, W.J., Harding, S.A., Lung, J., Popko, J.L., Ralph, J., Stokke, D.D., Tsai, C.J., Chiang, V.L., 1999. Repression of lignin biosynthesis promotes cellulose accumulation and growth in transgenic trees. Nat. Biotechnol. 17, 808–812.

Huntley, S.K., Ellis, D., Gilbert, M., Chapple, C., Mansfield, S.D., 2003. Significant increases in pulping efficiency in C4H-F5H-transformed poplars: Improved chemical savings and reduced environmental toxins. J. Agric. Food Chem. 51, 6178–6183.

Hussey, S.G., Mizrachi, E., Creux, N.M., Myburg, A.A., 2013. Navigating the transcriptional roadmap regulating plant secondary cell wall deposition. Front. Plant Sci. 29 (4), 325.

Iliev, I., Savidge, R., 1999. Proteolytic activity in relation to seasonal cambial growth and xylogenesis in *Pinus banksiana*. Phytochemistry 50 (6), 953–960.

Isenberg, I.H., 1980, Pulpwoods of the United States and Canada. Hard-woods, vol 2, third ed.The Institute of Paper Chemistry, Appleton, WI, 168p.

Jiao, Y., Lau, O.S., Deng, X.W., 2007. Light-regulated transcriptional networks in higher plants. Nat. Rev. Genet. 8 (3), 217–230.

Jin, H., Do, J., Moon, D., Noh, E.-W., Kim, W., Kwon, M., 2011. EST analysis of functional genes associated with cell wall biosynthesis and modification in the secondary xylem of the yellow poplar (*Liriodendron tulipifera*) stem during early stage of tension wood formation. Planta 234 (5), 959–977.

Joshi, C.P., Bhandari, S., Ranjan, P., Kalluri, U.C., Liang, X., Fujino, T., Samuga, A., 2004. Genomics of cellulose biosynthesis in poplars. New Phytol. 164, 53–61.

Karpinska, B., Karlsson, M., Srivastava, M., Stenberg, A., Schrader, J., Sterky, F., Bhalera, R., Wingsle, G., 2004. MYB transcription factors are differentially expressed and regulated during secondary vascular tissue development in hybrid aspen. Plant Mol. Biol. 56, 255–270.

Khanna, R., Huq, E., Kikis, E.A., Al-Sady, B., Lanzatella, C., Quail, P.H., 2004. A novel molecular recognition motif necessary for targeting photoactivated phytochrome signaling to specific basic helix-loop-helix transcription factors. Plant Cell 16 (11), 3033–3044.

Kim, W.-C., Ko, J.-H., Kim, J.Y., Kim, J.-M., Bae, H.-J., Han, K.-H., 2013a. MYB46 directly regulates the gene expression of secondary wall-associated cellulose synthases in *Arabidopsis*. Plant J. 73, 26–36.

Kim, W.-C., Kim, J.Y., Ko, J.-H., Kim, J., Han, K.-H., 2013b. Transcription factor MYB46 is an obligate component of the transcriptional regulatory complex for functional expression of secondary wall-associated cellulose synthases in *Arabidopsis thaliana*. J. Plant Physiol. 170 (15), 1374–1378.

Kim, W.-C., Reca, I.-B., Kim, Y., Park, S., Thomashow, M.F., Keegstra, K., Han, K.-H., 2014a. Transcription factors that directly regulate the expression of CSLA9 encoding mannan synthase in *Arabidopsis thaliana*. Plant Mol. Biol. 84, 577–587.

Kim, W.-C., Kim, J.Y., Ko, J.-H., Kang, H., Han, K.-H., 2014b. Identification of direct targets of transcription factor MYB46 provides insights into the transcriptional regulation of secondary wall biosynthesis. Plant Mol. Biol. 85, 589–599.

Klee, H.J., Horsch, R.B., Hinchee, M.A., Hein, M.B., Hoffmann, N.L., 1987. The effects of overproduction of two *Agrobacterium tumefaciens* T-DNA auxin biosynthetic gene products in transgenic petunia plants. Genes Dev. 1, 86–96.

Ko, J.H., Han, K.H., 2004. *Arabidopsis* whole-transcriptome profiling defines the features of coordinated regulations that occur during secondary growth. Plant Mol. Biol. 55, 433–453.

Ko, J.H., Han, K.H., Park, S., Yang, J., 2004. Plant body weight-induced secondary growth in *Arabidopsis* and its transcription phenotype revealed by whole-transcriptome profiling. Plant Physiol. 135, 1069–1083.

Ko, J.H., Beers, E.P., Han, K.H., 2006a. Global comparative transcriptome analysis identifies gene network regulating secondary xylem development in *Arabidopsis thaliana*. Mol. Genet. Genome 276, 517–531.

Ko, J.H., Kim, W.C., Han, K.H., 2009. Ectopic expression of MYB46 identifies transcriptional regulatory genes involved in secondary wall biosynthesis in *Arabidopsis*. Plant J. 60 (4), 649–665.

Ko, J.-H., Prassinos, C., Keathley, D., Han, K.-H., 2011. Novel aspects of transcriptional regulation in the winter survival and maintenance mechanism of poplar. Tree Physiol. 31 (2), 208–225.

Ko, J.H., Kim, W.C., Kim, J.Y., Ahn, S.J., Han, K.H., 2012. MYB46-mediated transcriptional regulation of secondary wall biosynthesis. Mol. Plant 5 (5), 961–963.

Ko, J.-H., Jeon, H.-W., Kim, W.-C., Kim, J.-Y., Han, K.-H., 2014. MYB46/MYB83-mediated transcriptional regulatory program is gatekeeper of secondary wall biosynthesis. Ann. Bot. 114 (6), 1099–1107.

Kubo, M., Udagawa, M., Nishikubo, N., Horiguchi, G., Yamaguchi, M., Ito, J., Mimura, T, Fukuda, H., Demura, T., 2005. Transcription switches for protoxylem and metaxylem vessel formation. Genes Dev. 19, 1855–1860.

Kumar, M., Thammannagowda, S., Bulone, V., Chiang, V., Han, K.-H., Joshi, C.P., Mansfield, S.D., Mellerowicz, E., Sundberg, B., Teeri, T., Ellis, B.E., 2009. An update on the nomenclature for the cellulose synthase genes in *Populus*. Trends Plant Sci. 14 (5), 248–254.

Larson, P.R., 1963. Stem form and development of forest trees. For. Sci. Monogr. 5, 1–41.

Larson, P.R., 1994. The Vascular Cambium. Springer-Verlag, Berlin.

Lee, C., Teng, Huang, W., Zhong, R., Ye, Z.H., 2009. Down-regulation of PoGT47C expression in poplar results in a reduced glucuronoxylan content and an increased wood digestibility by cellulase. Plant Cell Physiol. 50, 1075–1089.

Leivar, P., Quail, P.H., 2011. PIFs: pivotal components in a cellular signaling hub. Trends Plant Sci. 16 (1), 19–28.

Lerouxel, O., Cavalier, D.M., Liepman, A.H., Keegstra, K., 2006. Biosynthesis of plant cell wall polysaccharides: a complex process. Curr. Opin. Plant Biol. 9, 621–630.

Lev-Yadun, S., 1994. Induction of sclereid differentiation in the pith of *Arabidopsis thaliana* (L.) Heynh. J. Exp. Bot. 45, 1845–1849.

Lev-Yadun, S., Aloni, R., 1995. Differentiation of the ray system in woody plants. Bot. Rev. 61, 45–84.

Leyser, O., 2006. Dynamic integration of auxin transport and signalling. Curr. Biol. 16, R424–R433.

Li, Y., Zhou, Y., Cheng, X., Sun, J., Marita, J.M., Ralph, J., Chiang, V.L., 2003. Combinatorial modification of multiple lignin traits in trees through multigene cotransformation. Proc. Natl. Acad. Sci. USA 100, 4939–4944.

Li, Y., Wang, F., Lee, J.-A., Gao, F.-B., 2006. MicroRNA-9a ensures the precise specification of sensory organ precursors in *Drosophila*. Genes Dev. 20, 2769–2772.

Liepman, A.H., Wilkerson, C.G., Keegstra, K., 2005. Expression of cellulose synthase-like (Csl) genes in insect cells reveals that CslA family members encode mannan synthases. Proc. Natl. Acad. Sci. USA 102, 2221–2226.

Little, C.H.A., Sundberg, B., 1991. Tracheid production in response to indole-3-acetic-acid varies with internode age in *Pinus sylvestris* stems. Trees 5, 101–106.

Liu, L., Filkov, V., Groover, A., 2013. Modeling transcriptional networks regulating secondary growth and wood formation in forest trees. Physiol. Plant 151 (2), 156–163.

Lu, S., Li, L., Yi, X., Joshi, C.P., Chiang, V.L., 2008. Differential expression of three eucalyptus secondary cell wall-related cellulose synthase genes in response to tension stress. J. Exp. Bot. 59 (3), 681–695.

Mauseth, J., 1998. Botany: An Introduction to Plant Biology. Jones and Bartlett Publishers, Sudbury, Massachusetts.

McDougall, G.J., Morrison, I.M., Stewart, D., Weyers, J.D.B., Hillman, J.R., 1993. Plant fibres: botany, chemistry and processing for industrial use. J. Sci. Food Agric. 62, 1–20.

Mellerowicz, E.J., Coleman, W.K., Riding, R.T., Little, C.H.A., 1992a. Periodicity of cambial activity in *Abies balsamea*.1. Effects of temperature and photoperiod on cambial dormancy and frost hardiness. Physiol. Plant 85, 515–525.

Mellerowicz, E.J., Riding, R.T., Little, C.H.A., 1992b. Periodicity of cambial activity in *Abies balsamea*. 2. Effects of temperature and photoperiod on the size of the nuclear genome in fusiform cambial cells. Physiol. Plant 85, 526–530.

Mellerowicz, E.J., Baucher, M., Sundberg, B., Boerjan, W., 2001. Unravelling cell wall formation in the woody dicot stem. Plant Mol. Biol. 47, 239–274.

Mellerowicz, E.J., Immerzeel, P., Hayashi, T., 2008. Xyloglucan: the molecular muscle of trees. Ann. Bot. 102 (5), 659–665.

Myburg, A.A., Grattapaglia, D., Tuskan, G.A., Hellsten, U., Hayes, R.D., Grimwood, J., et al., 2014. The genome of *Eucalyptus grandis*. Nature 510 (7505), 356–362.

Nishikubo, N., Awano, T., Banasiak, A., Bourquin, V., Ibatullin, F., Funada, R., Brumer, H., Teeri, T.T., Hayashi, T., Sundberg, B., Mellerowicz, E.J., 2007. Xyloglucan endo-transglycosylase (XET) functions in gelatinous layers of tension wood fibers in poplar – a glimpse into the mechanism of the balancing act of trees. Plant Cell Physiol. 48 (6), 843–855.

Oh, S., Park, S., Han, K.-H., 2003. Transcriptional regulation of secondary growth in *Arabidopsis thaliana*. J. Exp. Bot. 54 (393), 2709–2722.

Ohashi-Ito, K., Fukuda, H., 2003. HD-zip III homeobox genes that include a novel member, ZeHB-13 (*Zinnia*)/ATHB-15 (*Arabidopsis*), are involved in procambium and xylem cell differentiation. Plant Cell Physiol. 44, 1350–1358.

Ohtani, M., Nishikubo, N., Xu, B., Yamaguchi, M., Mitsuda, N., Goué, N., Shi, F., Ohme-Takagi, M., Demura, T., 2011. A NAC domain protein family contributing to the regulation of wood formation in poplar. Plant J. 67 (3), 499–512.

Olsen, J., Junttila, O., Nilson, J., Eriksson, M., Martinussen, I., Olsson, O., Sandberg, G., Moritz, T., 1997. Ectopic expression of oat phytochrome A in hybrid aspen changes critical day-length for growth and prevents cold acclimatization. Plant J. 12 (6), 1339–1350.

Olsen, A.N., Ernst, H.A., Leggio, L.L., Skriver, K., 2005. NAC transcription factors: structurally distinct, functionally diverse. Trends Plant Sci. 10, 79–87.

Park, S., Keathley, D.E., Han, K.-H., 2008. Transcriptional profiles of the annual growth cycle in *Populus deltoides*. Tree Physiol. 28 (3), 321–329.

Paux, X., Carocha, V., Marques, C., Mendes de Sousa, A., Borralho, N., Sivadon, P., Grima-Pettenati, J., 2005. Transcript profiling of *Eucalyptus* xylem genes during tension wood formation. New Phytol. 167 (1), 89–100.

Pear, J.R., Kawagoe, Y., Schreckengost, W.E., Delmer, D.P., Stalker, D.M., 1996. Higher plants contain homologs of the bacterial celA genes encoding the catalytic subunit of cellulose synthase. Proc. Natl. Acad. Sci. USA 93, 12637–12642.

Pilate, G., Dejardin, A., Laurans, F., Leple, J.-C., 2004. Tension wood as a model for functional genomics of wood formation. New Phytol. 164, 63–72.

Pimrote, K., Tian, Y., Lu, X., 2012. Transcriptional regulatory network controlling secondary cell wall biosynthesis and biomass production in vascular plants. Afr. J. Biotechnol. 11, 13928–13937.

Plomion, C., Leprovost, G., Stokes, A., 2001. Wood formation in trees. Plant Physiol. 127, 1513–1523.

Qiu, D., Wilson, I.W., Gan, S., Washusen, R., Moran, G.F., Southerton, S.G., 2008. Gene expression in *Eucalyptus* branch wood with marked variation in cellulose microfibril orientation and lacking G-layers. New Phytol. 179 (1), 94–103.

Quail, P.H., 2002. Photosensory perception and signalling in plant cells: new paradigms? Curr. Opin. Cell Biol. 14 (2), 180–188.

Raes, J., Rohde, A., Christensen, J.H., Van de Peer, Y., Boerjan, W., 2003. Plant Physiol. 133, 1051–1071.

Ragni, L., Nieminen, K., Pacheco-Villalobos, D., Sibout, R., Schechheimer, C., Hardtke, C.S., 2011. Mobile gibberellin directly stimulates *Arabidopsis* hypocotyl xylem expansion. Plant Cell 23 (4), 1322–1336.

Rengel, D., San Clemente, H., Servant, F., Ladouce, N., Paux, E., Wincker, P., Couloux, A., Sivadon, P., Grima-Pettenati, J., 2009. A new genomic resource dedicated to wood formation in Eucalyptus. BMC Plant Biol. 9, 36.

Rockwell, N.C., Su, Y.-S., Lagarias, J.C., 2006. Phytochrome structure and signaling mechanisms. Annu. Rev. Plant Biol. 57 (1), 837–858.

Rodriguez-A, J., Sherman, W.B., Scorza, R., Wisniewski, M., Okie, W.R., 1994. 'Evergreen' peach, its inheritance and dormant behavior. J. Am. Soc. Hort. Sci. 119 (4), 789–792.

Roger, L.A., Campbell, M.M., 2004. The genetic control of lignin deposition during plant growth and development. New Phytol. 164, 17–30.

Rohde, A., Bhalerao, R.P., 2007. Plant dormancy in the perennial context. Trends Plant Sci. 12 (5), 217–223.

Sachs, T., 1981. The control of the patterned differentiation of vascular tissues. Adv. Bot. Res. 9, 151–262.

Sachs, T., 2000. Integrating cellular and organismic aspects of vascular differentiation. Plant Cell Physiol. 41, 649–656.

Saxena, I.M., Brown, Jr., R.M., 2005. Cellulose biosynthesis: current views and evolving concepts. Ann. Bot. 96, 9–21.

Schuetz, M., Smith, R., Ellis, B., 2013. Xylem tissue specification, patterning, and differentiation mechanisms. J. Exp. Bot. 64, 11–31.

Sederoff, R.R., MacKay, J.J., Ralph, J., Hatfield, R.D., 1999. Unexpected variation in lignin. Curr. Opin. Plant Biol. 2, 145–152.

Somerville, C., 2006. Cellulose synthesis in higher plants. Annu. Rev. Cell Dev. Bio. 22, 53–78.

Souer, E., van Houwelingen, A., Kloos, D., Mol, J., Koes, R., 1996. The no apical meristem gene of *Petunia* is required for pattern formation in embryos and flowers and is expressed at meristem and primordia boundaries. Cell 85, 159–170.

Stewart, G.H., Ignatieva, M.E., Meurk, C.D., Earl, R.D., 2004. The re-emergence of indigenous forest in an urban environment, Christchurch, New Zealand. Urban For Urban Greening 2, 149–158.

Suzuki, S., Li, L., Sun, Y.H., Chiang, V.L., 2006. The cellulose synthase gene superfamily and biochemical functions of xylem-specific cellulose synthase-like genes in *Populus trichocarpa*. Plant Physiol. 142, 1233–1245.

Tanaka, K., Murata, K., Yamazaki, M., Onosato, K., Miyao, A., Hirochika, H., 2003. Three distinct rice cellulose synthase catalytic subunit genes required for cellulose synthesis in the secondary wall. Plant Physiol. 133, 73–83.

Taylor, N.G., Scheible, W.R., Cutler, S., Somerville, C.R., Turne, S.R., 1999. The irregular xylem3 locus of *Arabidopsis* encodes a cellulose synthase required for secondary cell wall synthesis. Plant Cell 11, 769–779.

Taylor, N.G., Howells, R.M., Huttly, A.K., Vickers, K., Turner, S.R., 2003. Interactions among three distinct CesA proteins essential for cellulose synthesis. Proc. Natl. Acad. Sci. USA 100, 1450–1455.

Timell, T.E., 1969. The chemical composition of tension wood. Svensk Papperstidning 72, 173–181.

Timell, T.E., 1986. Compression Wood in Gymnosperms, vol. 2. Springer-Verlag, Heidelberg, pp. 259–262.

Turner, S.R., Taylor, N., Jones, L., 2001. Mutations of the secondary cell wall. Plant Mol. Biol. 47, 209–219.

Turner, S., Gallois, P., Brown, D., 2007. Tracheary element differentiation. Annu. Rev. Plant Biol. 58, 407–433.

Tuskan, G.A., DiFazio, S., Jansson, S., Bohlmann, J., Grigoriev, I., Hellsten, U., et al., 2006. The genome of black cottonwood *Populus trichocarpa* (Torr. & Gray). Science 313 (5793), 1596–1604.

Uggla, C., Moritz, T., Sandberg, G., Sundberg, B., 1996. Auxin as a positional signal in pattern formation in plants. Proc. Natl. Acad. Sci. USA 93, 9282–9286.

Uggla, C., Mellerowicz, E.J., Sundberg, B., 1998. Indole-3-acetic acid controls cambial growth in Scots pine by positional signaling. Plant Physiol. 117, 113–121.

Vanholme, R., Morreel, K., Ralph, J., Boerjan, W., 2008. Lignin engineering. Curr. Opin. Plant Biol. 11, 278–285.

Verde, I., Abbott, A.G., Scalabrin, S., Jung, S., Shu, S., Marroni, F., et al., 2013. The high-quality draft genome of peach (*Prunus persica*) identifies unique patterns of genetic diversity, domestication and genome evolution. Nat. Genet. 45 (5), 487–494.

Wang, H.Z., Dixon, R.A., 2012. On-off switches for secondary cell wall biosynthesis. Mol. Plant 5, 297–303.

Wang, Q., Little, C.H.A., Oden, P.C., 1995. Effect of laterally applied gibberellin $A_{4/7}$ on cambial growth and the level of indole-3-acetic-acid in *Pinus sylvestris* shoots. Physiol. Plant 95, 187–194.

Wang, Y., Georgi, L.L., Reighard, G.L., Scorza, R., Abbott, A.G., 2002. Genetic mapping of the evergrowing gene in peach [*Prunus persica* (L.) Batsch]. J. Hered. 93 (5), 352–358.

Wang, C., Zhang, N., Gao, C., Cui, Z., Sun, D., Chang, C., Wang, Y., 2014. Comprehensive transcriptome analysis of developing xylem responding to artificial bending and gravitational stimuli in *Betula platyphylla*. PLoS One 9 (2), e87556.

Welling, A., Moritz, T., Palva, E.T., Junttila, O., 2002. Independent activation of cold acclimation by low temperature and short photoperiod in hybrid aspen. Plant Physiol. 129 (4), 1633–1641.

Wolpert, L., 1996. One hundred years of positional information. Trends Genet. 12, 359–364.

Wu, L., Joshi, C.P., Chiang, V.L., 2000. A xylem-specific cellulose synthase gene from aspen (*Populus tremuloides*) is responsive to mechanical stress. Plant J. 22, 495–502.

Wu, A.M., Rihouey, C., Seveno, M., Hörnblad, E., Singh, S.K., Matsunaga, T., Ishii, T., Lerouge, P., Marchant, A., 2009. The *Arabidopsis* IRX10 and IRX10-LIKE glycosyltransferases are critical for glucuronoxylan biosynthesis during secondary cell wall formation. Plant 57, 718–731.

Xie, Q., Frugis, G., Colgan, D., Chua, N.H., 2000. *Arabidopsis* NAC1 transduces auxin signal downstream of TIR1 to promote lateral root development. Genes Dev. 14, 3024–3036.

Yamaguchi, M., Demura, T., 2010. Transcriptional regulation of secondary wall formation controlled by NAC domain proteins. Plant Biotechnol. 27 (3), 237–242.

Yamamoto, H., Abe, K., Arakawa, Y., Okuyama, T., Grill, J., 2005. Role of the gelatinous layer (G-layer) on the origin of the physical properties of the tension wood of *Acer sieboldianum*. J. Wood Sci. 51, 222–233.

Ye, Z.H., 2002. Vascular tissue differentiation and pattern formation in plants. Annu. Rev. Plant Biol. 53, 183–202.

Zhang, J., Elo, A., Helariutta, Y., 2010. *Arabidopsis* as a model for wood formation. Curr. Opin. Biotechnol. 22, 1–7.

Zhang, Q., Chen, W., Sun, L., Zhao, F., Huang, B., Yang, W., Tao, Y., Wang, J., Yuan, Z., Fan, G., Xing, Z., Han, C., Pan, H., Zhong, X., Shi, W., Liang, X., Du, D., Sun, F., Xu, Z., Hao, R., Lv, Y., Zheng, Z., Sun, M., Luo, L., Cai, M., Gao, Y., Wang, J., Yin, Y., Xu, X., Cheng, T., Wang, J., 2012. The genome of *Prunus mume*. Nat. Commun. 3, 1318.

Zhao, Q., Dixon, R.A., 2011. Transcriptional networks for lignin biosynthesis: more complex than we thought? Trends Plant Sci. 16, 227–233.

Zhong, R.Q., Ye, Z.H., 1999. IFL1, a gene regulating interfascicular fiber differentiation in *Arabidopsis*, encodes a homeodomain-leucine zipper protein. Plant Cell 11, 2139–2152.

Zhong, R., Ye, Z.H., 2001. Alteration of auxin polar transport in the *Arabidopsis ifl1* mutants. Plant Physiol. 126, 549–563.

Zhong, R., Ye, Z.H., 2007. Regulation of cell wall biosynthesis. Curr. Opin. Plant Biol. 10 (6), 564–572.

Zhong, Y., Mellerowicz, E.J., Lloyd, A.D., Leinhos, V., Riding, R.T., Little, C.H.A., 1995. Seasonal variation in the nuclear genome size of ray cells in the vascular cambium of *Fraxinus americana*. Physiol. Plant. 93, 305–311.

Zhong, R., Pena, M., Zhou, G.K., Nairn, C.J., Wood-Jones, A., Richardson, E.A., Morrison, W.H., Darvill, A.G., York, W.S., Ye, Z.H., 2005. *Arabidopsis Fragile Fiber8*, which encodes a putative glucuronyltransferase, is essential for normal secondary wall synthesis. Plant Cell 17, 3390–3408.

Zhong, R., Lee, C., Ye, Z.H., 2010a. Evolutionary conservation of the transcriptional network regulating secondary cell wall biosynthesis. Trends Plant Sci. 15, 625–632.

Zimmermann, M.H., Brown, C.L., 1971. Tree Structure and Function. Springer-Verlag, Berlin and New York.

Zobel, B.J., Jett, J.B., 1995. Genetics of Wood Production. Springer-Verlag, Berlin.

Chapter 11

Secondary Xylem for Bioconversion

Shiro Saka*, Hyeun-Jong Bae**

*Graduate School of Energy Science, Department of Socio-Environmental Energy Science, Kyoto University, Yoshida-honmachi, Sakyo-ku, Kyoto, Japan; **Department of Bioenergy Science & Technology, Chonnam National University, Buk-gu, Gwangju, South Korea

Chapter Outline

INTRODUCTION

Environmental issues in energy generation and other technological processes are of major concern because of the alarming rate of global warming. Efforts are increasingly intensifying in developing biofuels and biochemicals from biomass, which are steadily replacing fossil resources. Biomass is not only a clean, renewable material but also abundantly available, and its bioconversion to liquids and gases is an attractive approach for generating biofuels and biochemicals.

Wood biomass is the most abundant renewable resource on earth. In trees, the secondary xylem or wood originates from the activity of the vascular cambium, a lateral meristem. The mass of the secondary xylem on the inside of the cambium is considerably greater than the secondary phloem produced on the outside of the cambium, and thus trees constitute an excellent source of convertible biomass. The wood from trees consists of several types of tissues. In softwoods, tracheids and parenchyma are the main tissues, whereas hardwoods consist of fibers, vessels, and parenchyma. Since tracheids and fibers constitute

Secondary Xylem Biology. http://dx.doi.org/10.1016/B978-0-12-802185-9.00011-5

the bulk of softwoods and hardwoods, respectively, they are the main contributors to the physical and chemical properties of wood (Saka, 2001a).

Cellulose, hemicellulose, and lignin are the major constituents of wood. Natural cellulose is a bundle of linear β-1,4-linked glucan chains and these chains are packed tightly by inter- and intrahydrogen bonds. Cellulose produced by woody plants is the most abundant biopolymer on earth, accounting for 40–60% of plant cell walls.

In addition to the traditional uses of cellulose in the pulp and paper, food, and textile industries, the new concept of bioethanol production from lignocellulosic biomass is considered a promising route to sustainable energy production. However, biomass resources, particularly woods, are natural composite materials composed of cellulose, hemicelluloses, and lignin, in which the crystalline cellulose microfibrils are associated with amorphous hemicelluloses encrusted with lignin dehydrogenatively polymerized (Higuchi, 1997). Such a complex structure of the lignin-encrusting cell walls restricts access of enzymes and microorganisms to hemicelluloses and cellulose, and therefore chemical pretreatments prior to biological approach are needed for efficient conversion of wood biomass. Lignocellulosic material is intrinsically recalcitrant to chemical and enzymatic breakdown into simple sugars that can be fermented to liquid biofuels. A deeper understanding of biomass recalcitrance is required for the potential of lignocellulosic bioconversion to be fully realized (Wang et al., 2011; Ragauskas et al., 2006).

In this chapter, chemical pretreatments for the bioconversion of woody biomass are first briefly described, followed by a detailed description of biological treatment processes to provide a more comprehensive knowledge of the treatment processes for the conversion of woody biomass. The correlative information also promotes a deeper understanding of bioconversion characteristics.

BIOCONVERSION OF WOODY BIOMASS BY CHEMICAL PROCEDURES

For bioconversion, woody biomass is often pretreated to enhance saccharification of cellulose and hemicelluloses by the subsequent enzymatic treatment. Among various pretreatments, delignification of lignin and decrystallization of cellulose are most effective. However, it is also important to avoid the loss of hemicelluloses as well as their conversion into inhibitory by-products, which can negatively affect enzymatic hydrolysis and fermentation (Himmel et al., 2007). With these in mind, conventional chemical and hydrothermal processes have been employed for pretreating woody biomass. A description of these processes is provided, introducing the current progress being made.

Acid Hydrolysis

Several types of acids, such as sulfuric, hydrochloric, hydrofluoric, phosphoric, and nitric acids, can be used for acid hydrolysis of wood biomass. The earliest reported use of concentrated sulfuric acid for saccharification of cellulose dates back to 1819 (Braconnot, 1819). Since then, many trials have been made to hydrolyze cellulose and hemicelluloses by either concentrated or dilute acids.

A major advantage of the use of concentrated acids is that hydrolysis can be carried out at low temperatures, with high yields of sugars, such as glucose from cellulose. However, due to the intrinsic properties of crystalline cellulose and amorphous hemicelluloses, hydrolysis rate is much higher for hemicelluloses than cellulose, thus hemicellulose-derived sugars, pentoses and hexoses, are more extensively degraded into furfural and 5-hydroxymethyl furfural (5-HMF), and so on. These products inhibit subsequent fermentation of the sugars, as observed in dilute acid hydrolysis (Larsson et al., 1999). Other drawbacks are corrosion of the equipment, high consumption of the acid, toxicity to the environment, and energy demand for acid recovery.

Dilute acid hydrolysis has also been employed to pretreat woody biomass. The advantage for the use of dilute acid is the low consumption of the acid. However, compared to the concentrated acid, higher temperature is required for the hydrolysis reaction, which causes corrosion of equipment. In achieving a reasonable yield of glucose from crystalline cellulose, amorphous hemicelluloses are extensively degraded.

To remedy the problem, a two-stage process has been developed for concentrated and dilute acid hydrolyses. One example of the concentrated sulfuric acid process (Taneda, 2006) involves two-stage treatments, first using about 70% sulfuric acid at 30–40°C to hydrolyze hemicelluloses and to decrystallize cellulose. In the first stage, hemicellulose-derived sugars are separated. In the second stage, sulfuric acid is diluted to 30–40% with hot water, raising the temperature to 90–95°C to hydrolyze decrystallized cellulose. The hydrolysates are then separated into sugars and sulfuric acid through an ion exchange column. Resulting sugars are then fermented by genetically modified microorganisms, such as yeast, under anaerobic conditions to obtain bioethanol. The sulfuric acid, on the other hand, is recycled for further acid hydrolysis (Saka, 2009).

Alkali Hydrolysis

For pretreatment of wood, some bases can also be used (alkaline hydrolysis) to enhance enzymatic hydrolysis (Karr and Holzapple, 2000). The alkaline hydrolysis is believed to cause saponification of intermolecular ester bonds between lignin and carbohydrates. The main reagents used for the pretreatment are sodium hydroxide (NaOH), ammonia, calcium hydroxide, and oxidative alkali. Dilute NaOH treatment of wood, as in soda pulping, causes swelling of cell walls, resulting in an increase in internal surface area. At the same time, molecular linkages between lignin and carbohydrates are cleaved and the cell walls of the fibers are delignified (McMillan, 1994).

The digestibility of NaOH-treated hardwood is reported to increase with a decrease in lignin content. However, no effect was observed for softwoods with lignin content higher than 26% (Millet et al., 1976). This is not only due to higher content of lignin but also its intrinsic properties in softwoods (Rabemanolontsoa and Saka, 2013). In alkaline pretreatment, ammonia-recycled percolation process is also used (Iyer et al., 1996) at 170°C for 1 h with its concentration of 2.5–20% for mixture of corn cob and stover as well as switchgrass. The resulting delignification was 60–80% for the mixture, while 65–85% for the switchgrass.

Ionic Liquid

Ionic liquids mainly composed of imidazolium cation and chloride anion, such as 1-ethyl-3-methylimidazolium chloride, 1-butyl-3-methylimidazolium chloride, and 1-allyl-3-methylimidazolium chloride, are new organic salts that have a melting point around ambient condition, and thus generally exist as liquids at a relatively low temperature. They are chemically and thermally stable, nonflammable, and have low volatility (Zhu et al., 2006). In addition, ionic liquids have good solvation power for a wide variety of compounds and some can be recycled several times without any negative effects (Zulfiqar and Kitazume, 2000). The use of ionic liquids is considered to be "green chemistry" for various chemical reactions (Zhu et al., 2006).

Hydrothermal Treatment

The hydrothermal treatment is defined here as "reactions occurring under the conditions of high temperature and high pressure in aqueous solutions in a closed system." Hydrothermal treatment includes supercritical and subcritical water treatments, depending on the conditions of temperature and pressure. Water is one of the most important solvents and abundantly present in nature. It is environmentally benign, nontoxic, nonflammable, noncarcinogenic, nonmutagenic, and thermodynamically stable. Furthermore, water is also volatile and thus can be readily removed from the products.

These characteristic features depend on the conditions of temperature and pressure because water exists in gas, liquid, and solid phases. The supercritical fluid is one of high density that exceeds its critical pressure (P_c) and critical temperature (T_c), being a noncondensable gas-like water but not liquefied anymore even when the pressure is increased. The T_c and P_c of water are 374°C, and 22.1 MPa, respectively. If temperature and pressure exceed T_c and P_c, water turns into supercritical water. When temperature and/or pressure do not exceed T_c and/or P_c, water is called subcritical water. The density of supercritical water is half to a third that of water at ambient condition. Although its density is several hundred times greater than that of water vapor, its viscosity coefficient is similar to that of water vapor, and its diffusion coefficient lies between that of liquid and gas forms. In addition, the ionic product of water is drastically increased in the subcritical and supercritical state of water (Marshall and Franck, 1981). Due to these properties, the supercritical or subcritical water is highly active and its reaction rate is expected to be very high.

The important parameters in the chemical reaction field of supercritical and subcritical water, dielectric constant and ionic product, can be largely controlled by temperature and pressure. As a result, properties of water can be continuously changed from aqueous solution to a nonaqueous one. In the supercritical and subcritical state, the ionic product of water increases, and hydrolytic degradation capacity is added (Marshall and Franck, 1981). Hydrolysis reaction field can be achieved in the supercritical and subcritical conditions (Saka, 2001b). Although the supercritical water is clearly defined, subcritical state of water is not well defined with respect to temperature and pressure. Water is called as subcritical water only in the vicinity of the critical point.

Supercritical and Subcritical Water Treatment

The linkages of ester and ether in chemical constituents of wood can be hydrolyzed using subcritical or supercritical water without any catalyst. Supercritical or subcritical water may be employed to obtain useful low-molecular-weight chemicals. Figure 11.1 shows such decomposition pathway of cellulose as treated by flow-type supercritical water at 380°C/40 MPa/0.12–0.48 s (Ehara and Saka, 2002). Cellulose is first hydrolyzed to polysaccharides, with a degree of polymerization (DP) in the 13–100 range. Polysaccharides are further hydrolyzed to oligosaccharides with DP between 2 and 12, whose reducing end was found to be dehydrated and fragmented to levoglucosan and erythrose/glycolaldehyde, respectively, by matrix-assisted laser desorption/ionization-time-of-flight mass spectrometry analysis (Ehara and Saka, 2002). These oligosaccharides are, further, hydrolyzed to glucose, which is, to some extent, isomerized to fructose.

If the treatment is prolonged, the resultant hexoses are further decomposed to levoglucosan, 5-HMF, erythrose, glycolaldehyde, methylglyoxal, and dihydroxyacetone by way of dehydration or fragmentation (Kabyemela et al., 1999; Ehara and Saka, 2002). The dehydrated and fragmented products are further oxidized to low-molecular-weight organic acids such as pyruvic, lactic, formic, and acetic acids.

Such hydrolysis reaction of cellulose is very fast, in the order of seconds. Recently, instantaneous heating of cellulose suspension was established by mixing it with supercritical water, followed by instantaneous cooling with decompression by a sudden expansion microreactor (Cantero, 2014). It was demonstrated that using this method 100% hydrolysis of crystalline cellulose can be achieved in a residence time of only 0.02 s. Even shorter residence time is required for amorphous hemicelluloses. Thus, hemicellulose hydrolysis by supercritical water is almost uncontrollable.

In woody biomass, the decomposition mechanism of lignin is also important for biomass conversion. To simulate the reaction of lignin, dimeric β-O-4 noncondensed and biphenyl 5–5′ condensed types of lignin model compounds

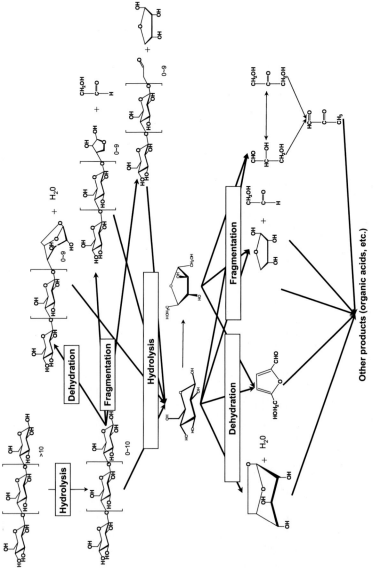

FIGURE 11.1 Proposed pathway of cellulose decomposition as treated by a flow-type supercritical water (Ehara and Saka, 2002).

were treated with supercritical water (400°C/115 MPa/8 s) for phenolic and nonphenolic structures. β-O-4 model compounds were readily cleaved in their ether linkages, whereas biphenyl-type compounds were stable for phenolic and nonphenolic compounds. These lines of evidence clearly indicate that the ether linkages of lignin are easily cleaved, whereas the condensed type linkages of lignin are rather stable under supercritical water treatment (Ehara et al., 2002). Keeping these results in mind, Japanese cedar (*Cryptomeria japonica*) was treated by supercritical water in a similar manner, and oily substances as lignin-derived products could be collected as the methanol-soluble portion.

Figure 11.2 shows the total ion chromatogram of the methanol-soluble portion. All these products have guaiacyl nuclei (2-methoxyphenol), except for 5-HMF, a contaminant from cellulose. The obtained phenylpropane units (C_6–C_3) are composed of eugenol, propylguaiacol, *cis*-iso-eugenol, *trans*-iso-eugenol, propioguaiacone, guaiacylacetone, 2-methoxy-4-(1-hydroxypropyl)-phenol, 2-methoxy-4-(prop-1-en-3-one)phenol, *trans*-coniferylaldehyde, and ferulic acid. These products would be derived through the cleavage of ether linkages of lignin, and will be useful as alternatives to aromatic chemicals from fossil resources (Ehara et al., 2002; Takada et al., 2004; Ehara and Saka, 2005).

In addition to C_6–C_3 units, C_6–C_2 and C_6–C_1 units of products were also detected. The C_6–C_2 units were ethylguaiacol, vinylguaiacol, homovanillin, acetoguaiacone, and homovanillic acid, while the C_6–C_1 units were methylguaiacol and vanillin. The cleavage between C_β and C_γ (C_β/C_γ) bond in the C_6–C_3 unit was observed (Ehara et al., 2002). The existence of these products suggests that the cleavage between C_β/C_γ and C_α/C_β linkages of lignin takes place in supercritical water. It is also reported that the dealkylation of the propyl chain of alkyl phenols takes place in supercritical water (Sato et al., 2002).

For dimeric lignin-derived products, biphenyl-type (5–5), diphenylethane-type (β-1), stilbene-type (β-1), and phenylcoumaran-type (β-5) products were identified. These condensed type products are considered to be more stable than those with ether linkages such as β-O-4 and α-O-4 during supercritical water treatment (Takada et al., 2004).

Steam Explosion

Steam explosion is the most commonly used method for hydrothermal pretreatment of woody biomass (McMillan, 1994). Wet wood chips are treated with high-pressure saturated steam at about 0.69–4.83 MPa with temperature ranging between 160°C and 260°C for several seconds to a few minutes. Subsequently, the pressure is swiftly reduced down to atmospheric condition. Steam that penetrates into the chips is expanded and the cell walls of fibers exploded being fibrillated and decomposed (Sun and Cheng, 2002). During this treatment, acetic acid generated from the acetyl residues of xylan hemicellulose rich in hardwoods can catalyze its hydrolysis. The steam explosion is considered to be an autohydrolysis process (Lora and Wayman, 1978), which leads to efficient enzymatic hydrolysis afterward.

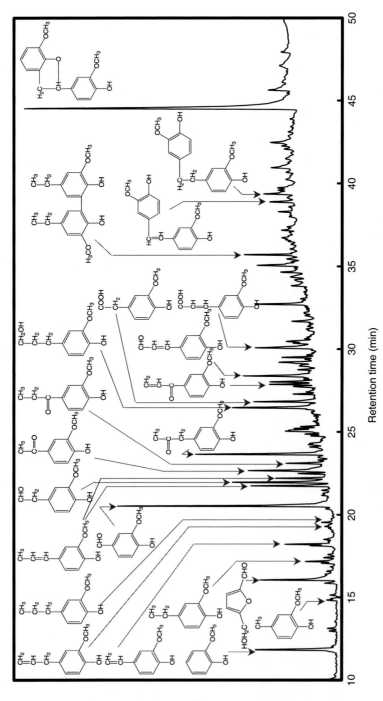

FIGURE 11.2 Various aromatic compounds from Japanese cedar lignin collected as methanol-soluble portion by supercritical water (380°C/100 MPa/8 s, Batch-type system) (Ehara and Saka, 2002).

Steam explosion is, thus, classified as one of the most cost-effective pretreatment processes. For example, the energy requirement with this process is considerably low than mechanical milling, which requires 70% more energy to achieve the same size results (Holtzapple et al., 1989). However, the drawbacks of steam explosion are lower recovery efficiency of products from hemicellulose, incomplete disruption of the lignin-carbohydrate matrix, and production of compounds inhibitory to microorganisms used in downstream processes (Mackie et al., 1985). Therefore, removal of inhibitory products is required for subsequent enzymatic hydrolysis (McMillan, 1994).

BIOCONVERSION OF WOODY BIOMASS

Introduction

In native plant cell walls, cellulose exists as nanometer-scale microfibril networks embedded in matrices of other biopolymers such as hemicelluloses, pectins, and lignins. Chemical pretreatment processes are often required to remove or relocate these "other" matrix polymers, particularly lignin, thereby exposing the cellulose to subsequent enzymatic hydrolysis to produce glucose.

Biochemical bioconversion entails breaking down biomass to make the carbohydrates available for processing into monomer sugars, which can be converted into bioethanol and biochemicals through the use of microorganism and biocatalysts. Each of the lignocellulose components is degraded by a variety of microorganisms, which produce hydrolytic and oxidative enzymes. A diverse spectrum of lignocellulose degrading microorganisms, mainly fungi and bacteria, have been isolated and identified over the years. Despite extensive studies and collected research information, most of the functions of lignocellulolytic microorganisms are involved in various enzymes and their mechanisms. Thus, conversion of cellulose and hemicelluloses of the biomass into monomer sugars requires biocatalysts, such as enzymes. Conversion is mainly explained based on the action of enzymes (or other biocatalysts) that enables the sugars within cellulose and hemicellulose in the pretreated material to be separated and released over a period of several days.

Cellulosic Conversion

Enzymatic hydrolysis of cellulose and hemicelluloses can be carried out by highly specific cellulase and hemicellulase enzymes. Cellulase is one of the hydrolytic enzymes that degrade the cellulosic material. It is mainly produced by microorganisms such as fungi, bacteria, actinomycetes, and plant pathogens. Microbial degradation of cellulosic materials is accomplished by a synergistic action of several enzymes, the most prominent of which is the cellulose. Cellulases are three different types according to the reaction mechanism and structural properties: (1) endoglucanase (EC 3.2.1.4), (2) exoglucanase (EC 3.2.1.91), and (3) β-glucosidase (EC 3.2.1.21) (Lynd et al., 2002; Taherzadeh and Karimi, 2007; Jalak et al., 2012).

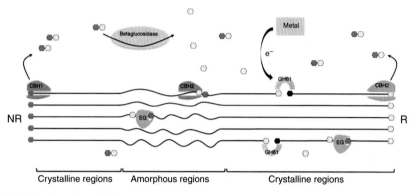

FIGURE 11.3   Simplified schematic of the classical enzyme system in the degradation of cellulose.

Generally, cellulase catalyzes the hydrolysis of 1,4-β-D-glycosidic link-ages to produce glucose, cellobiose, and cello-oligosaccharides from cellulose and other cellulosic materials. However, each cellulase has a different function and structure. Endoglucanase randomly cleaves internal glucosidic bonds at amorphous sites of cellulose, resulting in oligosaccharides of various lengths and new chain ends; whereas exoglucanase releases cellobioses from the ends of the exposed chains of cellulose. Exoglucanases are classified into two main types: (1) cellobiohydrolases I (CBHI) and (2) cellobiohydrolases II (CBH II). The CBHI mainly works from the reducing end of cellulose chain, whereas CBHII mainly works from the nonreducing end of cellulose; β-glucosidase hydrolyses the cellobiose to glucose from the nonreducing end (Fig. 11.3).

However, it has been recognized that the classification of cellulase into en-doglucanase and cellobiohydrolase is not as simple as has been considered. Cellulases have evolved to a continuum of overlapping modes of actions rang-ing from totally random endoglucanases through processive endoglucanase to strictly exo-acting highly processive cellobiohydrolases I and II (Terri, 1997). The detailed roles of individual enzymes with different degrees of processivity and endoglucanase activity in cellulose degradation are not fully understood. When we compare the DNA sequences of endoglucanase, a relatively large number of endo-type cellulase are distributed throughout various glycoside hy-drolase families (GH families 5, 6, 7, 12, and 45) but cellobiohydrolases belong to a relatively low number of GH family. Endoglucanase is, in general, a mono-meric protein with a molecular weight of approximately 20–50 kDa. The opti-mal pH range is 4–6 and temperature range is 40–70°C. Fungi endoglucanases are found in GH5, GH6, GH7, GH12, GH45, and GH74 (Table 11.1).

Cellobiohydrolase II, also known as exoglucanase, is only found in the GH6, while cellobiohydrolase I is found in the GH7 family. Cellobiohydrolases are among the most important cellulolytic enzymes in the bioconversion of ligno-cellulose. The soft-rot fungus *Trichoderma reesei* is the most extensively studied

TABLE 11.1 GH Families Related to Cellulose Degradation

GH family	Clan	Mechanism	Catalytic nucleophile/ catalytic proton donor	Structure	Enzyme activities
GH1	GH-A	Retaining	Glu/Glu	$(\beta/\alpha)8$	β-Glucosidase
GH3	—	Retaining	Asp/Glu	—	β-Glucosidase
GH5	GH-A	Retaining	Glu/Glu	$(\beta/\alpha)8$	Endo-β-1,4-glucanase
GH6	—	Inverting	Asp/Asp	—	Endoglucanase, cellobiohydrolase
GH7	GH-B	Retaining	Glu/Glu	β-Jelly roll	Endo-β-1,4-glucanase, reducing end-acting cellobiohydrolase
GH12	GH-C	Retaining	Glu/Glu	β-Jelly roll	Endoglucanase
GH45	—	Inverting	Asp/Asp	—	Endoglucanase
GH61	—	—	—	—	Cellulase-enhancing protein
GH74	—	Inverting	Asp/Asp	Seven-fold β-propeller	Endoglucanase, oligoxyloglucan reducing end-specific cellobiohydrolase

cellulolytic organism. Overall, the GH family 7 CBHI is a major component making up 60% and GH family 6 CBHII making up 15% of the protein mass of the cellulolytic system of *T. reesei*. In addition, when CBHI and CBHII are removed from the overall cellulase system, the enzymatic activity on crystalline cellulose is reduced by 70%, which is an interesting finding to understand in relation to bioconversion of woody biomass.

Cellulose is a linear monopolysaccharide of linked β-1,4-glucopyranose residues, and is the main component of lignocellulosic materials, typically accounting for 35–50% of the plant cell wall. Structural analysis of cellulose has shown that cellulose microfibrils are differentiated into highly ordered crystalline and noncrystalline (amorphous) regions. The crystallinity of cellulose hinders accessibility and adsorption of cellulase to cellulose surfaces, which leads to retardation of the initial rate of hydrolysis and a reduction in hydrolysis efficiency (Igarashi et al., 2007; Jeoh et al., 2007; Ishizawa et al., 2009). Although the correlation between crystallinity and the rate of enzymatic hydrolysis

(a)

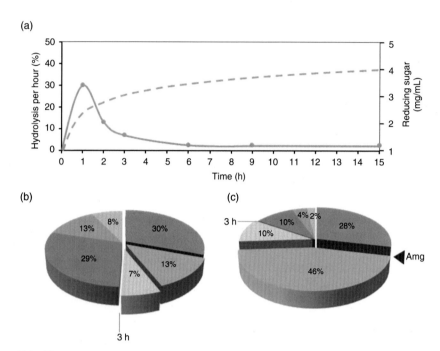

(b)   (c)

FIGURE 11.4   **Enzymatic hydrolysis rate of pretreated lignocellulose (PL).** (a) Solid line indicates hydrolysis rate (%) per hour, measuring reducing sugar concentration after rehydrolysis process of 1% PL at each time. Dashed line presents typical enzymatic hydrolysis pattern of 1% PL in time course. (b, c) Structural recalcitrance of cellulose during enzymatic hydrolysis of 1% PL. Rehydrolysis process was repetitively carried out by incubating with fresh cellulase after washing at each step (1 h × 3 times, 12 h × 1 time, 24 h × 2 times, sequentially) until no detection of reducing sugar. Amorphogenesis was carried out by treating 81% phosphoric acid after prehydrolysis for 1 h. Amg, amorphogenesis.

remains controversial, Hall et al. (2010) provided evidence for an important role of crystallinity in the enzymatic hydrolysis. Lee et al. (2014) reported that the cellulose structure has a considerable effect on hydrolysis efficiency and the rate of enzymatic hydrolysis by amorphogenesis of lignocellulose (Fig. 11.4). Reduction in the structural inhibition of cellulose by amorphogenesis results in considerably higher hydrolysis yield.

Lignin hinders enzymatic hydrolysis of lignocellulose. Lignin is a highly oxygenated aromatic polymer and complexes tightly with cellulose fibers, which constitute the major part of the available fermentable sugar-containing component of lignocellulosic biomass. Lignin physically blocks the access of cellulase to cellulose in cell walls and causes nonspecific enzyme adsorption (unproductive binding), reducing the enzymatic hydrolysis of lignocellulose (Ximenes et al., 2011). The shielding effects as well as enzyme binding onto lignin are considered to be the most influential inhibitory mechanisms (Nakagame et al., 2011). The extent of lignin-derived inhibition is dependent on the woody plant origin as well as on the pretreatment applied to the lignocellulosic biomass because both factors affect localization and the chemical properties of lignin. For example, localization of lignin in the woody plant tissue significantly differs in hardwood and softwood. Wood-based lignocellulosic materials must first be properly pretreated to improve enzymatic hydrolysis, while at the same time ensuring lignin removal. However, there is insufficient information on the effect of pretreatment on nonproductive adsorption of cellulases.

In addition, it has recently become very clear that there are synergistic proteins (called "auxiliary enzymes") that are now known to strongly and synergistically raise the activity of the cellulases and hemicellulases but the role and mechanism of these auxiliary enzymes in cellulose hydrolysis are still not well understood. The auxiliary enzyme classified as glycosyl hydrolase GH61 cellulase acts as auxiliary to hydrolytic cellulases. Other nonhydrolytic proteins may also play a role in enhancing cellulose hydrolysis but do not exhibit any enzymatic activity on cellulose themselves (such as expansin, and swollenin) (Saloheimo et al., 2002; Harris et al., 2010). Synergistic proteins exert their activity by inducing structural modifications in cellulose. Synergistic proteins from various biological sources, including bacteria, fungi, and plants, were identified (Kim et al., 2014b).

Taken together, the results suggest that lignocellulose hydrolysis is the rate-limiting step in the generation of biofuels from lignocellulosic substrates. The form that the cellulose assumes also greatly affects the rate of hydrolysis downstream. Pretreatment of cellulosic materials is very important for increasing the rate of hydrolysis as well as overall conversion rate. Also, the action of the endoglucanase becomes increasingly important as the number of free cellulose chain ends are increased for continued cellobiohydrolase hydrolysis. Cellobiohydrolase is considered a key enzyme in mediating cellulose decomposition. Also, unlike the classical cellulose degradation, synergistic proteins and accessory enzymes can improve the synergistic effect of hydrolytic enzymes (endoglucanase, exoglucanase, and beta-glucosidase) for the efficient hydrolysis of lignocellulose.

Hemicellulosic Conversion

Hemicelluloses are the second most abundant polysaccharides in nature and are a major component of cell walls. Hemicelluloses consist of heteropolymers such as xylan, xyloglucan, glucomannan, galactoglucomannan, and arabinogalactan. Of these, xylan comprises a large proportion of plant cell wall constituents (Motta et al., 2013). Xylan is found in significant quantities in the cell walls of hardwoods (15–30%), softwoods (7–10%), and annual plants (<30%) (Prade, 1996). Xylan is a complex polysaccharide with a backbone consisting of a β-1,4-D-linked xylose residue. The degree of polymerization of xylans is also variable in plants. For example, hardwood and softwood xylans generally consist of 150–200 DP and 70–130 DP β-xylose residues, respectively

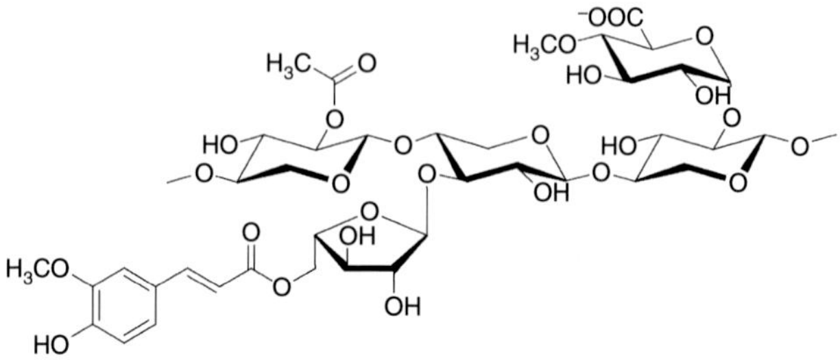

FIGURE 11.5   Substituted xylose backbone.

(Kulkarni et al., 1999). In addition, xylans are often partially substituted with O-acetyl, α-L-arabinofuranose and 4-O-methyl-D-glucuronic acid on the xylose backbone (Fig. 11.5).

The second most abundant hemicelluloses are galactoglucomannans, which are the main hemicelluloses in softwoods with a structure of randomly distributed β-1,4-D-linked mannopyranosyl and β-1,4-D-linked glucopyranosyl backbone substituted with α-galactopyranose, and xyloglucans that are present in primary walls and consist of a β-1,4-D-linked glucose backbone substituted with α-linked xylose (de Vries and Visser, 2001).

For the hydrolysis of hemicelluloses, all substituted residues have to be first released on the hemicellulose backbone. The release of substituted residues requires cooperative action of at least nine enzymes, such as endo-β-1,4-xylanase (EC 3.2.1.8), β-xylosidase (EC 3.2.1.37), α-L-arabinofuranosidase (EC 3.2.55), feruloyl esterase (EC 3.1.1.73), α-glucuronidase (EC 3.2.1.139), and acetyl xylan esterase (EC 3.1.1.72) (Table 11.1). After the release of substituted residues, hemicellulose backbones can be degraded by endo-1, 4-β-xylanase (Fig. 11.6).

Degradation of Xylan

Xylan degradation requires the endo-1,4-β-xylanase. Endo-1,4-β-xylanase (E.C.3.2.1.8) is a glycoside hydrolase, which mainly catalyzes random hydrolysis of β-1,4-D-xylosidic linkage in xylan (Polizeli et al., 2005). Generally, xylanases are synthesized by microbes and filamentous fungi, and the xylanase family is very diverse. According to the CAZy classification system, xylanases may belong to GH family 5, 8, 10, 11, and 43 with GH10 and 11 as the major components of endo-1, 4-β-xylanase group (Lombard et al., 2014).

Glycoside hydrolase 10 family consists of a few endo-1, 3-β-xylanase (EC 3.2.1.32) and majority endo-1, 4-β-xylanase. Its major acting is endo-1, 4-β-xylanase. But, some GH10 not only shows xylanase activity but also cellulase activity (Biely, 2003). Generally, GH10 has a large molecular mass (over 30 kDa), and a low pI (above 5.0) value (Collins et al., 2005). This family has a small substrate-binding site, and it is highly active on short-chain xylo-oligomer, -polymer. According to the kinetic and crystal structure analysis, GH10 has four

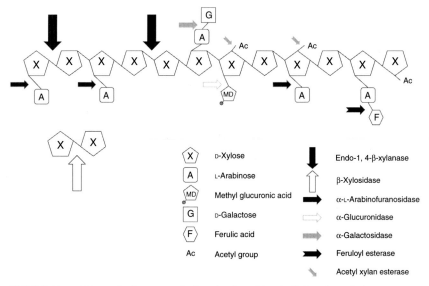

FIGURE 11.6   Schematic view on a substituted xylan of hemicellulose degradation system.

to five substrate-binding sites in the catalytic domain. In addition, GH10 family xylanases can hydrolyze substituted xylan backbone (Kim et al., 2014a).

In contrast to other GH families, GH11 appears to be active on xylan substrate – no other activities have been reported. Therefore, GH11 is called the "true xylanase." Family GH11 xylanases also retain enzymes like GH10 xylanase, but the properties of this family are different from GH10; GH 11 xylanases are generally characterized by a low molecular weight (under 30 kDa) and a high pI (above 9.0). These families are most active on long-chain xylo-oligomer-polymer. It has been reported that they have a larger substrate binding site, with at least seven subsites (Bray and Clarke, 1992). Besides, GH11 xylanases are not able to attack substituted xylan such as arabinoxylan (Song et al. 2013). GH11 xylanases require other accessory enzymes to hydrolyze substituted xylan.

Degradation of Substituted Residue on Backbone

Due to the heterogeneity and complexity of substituted residue on the hemicellulose backbone, its complete hydrolysis requires various accessory enzymes, such as acetyl xylan esterases (EC 3.1.1.72), α-D-glucuronidases (EC 3.2.1.139), α-L-arabinofuranosidases (EC 3.2.1.55), and so on (Subramaniyan and Prema, 2002). α-L-Arabinofuranosidases (EC 3.2.1.55) catalyze the hydrolysis of α-L-1-2, α-L-1-3 arabinofuranosyl residues attached to xylose units of hemicelluloses such as arabinoxylan and other L-arabinose-containing polysaccharides (Lee et al., 2011). Acetyl xylan esterase (EC 3.1.1.72) attacks O-acetyl groups from positions 2 and/or 3 on the β-D-xylopyranosyl residues of acetyl xylan. Acetyl xylan plays an important role in the hydrolysis of xylan, since the acetyl side substitute on acetyl xylan can interfere with the binding of endo-xylanases, which cleave the backbone (Caufrier et al., 2003). α-Glucuronidase (EC 3.2.1.131) hydrolyzes the α-1,2 bonds between the glucuronic acid residues on the β-D-xylopyranosyl backbone in glucuronoxylan. The GH67 enzymes attack the glucuronic acid substituted to the C2–OH of the β-D-xylopyranosyl residues of glucuronoxylan and display a preference for MeGlcA side chains. Glycoside hydrolases of GH115 catalyze the cleavage of 4-O-methyl D-glucuronic acid side chains from native xylan polysaccharides. It has also been noted that acetyl groups close to the glucuronosyl substituents can partially hinder the α-glucuronidase activity (Harris and Ramalingam, 2010).

Complete hydrolysis of hemicellulose is very important because hemicellulose plays a role as a cross-linking agent, connecting cellulose to lignin in plant cell walls (Subramaniyan and Prema, 2002). Therefore, hemicellulose degradation is the first step in lignocellulose degradation, which enhances cellulase accessibility to cellulose fibrils (Hu et al., 2011).

CONCLUDING REMARKS

A better understanding of the bioconversion of woody biomass using enzymes and microorganisms is essential for achieving high yield of bio-based chemicals and fuels. However, due to the structural complexity of secondary xylem cell walls, biological decomposition of woody biomass is difficult, and therefore wood is chemically pretreated prior to biological treatments. Chemical procedures and biological treatment in combination can lead to efficient conversion of wood biomass into biofuels and novel chemicals, thus minimizing negative impacts on the environment.

REFERENCES

Biely, P., 2003. Diversity of microbial endo-β-1,4-xylanases. In: Mansfield, S.D., Saddler, J.N. (Eds.), Applications of Enzymes to Lignocellulosics. American Chemical Society, Washington DC, pp. 361–380.

Braconnot, H., 1819. Verwandlungen des Holzstoffs mittelst schwefelsäure in Gummi, Zucker und eine eigne Säure, und mittelst Kali in Ulmin. Annalen der Physik 63, 347–371.

Bray, M.R., Clarke, A.J., 1992. Action pattern of xylo-oligosaccharide hydrolysis by *Schizophyllum commune* xylanase A. Eur. J. Biochem. 204, 191–196.

Cantero, D.A., 2014. Intensification of cellulose hydrolysis process by supercritical water. Obtaining of added value products. PhD thesis. Universidad de Valladolid. pp. 30–31.

Caufrier, F., Martinou, A., Dupont, C., Bouriotis, V., 2003. Carbohydrate esterase family 4 enzymes: substrate specificity. Carbohydr. Res. 338, 687–692.

Collins, T., Gerday, C., Feller, G., 2005. Xylanases, xylanase families and extremophilic xylanases. FEMS Microbiol. Rev. 29, 3–23.

de Vries, R.P., Visser, J., 2001. *Aspergillus* enzymes involved in degradation of plant cell wall polysaccharides. Microbiol. Mol. Biol. Rev. 65, 497–522.

Ehara, K., Saka, S., 2002. A comparative study on chemical conversion of cellulose between the batch-type and flow-type systems in supercritical water. Cellulose 9, 301–311.

Ehara, K., Saka, S., 2005. Decomposition behavior of cellulose in supercritical water, subcritical water, and their combined treatments. J. Wood Sci. 51, 148–153.

Ehara, K., Saka, S., Kawamoto, H., 2002. Characterization of the lignin-derived products from wood as treated in supercritical water. J. Wood Sci. 48, 320–325.

Hall, M., Bansal, P.J., Lee, H., Realff, M.J., Bommarius, A.S., 2010. Cellulose crystallinity – A key predictor of the enzymatic hydrolysis rate. FEBS J. 277, 1571–1582.

Harris, A.D., Ramalingam, C., 2010. Xylanases and its application in food industry: a review. J. Exp. Sci. 7, 1–11.

Harris, P.V., Welner, D., McFarland, K.C., Re, E., Navarro Poulsen, J.C., Brown, K., Salbo, R., Ding, H., Vlasenko, E., Merino, S., Xu, F., Cherry, J., Larsen, S., Lo Leggio, L., 2010. Stimulation of lignocellulosic biomass hydrolysis by proteins of glycoside hydrolase family 61: structure and function of a large, enigmatic family. Biochemistry 49, 3305–3316.

Higuchi, T., 1997. Biochemistry and Molecular Biology of Wood. Springer-Verlag, New York.

Himmel, M.E., et al., 2007. Biomass recalcitrance: engineering plant and enzymes for biofuels production. Science 315, 804–807.

Holtzapple, M.T., Humphrey, A.E., Taylor, J.D., 1989. Energy requirements for the size reduction of poplar and aspen wood. Biotechnol. Bioeng. 33, 207–210.

Hu, J., Arantes, V., Saddler, J.N., 2011. The enhancement of enzymatic hydrolysis of lignocellulosic substrates by the addition of accessory enzymes such as xylanase: is it an additive or synergistic effect? Biotechnol. Biofuels 4, 36.

Igarashi, K., Wada, M., Samejima, M., 2007. Activation of crystalline cellulose to cellulose III_1 results in efficient hydrolysis by cellobiohydrolase. FEBS J. 274, 1785–1792.

Ishizawa, C.I., Jeoh, T., Adney, W.S., Himmel, M.E., Johnson, D.K., Davis, M.F., 2009. Can delignification decrease cellulose digestibility in acid pretreated corn stover? Cellulose 16, 677–686.

Iyer, P.V., Wu, Z.-W., Kim, S.B., Lee, Y.Y., 1996. Ammonia recycled percolation process for pretreatment of herbaceous biomass. Appl. Biochem. Biotechnol. 57/58, 121–132.

Jalak, J., Kurašin, M., Teugjas, H., Väljamäe, P., 2012. Endo–exo synergism in cellulose hydrolysis revisited. J. Biol. Chem. 287 (34), 28802–28815.

Jeoh, T., Ishizawa, C.I., Davis, M.F., Himmel, M.E., Adney, W.S., Johnson, D.K., 2007. Cellulase digestibility of pretreated biomass is limited by cellulose accessibility. Biotechnol. Bioeng. 98, 112–122.

Kabyemela, B.M., Adschiri, T., Malaluan, R., Arai, K., 1999. Glucose and fructose decomposition in subcritical and supercritical water: detailed reaction pathway, mechanisms, and kinetics. Ind. Eng. Chem. Res. 38, 2888–2895.

Karr, W.E., Holzapple, M.T., 2000. Using lime pretreatment to facilitate the enzymatic hydrolysis of corn stover. Biomass Bioenerg. 18, 189–199.

Kim, H.M., Lee, K.H., Kim, K.H., Lee, D.S., Nguyen, Q.A., Bae, H.J., 2014a. Efficient function and characterization of GH10 xylanase (Xyl10g) from *Gloeophyllum trabeum* in lignocellulose degradation. J. Biotechnol. 172, 38–45.

Kim, I.J., Lee, H.J., Choi, I.G., Kim, K.H., 2014b. Synergistic proteins for the enhanced enzymatic hydrolysis of cellulose by cellulase. Appl. Microbiol. Biotechnol. 98 (20), 8469–8480.

Kulkarni, N., Shendye, A., Rao, M., 1999. Molecular and biotechnological aspects of xylanases. FEMS Microbiol. Rev. 23, 411–456.

Larsson, S., Palmqvist, E., Hahn-Hägerdal, B., Tengborg, C., Stenberg, K., Zacchi, G., Nilvebrant, N.-O., 1999. The generation of fermentation inhibitors during dilute acid hydrolysis of softwood. Enzyme Microb. Technol. 24, 151–159.

Lee, D.S., Wi, S.G., Lee, Y.G., Cho, E.J., Chung, B.Y., Bae, H.J., 2011. Characterization of a new α-L-Arabinofuranosidase from *Penicillium* sp. LYG 0704, and their application in lignocelluloses degradation. Mol. Biotechnol. 49, 229–239.

Lee, D.S., Wi, S.G., Lee, S.J., Lee, Y.G., Kim, Y.S., Bae, H.J., 2014. Rapid saccharification for production of cellulosic biofuels. Bioresour. Technol. 158, 239–247.

Lombard, V., Golaconda Ramulu, H., Drula, E., Coutinho, P.M., Henrissat, B., 2014. The carbohydrate-active enzymes database (CAZy) in 2013. Nucl. Acids Res. 42, 490–495.

Lora, J.H., Wayman, M., 1978. Delignification of hardwoods by autohydrolysis and extraction. Tappi J. 61, 47–50.

Lynd, L.R., Weimer, P.J., Zyl, W.H., 2002. Microbial cellulose utilization: fundamentals and biotechnology. Microbiol. Mol. Biol. Rev. 66, 506–577.

Mackie, K.L., Brownell, H.H., West, K.L., Saddler, J., 1985. Effect of sulphur dioxide and sulphuric acid on steam explosion of aspenwood. J. Wood Chem. Technol. 5, 405–425.

Marshall, W.L., Franck, E.U., 1981. Ion product of water substance, 0–1000°C, 1–10,000 bars – New international formulation and its background. J. Phys. Chem. Ref. Data 10, 295–304.

McMillan, J.D., 1994. Pretreatment of lignocellulosic biomass. In: Himmel, M.E., Baker, J.O., Overend, R.P. (Eds.), Enzymatic Conversion of Biomass for Fuels Production. American Chemical Society, Washington DC, pp. 292–324.

Millet, M.A., Baker, A.J., Scatter, L.D., 1976. Physical and chemical pretreatment for enhancing cellulose saccharification. Biotechnol. Bioeng. Symp. 6, 125–153.

Motta, F.L., Andrade, C.C., Santana, M.H., 2013. In: Chandel, A.K., da Silva, S.S. (Eds.), A Review of Xylanase Production by the Fermentation of Xylan: Classification, Characterization and Applications. Intec, pp. 251–275.

Nakagame, S., Chandra, R.P., Saddler, J.N., 2011. In: Zhu, J, Zhang, X., Pan, X. (Eds.), The Influence of Lignin on the Enzymatic Hydrolysis of Pretreated Biomass Substrates. Sustainable Production of Fuels, Chemicals, and Fibers from Forest Biomass. American Chemical Society, Washington DC, pp. 145–167.

Polizeli, M.L., Rizzatti, A.C., Monti, R., Terenzi, H.F., Jorge, J.A., Amorim, D.S., 2005. Xylanases from fungi: properties and industrial applications. Appl. Microbiol. Biotechnol. 67, 577–591.

Prade, R.A., 1996. Xylanases: from biology to biotechnology. Biotechnol. Genet. Eng. Rev. 13, 101–131.

Rabemanolontsoa, H., Saka, S., 2013. Comparative study on chemical composition of various biomass species. RSC Adv. 3, 3946–3956.

Ragauskas, A., et al., 2006. The path forward for biofuels and biomaterials. Science 311, 484–489.

Saka, S., 2001a. Chemical composition and distribution. In: Hon, D.N.-S., Shiraishi, N. (Eds.), Wood and Cellulosic Chemistry. 2nd ed. Marcel Dekker, New York, pp. 51–81.

Saka, S., 2001b. Post-petrochemistry of woody biomass by supercritical water. Wood Ind. 56, 105–110.

Saka, S., 2009. Recent progress of biofuels in Japan. IEA Task 39 Newslett. 23, 2–10.

Saloheimo, M., Paloheimo, M., Hakola, S., Pere, J., Swanson, B., Nyyssonen, E., Bhatia, A., Ward, M., Penttila, M., 2002. Swollenin, a *Trichoderma reesei* protein with sequence similarity to the plant expansins, exhibits disruption activity on cellulosic materials. Eur. J. Biochem. 269, 4202–4211.

Sato, T., Sekiguchi, G., Saisu, M., Watanabe, M., Adschiri, T., Arai, K., 2002. Dealkylation and rearrangement kinetics of 2-isopropylphenol in supercritical water. Ind. Eng. Chem. Res. 41, 3124–3130.

Song, Y., Lee, Y.G., Choi, I.S., Lee, K.H., Cho, E.J., Bae, H.J., 2013. Heterologous expression of endo-1,4-β-xylanase A from *Schizophyllum commune* in *Pichia pastoris* and functional characterization of the recombinant enzyme. Enzyme Microb. Technol. 52, 170–176.

Subramaniyan, S., Prema, P., 2002. Biotechnology of microbial xylanases: enzymology, molecular biology, and application. Crit. Rev. Biotechnol. 22, 33–64.

Sun, Y., Cheng, J., 2002. Hydrolysis of lignocellulosic materials for ethanol production: a review. Bioresour. Technol. 83, 1–11.

Taherzadeh, M.J., Karimi, K., 2007. Enzyme-based hydrolysis processes for ethanol from lignocellulosic materials: a review. BioResources 2, 707–738.

Takada, D., Ehara, K., Saka, S., 2004. Gas chromatographic and mass spectrometric (GC-MS) analysis of lignin-derived products from *Cryptomeria japonica* treated in supercritical water. J. Wood Sci. 50, 253–259.

Taneda, D., 2006. Concentrated sulfuric acid biomass ethanol process. Cellulose Commun. 13, 49–52.

Terri, T.T., 1997. Crystalline cellulose degradation: new insight into the function of cellobiohydrolases. Trends Biotechnol. 15, 160–167.

Wang, Z., Xu, J., Cheng, J.J., 2011. Modeling biochemical conversion of lignocellulosic materials for sugar production: a review. BioResources 6, 5282–5306.

Ximenes, E., Kim, Y., Mosier, N., Dien, B., Ladisch, M., 2011. Deactivation of cellulases by phenols. Enzyme Microb. Technol. 48, 54–60.

Zhu, S., Wu, Y., Chen, Q., Yu, Z., Wang, C., Jin, S., Ding, Y., Wu, G., 2006. Dissolution of cellulose with ionic liquids and its application: a mini-review. Green Chem. 8, 325–327.

Zulfiqar, F., Kitazume, T., 2000. One-pot aza-Diels–Alder reaction in ionic liquids. Green Chem. 2, 137–139.

Chapter 12

Wood as Cultural Heritage Material and its Deterioration by Biotic and Abiotic Agents

Yoon Soo Kim*, Adya P. Singh†

*Department of Wood Science and Engineering, Chonnam National University, Gwangju, South Korea; †Manufacturing and Bioproducts Group, Scion (New Zealand Forest Research Institute), Rotorua, New Zealand

Chapter Outline

WOODEN CULTURAL HERITAGES AND THEIR PROPERTY DIAGNOSIS

Wooden Cultural Heritages

Wood has been widely used in structural applications as well as for decorative, ritual, and religious purposes. Intimate human links to wood over thousands of years are embedded in wooden cultural heritage (WCH) representing human life and values. Stone, bronze, and iron have their own epoch of utilization but

Secondary Xylem Biology. http://dx.doi.org/10.1016/B978-0-12-802185-9.00012-7

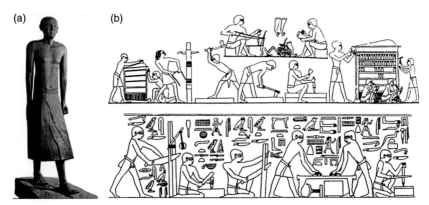

FIGURE 12.1   (a) Statue of the Chancellor Nakt, Egyptian middle kingdom at the time of 13th dynasty (Louvre Museum) and (b) Relief on the Tombs of Tebe (1400 BC) and Saqquara (2500 BC) showing the utilization of wood by the Egyptians. *(Wooden statue from "Masterpieces of the Louvre and the Jeu de Paume", Casa Editrice Bonechi, Florence, 1974 and Relief on the tomb from Kuenhn H, second. ed. "Erhalting und Pflege von Kunstwerken und Antiquitaeten". Keyer, Munich, 1981.)*

wood has been continuously used throughout history. There is no indication of the direct use of wood by the earliest hominids, although Neanderthals exhibited many complex behaviors such as funerary practices (d'Errico et al., 2012). Cultural uses of wood appear to be related to the activity of humans rather than hominids (Bamford, 2010) (Fig. 12.1).

Any wooden artifact that provides us information about human life and culture, and that is considered worthy of preservation for the future can be defined as wooden cultural heritage (Rowell and Barbour, 1990). Vocational, technical artifacts, creative activities, and cannons of craftsmanship are embedded in WCH, reflecting past human culture, ideals, and symbols. Thus, the technological and sociological aspects of human activities can be understood from excavating WCH objects or remains. Consequently, a holistic approach comprising of natural sciences, humanities, and the arts is needed for a meaningful and informative study of WCH.

WCH as tangible objects can be generally classified into three groups: (1) moveable, (2) immoveable, and (3) underwater (according to UNESCO). Moveable WCH include musical instruments, painted panels, furniture, Italian intarsia, sculptures, iconic altar, Tripitaka-plates, coffins, and masks. In general, moveable WCH are displayed and/or kept indoors with air-conditioning. Immoveable WCH include Buddhist temples, churches, chapels, royal palaces, wooden pagodas, watermills, windmills, and wooden bridges of historic value. Most immoveable WCH are situated under outdoor conditions without air-conditioning and are exposed to fluctuations in temperature and humidity (Fig. 12.2). In contrast, underwater WCH include shipwrecks, foundation piles,

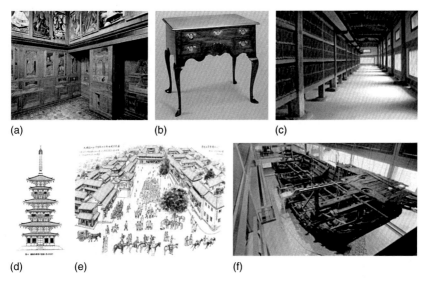

(a)   (b)   (c)

(d)   (e)   (f)

FIGURE 12.2   Typical wooden cultural heritages as movable, immoveable, and underwater WCH. (a) Studiolo in Florence, (b) dressing table in 1740–1750 in the United States, (c) Tripitaka in Korea as moveable WCH, (d) wooden pagoda in Japan, (e) wooden house village in China as immoveable WCH, (f) sunken ship of Kogge in Germany as underwater WCH. *(Studiolo in Florence and dressing table from "The Metropolitan Museum of Art Bullentin," Fall, 1995; Tripitaka in Korea from Kim, Y.S. et al. (Eds.) "Wood Preservation Science," Chonnam Nat'l Univ Press, Korea, 2004; wood pagoda from Itoh, T. (Ed.), "Wood Culture and Sciene," Kaiseisha Press, Japan, 2008; wooden house village from Liu, P. and Liu, Y. "Yunnan Ancient Architecture," China 2012; and Kogge ship from "Die Kogge," Convent, Hamburg 2003.)*

wooden cargo or contents with a cultural, historical, or archaeological character, which were partially or totally underwater periodically or continuously for at least 100 years.

Property Diagnosis of WCH

Virtually all WCH are in a constant state of chemical transformation and can deteriorate over time. Sometimes deterioration of WCH is relatively slow, either because wood species are quite durable or exposure conditions have been mild and nonaggressive (Koestler et al., 2010). However, many WCH are disintegrating and disappearing at a distressing rate simply from neglect and unintended abuse. The extent of deterioration of WCH varies from near normal appearance to disintegrated and extensively changed wood, depending on the agents of deterioration, wood species, and environmental conditions. Various abiotic and biotic agents can deteriorate WCH (Table 12.1). Exact diagnostic data on the existing state of recovered WCH are crucial not only for their conservation but also for understanding the degradation processes.

TABLE 12.1 Deterioration of WCH by Biotic and Abiotic Agents

Types of WCH	Biotic agents	Abiotic agents
Dry-type WCH	Fungi Insects Termites	Moisture fluctuation Weathering
Wet-type WCH	Fungi Bacteria Marine borers	Hydrolysis

Application of Wood Science in WCH Studies

Modern wood science knowledge can contribute to the diagnosis and conservation of WCH by understanding their anatomical, biological, physical, and chemical characteristics. Identification of wood species, strength analysis, monitoring physical and dimensional changes, and evaluation of the degree of deterioration by microorganisms and insects are the essential tasks for proper evaluation and diagnosis of WCH by wood scientists (Itoh and Yamada, 2012). Furthermore, knowledge of wood science can aid in the implementation of nondestructive methods and guidelines for the conservation of WCH, although the methodologies and techniques employed by wood scientists are not always welcomed by curators, museum authorities, and conservators, who have a different mindset to understanding and conserving WCH. They are concerned with the loss of the originality and integrity of WCH during analytical investigation because the diagnostic methods applied by wood scientists are generally of an invasive or destructive nature. A rational integration of science and art should be developed for anthropological interpretation without damaging the historical value and beauty of WCH.

Identification of wood species is a compulsory step prior to any investigation of WCH because identification of wood species provides information on the human control over the wood and also on the floral ecosystem (Itoh, 2008). For example, identification of wood species by our ancestors contributed to the knowledge of wood durability. The ancient Chinese used durable woods for coffins of their emperors and noble classes (Wang, 2010). Proietti et al. (2011) found that the main wood species of Egyptian sarcophagus was yew (*Taxus* spp.), which was not native to Egypt. The wood was probably imported from Turkey or Northwest Africa because of its high durability, which guaranteed a better preservation and thus improved the mummy conservation of the sarcophagus. Identification of wood species is also essential for the conservation treatment of WCH in order to distinguish the original work from later restorations. The palette of the fifteenth century Florentine intarsia cutter consisted of local woods such as walnut (*Juglans* sp.), pear (*Pyrus* sp.), cherry (*Prunus* sp.), maple (*Acer* sp.), and oak (*Quercus* sp.). However, Wilmering (1996) found that the nineteenth century restorations were

TABLE 12.2 Currently Available Nondestructive Methods

Mechanical	Stiffness, bending, tension, compression, probes	
Sonic/ultrasonic	Wave propagation, acoustic emission	
Electrical	Electrical resistance, dielectric and piezoelectric properties	
Magnetic	Nuclear magnetic resonance	
Electromagnetic and nuclear radiation	Reflection	Visual assessment, IR/NIR spectroscope
	Diffraction	X-ray small-angle scattering
	Transmission	imaging and tomography with X-ray or neutrons; X-ray densitometry
	Emission	IR-thermography
Chemical	Composition	

Mannes (2009).

done mostly in rosewood (*Dalbergia* sp.), maple, artificially stained oak, syca-more (*Platanus* sp.), and fir (*Abies* sp.). Direct visual evaluation of wood species in WCH based on the characteristic macroscopic features is not always possible. Visual identification of WCH is not conclusive due to its biological degradation or an opaque finishing or painting. Although the microscopic approach is more reliable than visual examination, sampling of wood from the WCH for identifi-cation is not always permitted. In this regard, trained specialists as well as non-invasive methods should be developed to overcome these difficulties. Recently, the use of two-photon excited fluorescence (2PEF) microscopy produced highly contrasted 3D images at micrometer scale without any preparation or sampling of the artifacts, making *in situ* observations of historical violins possible with-out any damage to the instrument (Latour et al., 2012). X-ray microtomography was also applied for noninvasive identification of ancient wood objects (Steppe et al., 2004; Mizuno et al., 2010; Maierhofer et al., 2010).

Because most of WCH are rare or irreplaceable objects, the use of noninva-sive or nondestructive testing (NDT) is more favored than destructive methods. NDT methods in general do not affect the integrity of the examined objects. However, each NDT method has its own specific limitations, such as low con-trast, resolution, sensitivity for hydrogen (water), and accessible sample size (from millimeter to decimeter range). Usually a combination of several different examination or analytical procedures is involved for the study of a given WCH. NDT methods applied in the identification of the current state of WCH are sum-marized in Table 12.2 (Mannes, 2009).

PHYSICAL AND CHEMICAL CHARACTERISTICS OF WCH

Physical Characteristics of WCH

Largely, the physical and mechanical properties of moveable and immoveable WCH are likely to be the same as for recent wood because the dry conditions of WCH inhibit the activity of most agents of deterioration. Inspections have shown that some old wooden objects, such as Buddhist monasteries and temples, are similar to recent wood in density and mechanical characteristics. In some cases, immovable WCH showed higher density and mechanical strength than recent wood. Kohara (1958) found that the bending strength of 1800-year-old Japanese cypress (softwood) increased for the first 200 years and then gradually decreased to about the same strength as for recent wood, whereas *Zelkoba* spp. (hardwood) showed a gradual decrease in mechanical strength. Changes in the stiffness of softwoods and hardwoods over time can be explained by changes in the crystallinity of cellulose (Nakao et al., 1989) (Fig. 12.3). Recently, Yokoyama et al. (2009) found that aged softwood was brittled especially in the radial direction, where the linear elastic limit was not reached.

Mechanical properties of WCH are affected greatly by moisture content and temperature. In particular, damages resulting from moisture are costly and hazardous. Structural failure due to decay, physical deterioration of the building, loss of thermal insulation, warping, corrosion of metal components, blistering and peeling of protective coatings and finishes, growth of molds, and discoloration are the typical moisture-related systems (Fazio et al., 2010; Irbe et al., 2012; Ortiz et al., 2014; Strzelczyk, 2004). Immoveable WCH, such as

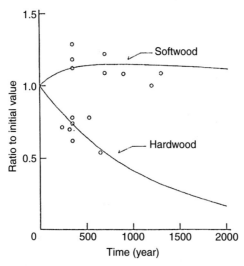

FIGURE 12.3   Changes in the degree of crystallinity and wood stiffness over time. *(Solid lines show degree of crystallinity and open circles denote values of the MOE (Nakao et al., 1989).)*

uninsulated outbuildings, can be severely affected even from a brief exposure to moisture, leading to cracks in musical instruments and veneer lifting. Moveable WCH, such as furniture and musical instruments, in a steam-heated environment (due to low relative humidity) will shrink and crack unless humidification and a long period of acclimatization are provided.

The physical properties and mechanical strength of underwater WCH are inferior to those of recent wood. However, the values of strength decreases are not directly proportional to the losses of woody mass because the deterioration of underwater WCH is not uniform and degradation is usually confined to the surface, particularly in large timbers. Furthermore, individual samples in the underwater WCH usually show great variations in wood properties depending upon the surrounding environmental and burial conditions. Thus, the age of underwater WCH is not a reliable indicator for estimating the degree of their deterioration (Schniwind, 1990). In many cases, the maximum moisture content of waterlogged archaeological wood has been used as a function of strength decreases and dimensional changes.

In general, tensile and bending strength are likely to be much more affected than compression strength parallel to the grain. Dimensional changes of archaeological woods are of great importance because significant shrinkage can occur during drying. The hygroscopicity of underwater WCH sometimes increases drastically. Equilibrium moisture content of the waterlogged ancient wood has been estimated to be more than 50%. Reduced cellulose crystallinity is also responsible for an increase in hygroscopicity and dimensional changes in underwater WCH (Schniwind, 1990) (Table 12.3). However, the degree of dimensional changes depends on the extent of deterioration of underwater WCH. Safe drying of underwater-waterlogged woods is one of the most challenging tasks in the conservation of archaeological woods.

Chemical Characteristics of WCH

It has been reported that usually chemical properties remain unchanged in the moveable and immovable WCH (Takei et al., 1997). In contrast, in underwater

TABLE 12.3 Density and Sorption of Buried Oak Wood

Age years	Residual density (%)	EMC in old wood (%)	EMC in recent wood (%)
570	94	52.0	32.0
1000	75	36.4	30.8
2500	42	60.0	30.0
4700	103	32.0	30.8

Data summarized from Schniwind (1990).

WCH, significant changes in their physical and chemical properties can occur. Chemical characteristics of WCH have been carried out mostly in the underwater WCH, such as waterlogged archaeological woods (Gelbrich et al., 2008; Passialis, 1997). Chemical data on the moveable and immovable WCH are rare because such objects are basically immune to microbial degradation due to their low moisture content.

Major chemical changes in waterlogged archelogical woods (WAW) under long-term waterlogging have been determined by pyrolytic-GC/MS (Tamburini et al., 2014), Fourier transform attenuated total reflectance (FT-MIR-ATR), FT-IR, FT-ATR, FT-NIR (Tsuchikawa et al., 2005; Pedersen et al., 2014), and solid-state nuclear magnetic resonance (NMR) by spectral comparison (Viel et al., 2004). Rapid and accurate estimation of the extent of degradation is important for optimizing restoration and conservation. Wet chemical analyses allow detailed interpretation of the degradation processes due to accuracy and high resolution. However, because of its destructive nature and long examination period, wet analysis is not ideally suited for examining archaeological wooden objects. Recent developments in nondestructive measurements, combining the field of optics (X-ray computed microtomography) and electronics, complement traditional wet chemical methods due to their simplicity of measurement, minimal sample preparation, possibility of repetitions, and highly enhanced reliability and accuracy of the data.

Chemical changes in underwater WCH occurred mainly in structural polysaccharides (Gelbrich et al., 2008; Giachi et al., 2003). Due to the preferential degradation of polysaccharides, the ratio of holocelluose to lignin in the underwater WCH was $0.3 \sim 1:1$, and in recent wood 3:1. Not only amorphous but also crystalline cellulose was decomposed during long-term waterlogging situation. Degradation of crystalline cellulose in underwater WCH can be confirmed by X-ray diffraction as well as polarized microscopy, the latter showing loss of the birefringence. Low molecular weight substances were also greatly affected during long-term underwater submersion of WCH. Hot water and 1% alkaline extractives showed a significant increase due to degradation of a large part of low-molecular hydrocarbons derived from hemicelluloses. In contrast, the amount of organic solvent soluble materials decreased. Chemical data of waterlogged bamboo exhibited the same trends as in the waterlogged wood (Table 12.4).

The ash content of the underwater wood and bamboo was greater than that of the recent wood samples partly because of migration of minerals from the sediments into the porous sites in the WAW and partly because of microbial activities (Fig. 12.4). Sulfur accumulation in WAW was induced by bacteria in a seawater environment. Fors et al. (2012) found the presence of sulfate-reducing bacteria (anaerobic) in diverse environments, which reduce sulfates to sulfides and utilize aromatic compounds including lignin and tannins. Some bacteria-degrading WAW in seawater promote the accumulation of sulfur as thiols in the lignin and also in iron sulfides, depending on the conditions.

TABLE 12.4 Chemical Characteristics of Recent and Waterlogged Archeological Wood of *Pinus* spp. and Bamboo

	Holocel-lulose	Lignin	Extractives			Ash
			Hot water	1% alkaline	Alcohol benzene	
Recent pine	73.2	26.85	2.8	12.67	3.0	0.3
Waterlogged *Pinus* spp.	29.47	64.18	7.8	18.73	0.7	7.6
Recent bamboo	75.2	24.3	6.6	25.1	4.4	1.3
Waterlogged bamboo	43.3	62.1	9.8	83.5	2.3	5.4

Kim (1990); Cha et al. (2014).

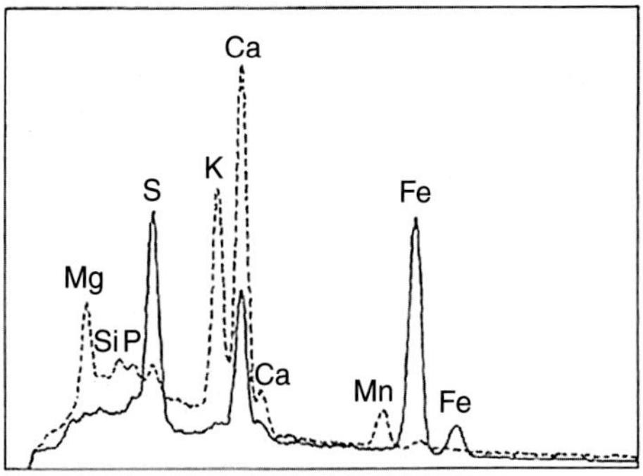

FIGURE 12.4   Diversity in ash present in the waterlogged archeological wood (solid line) in comparison with the recent wood of *Pinus massonia* (dotted line) (Kim, 1990).

Spectral and signal comparison in FT-IR and NMR indicated a significant decrease in polysaccharides and an increase in lignin in the underwater WCH. FT-IR analyses showed that band intensities assigned to xylan (1730 and 1320 cm^{-1}) and cellulose (1160, 1110, and 1060 cm^{-1}) decreased, whereas band intensities assigned to lignins (1220–1230, 1270 cm^{-1}) significantly increased (Fig. 12.5). Among the polysaccharides, the band assigned to hemicelluloses showed a greater decrease in comparison with cellulose, which suggests preferential degradation of hemicelluloses in the underwater WCH. In contrast, 10,000-year-old spruce lignin revealed no significant structural differences in comparison with the recent spruce wood by quantitative NMR analysis (Christiernin et al., 2009). The data from FT-IR and pyrolysis MS were consistent with results obtained from solid-state NMR (Bardet et al., 2009; Cha et al., 2014; Li et al., 2014).

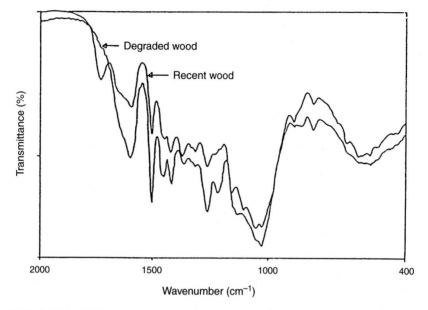

FIGURE 12.5   FT-IR spectra of recent and heavily degraded wood of *Pinus massoniana* (Kim, 1989).

Solid-state NMR analyses of waterlogged bamboo exhibited the same trends as for waterlogged archaeological woods. NMR spectra in the underwater bamboo showed the following: (1) complete disappearance of the signal at 21 ppm assigned to CH_3 of hemicelluloses; (2) a decrease in intensities and/or loss of signals at 64, 74, and 106 ppm, assigned to cellulose; and (3) stronger peak intensities at 105–160 ppm, assigned specifically to the aromatic compounds – the signals assigned to guaiacyl (G) lignin (84 and 115 ppm), syringyl (S) lignin (134, 148, and 152 ppm), and lignin methoxy carbon (56 ppm) showed significant increase. Significant increases in almost all the signals assigned to lignin polymer resulted from the suppression by cellulose and hemicelluloses, which were exposed in the WAW after degradation of structural carbohydrates against microbial attack under waterlogged conditions (Fig. 12.6).

All analyses showed that lignin in WAW is relatively more resistant than structural polysaccharides against microbial attack under waterlogged situations although depolymerization of lignin by cleavage of ether bonds (Salanti et al., 2010; Colombini et al., 2009; Pedersen et al., 2014) and demethoxylation of guaiacyl units was found (van Bergen et al., 2000). Dedic et al. (2014) found that the condition of lignin in the Vasa wood is similar to fresh oak and that potentially harmful tannins are not present in high amounts, suggesting that oxidative degradation mechanisms are not the primary route for cellulose depolymerization in the waterlogged archaeological wooden ships.

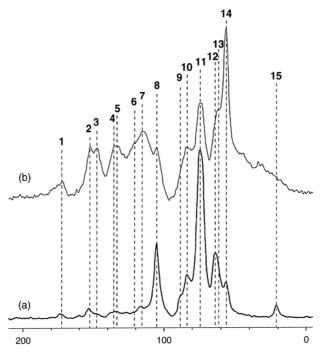

**FIGURE 12.6   Solid-state NMR of waterlogged ancient bamboo (b) and recent bamboo (a)
(Cha et al., 2014).** Resonance assignment: 1, carbohydrate; 2–5, syringyl and guaiacyl lignin; 6
and 7, guaiacyl lignin; 8 and 9, cellulose; 10 and 11, lignins; 12L, carbohydrates; 13, lignins; 14,
methoxyl in lignin; 15L, carbohydrates.

ABIOTIC AGENTS IN THE DETERIORATION OF MOVEABLE AND IMMOVEABLE WCH

When moveable and immoveable WCHs are exposed to natural weather conditions, the surface of dry-type WCH becomes rough, the grain rises, and
the wood develops checking, warping, and discoloration. Such deterioration
results from the process called weathering, which is mainly confined to the
surface. Three distinct alterations in the dry-type WCH by weathering are
the following: (1) loss of original color, (2) formation of cracks, and (3) erosion of wood mass (Fig. 12.7). Most alterations in the dry-type WCH occur
gradually on the surface; the roughened surface ultimately loses its surface co-
herence and corrugated appearance of the surface is caused by more rapid erosion of earlywood than latewood. The combined effect of light, precipitation,
heat, sand, wind, pollutants, and other factors bring about the degradation of
exposed WCH to natural weather conditions (Feist, 1990). Biotic agents such
as wood-decay fungi and molds are not considered to be primary agents of
weathering in dry-type WCH.

FIGURE 12.7   Wood plate in a Buddhist temple exposed to natural weathering in Japan. *(Unpublished file from Kim, Y.S.)*

Discoloration

WCH that are located outdoors in historic buildings and that do not have a surface coating or treatment, undergo rapid changes in color and a loss in their integrity. The surface color of WCH under outdoor conditions changes to brownish color due to oxidation and degradation of lignin. Later wood surface becomes covered with a spongy grayish layer due to the removal of lignin degradation products by rainfall. In contrast, the dry-type WCH stored under indoor conditions exhibit fading of colors by light; uncoated wood turning yellow or red is an indication of fading. The light-colored spruce of a soundboard will darken and the red mahogany of a piano lid will bleach when exposed to light. Discoloration is cumulative, irreversible, and insidious. Discoloration of the unprotected surface can result even from exposure to room light. Lee et al. (2009) found that about 70% of the original color of Japanese cedar (*Cryptomeria japonica*) wooden box was lost in the first year of indoor exposure. However, significantly greater color changes were not found after 5 years. Discoloration is accelerated by moisture content in the air, repetition of drying and wetting, and of hot and cold periods (Matsuo et al., 2008).

Cracks

Dry-type WCH exposed to indoor conditions can exhibit checks or cracks throughout the surface (Fig. 12.8). Cracking and distortion in the WCH derive from the moisture gradient along the wooden objects, producing a differential shrinkage perpendicular to the wood grain. Furthermore, repeated drying and wetting can lead to surface checks and cracks. Warping, tearing, and cracks on the surface of dry-type WCH can occur under very dry conditions. Paint layers, protective varnishes, and lacquers have often been present on dry-type WCH, such as old wooden musical instruments – pianos, harpsichords, and violins.

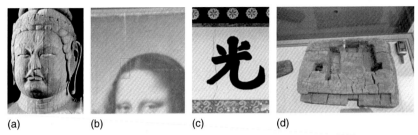

(a)   (b)   (c)   (d)

FIGURE 12.8   Cracks in various WCH. (a) Buddhist statue (Japan), (b) Mona Lisa panel (Louvre, France), (c) wooden plate (Korea), (d) waterlogged wooden pile (Korea). *(Buddhist statue from "Tokyo Nat'l Museum Bulletin"; Mona Lisa Panel from Uzielli, L., Proc. "The Wood Culture and Science Kyoto, Kyoto 2011"; wooden plate and waterlogged wooden pile from Kim, Y.S.s, unpublished files.)*

Wooden boxes and panels were fastened sometimes by metals. The dissimilar substances expand and shrink disparately with fluctuations in relative humidity and temperature, leading wood and coating materials to swell and shrink differently, resulting in cracks in the wooden part and in the flake-off of paints. Under low humidity and dry conditions, the layers of paints tend to crack more severely. The "Mona Lisa" painted about 500 years ago on one piece of a board of poplar (*Populus alba* L. 790 × 530 mm × 13 mm thick), showed an 11-cm-long crack running through the whole thickness (Fig. 12.8b) (Uzielli, 2011). Even the coating layers, surface treatments, and the original adhesives binding the decorative materials to wooden objects of WCH also deteriorate over time with environmental changes. Such alterations are difficult to repair without changing the visual appearance of the artifact.

Erosion of Wood Mass

It has been observed that erosion in WCH exposed to outdoor conditions occurs mainly in earlywood and that latewood is relatively resistant against weathering. Ultraviolet (UV) radiation in sunlight plays an important role in the disintegration of wood cell wall during the weathering process. Photodegradation in WCH is more or less localized on the surface because of shallow penetration of UV rays into wood. Kim et al. (2008) found that cell wall disintegration caused by weathering starts from the middle lamella, followed by the degradation of the S3 and S1 layers. The S2 layer was the most resistant of the secondary cell wall layers. However, even the S2 layer was degraded in later stages of weathering, caused by progressive detachment of microfibrils due to losses in the cell wall matrix materials, mainly the lignin. In the most advanced stages, the S2 layer also became defibrillated and the cell wall tended to disintegrate from peeling off of bundles of microfibrils (Fig. 12.9).

Erosion of wood surfaces during weathering is a slow process; on average the wood loss is 6 mm in a century, although the rate of surface erosion varies

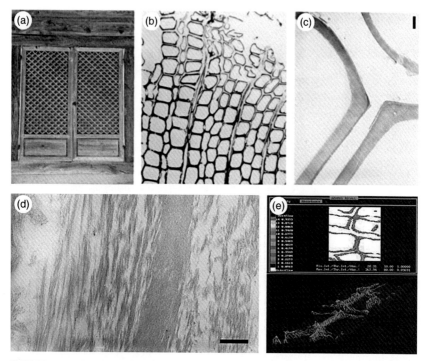

FIGURE 12.9 Disintegration of *Pinus* spp. in the windows of an old Buddhist Temple in Korea exposed to natural weathering (Kim et al., 2008).

between different species (Feist, 1990). In general, hardwoods are more resistant to weathering than softwoods mainly because of their greater density. When dry-type WCH is displayed in a location not directly exposed to sunlight, a much slower erosion of the surface occurs from delignification. Rainfall washes away water-soluble photodegradation products from the wood surfaces, resulting in surface erosion. Mechanical abrasion by wind and sand can aid in the removal of wood mass from degrading wood surfaces.

BIOTIC AGENTS IN THE DETERIORATION OF MOVEABLE AND IMMOVEABLE WCH

Deterioration of Moveable and Immoveable WCH by Insects

Regardless of their moisture content, WCH are prone to insect infestations. Nearly all wood and forest insects are now synanthropic species, which have expanded their niche to include buildings and processed wooden objects (Noldt, 2009; Sutter, 2002; Grosser, 1985). Most insects that attack dry-type WCH belong to the Coleoptera order, the members of which are specialized to attack dried wood with lower moisture content below the fiber saturation point.

(a)   (b)   (c)

FIGURE 12.10   Damages in indoor WCH by insects. (a) Common furniture beetle in the wooden sculpture in Basel Cathedral, (b) powder-post beetle in the xylarium of Nanjing Forest University, (c) termite attack on the pole of *Pinus* spp. in an old Buddhist temple in Korea. *(Unpublished files from Kim, Y.S.)*

Predominantly, wood-destroying insects in historic buildings in central Europe were found to be the common furniture beetle (*Anobium punctatum*), the deathwatch beetle (*Xestobium nifovillosum*), the anobiid species *Coelostethus pertinnax*, and the European house-borer (*Hylotrupes bajulus*) (Cymorek, 1985; Noldt, 2009; Kisternaya and Kozlov, 2012). The cigarette beetle (*Nicobium hirtum*) was found in a historic building of a Japanese wooden temple (Saito et al., 2008).

Dry-type WCH are also prone to insect infestations even under controlled relative humidity and temperature. Querner et al. (2013) found wood-destroying insects such as *Nicobium castaneum*, the common furniture beetle, *Hexathrum exiguum*, and powder-post beetle (*Lyctus brunneus*) in large museums in central Europe. It is interesting to note that powder-post beetle, which occurs frequently in modern buildings made from imported tropical hardwoods, was found even in the museums in Central Europe. Powder-post beetle is known to attack mainly those species that have large pores, such as oak, ash, walnut, elm, and hickory and bamboo culms containing adequate starch (Fig. 12.10). Softwoods and hardwoods with lower starch content are free from the attack of this beetle. Tropical hardwoods are susceptible to attack by this insect, particularly the light-colored species such as African mahogany, iroko, afzelia, Rhodesian teak, black wattle, ramin, and the many *Shorea* spp., such as seraya, meranti, lauan, and Eucalypts in Australia (Richardson, 1993). Imported household items made from tropical hardwoods, such as furniture, wooden floors, and panels, are often attacked by the powder-post beetle.

In addition to beetles, termite attack was found even in the wooden poles in an old Buddhist temple in South Korea (Fig. 12.10). Termite hazards are confined mainly to tropical and subtropical regions (Nunes, 2010) but are now encountered in New Zealand, Paris in France, and Hamburg in Germany. An increase in transportation and world trade has been responsible for the wide distribution of

TABLE 12.5 Characteristics of Damage by Wood-Boring Insects

	Preferable occurrence in wood species	Length of beetle (mm)	Diameter of exit hole (mm)	Full life cycle (years)
Furniture beetle	Softwood, hardwood, in particular pine, spruce, also poplar, elm, birch	3–6	1.6–2	2–3
Death-watch beetle	Softwood and heartwood primarily in oak wood previously damaged by fungi	6–8	2.5–7	1
Old house horn	Sapwood in softwoods such as pine, spruce and Douglas fir	10–25	10	3–11 or longer
Power post beetle	Sapwood only in hardwood with a certain amount of starch, tropical hardwoods, bamboo	0.8–1.5	0.8–1.6	1 or 2

Data Summarized from Cymorek (1985) and Richardson (1993).

termites. Subterranean termites belonging to genus *Reticulitermes*, *Heterotermes*, and *Coptotermes* (Formosan termites) in the Northern Hemisphere expanded their distribution beyond the latitude 40°N (Richardson, 1993). Subterranean termites attack only the softer earlywood and not the harder latewood, leaving a thin peripheral shell, which poses difficulties in visual inspection. Many foundations in the old Buddhist temples in East Asian countries, constructed with pine (*Pinus* sp.), are prone to termite attacks (Table 12.5).

Deterioration of Moveable and Immoveable WCH by Microorganisms

Dry-type WCHs are considered to be immune to microbial attacks due to their low moisture content. However, some fungal attacks have been observed even in the dry-type WCHs. Some workers (Nilsson and Daniel, 1990; Blanchette, 2010) observed some fungal hyphae in Egyptian coffins, indicating infection and growth of fungi within indoor WCH caused by occasional wetting, leaks, flooding, or other events at some stages. High humidity resulting from flooding and water leakage and poor ventilation in tombs and museums can provide an opportunity for fungal invasion in the dry-type archaeological wooden objects. In particular, molds developed quickly on the wood when wooden furniture, painted panels, and old books were exposed to flooding even though the duration of exposure was short. Fungal colonization over the Tripitaka-Koreana was found due to mistreatment of the plates after ink printing (Park et al., 1996). Dry-type WCHs in the outdoors have been more prone to a wide range of predators and deleterious factors than those in the indoor because they are frequently exposed to above 70% relative humidity, which can cause molds to readily grow.

When dry-type WCHs are dampened or are resting on a moisture-absorbing surface, such as basement floor, fungal growth and decay can occur. Soft-rot decay has been described frequently in historical and cultural wooden objects from construction timbers in great houses lived in by native Americans, in the ancient Egyptian wooden coffins, in the historic expedition huts in Antarctica (Blanchette, 2000; Blanchette et al., 2004), and in the wooden floor of an old Buddhist temple, which had been treated with fire retardants (Kim et al., 2008). It appears that the moisture-absorbing properties of the fire retardants induced the soft-rot decay in the wooden floors. The occurrence of soft-rot decay in water-logged archaeological woods has been well documented (Kim and Singh, 2000; Bjoerdal, 2012). Soft-rot fungi belonging to Deuteromycota and Ascomycota are known to be more tolerant to severe growth conditions such as high moisture content, pH, extreme temperature, and high preservative concentration than wood decaying basidiomycetes (Daniel and Nilsson, 1998).

In comparison with soft-rot fungi, other wood decay fungi have a relatively narrow spectrum of growth conditions. The optimum moisture content for growth is between 35% and 50%. A high moisture content of wood inhibits fungal growth due to oxygen restriction in the wood. However, wood decay fungi have been frequently observed in historic buildings exposed to outdoor conditions (Palanti et al., 2012; Pilt et al., 2009; Sterflinger, 2010). Brown- and white-rot fungi were observed in the historic buildings in Chilean wooden churches designated as a World Heritage Site by UNESCO (Ortiz et al., 2014). Kisternaya and Kozlov (2012) found that the most common destructors of historic timbers in northern Russia were brown-rot fungi, such as *Serpula lacrimans, Coniophora puteana, Coriolellus sinuosus, Fibuloporia vaillanti,* and *Paxillus paunmoides.* The dry-rot fungus *Serpular* spp. occurred frequently in poorly ventilated spaces such as on the backside of floorboards or ceilings in basement wood with moisture content of 20–30%. Limited period of exposure to wetting conditions in a certain season did not prevent the invasion of historic buildings by wood decay fungi. It should also be mentioned that in extreme cases, some wood decay fungal mycelium grew on wood with 17.4% moisture content. Stienen et al. (2014) observed that *Antrodia xantha, Coniophora* spp., *Donkioporia expansa,* and *Gloeophyllum abietinum* are able to degrade wood samples below fiber saturation over the temperature range of 10–25 °C, suggesting that dry-type WCH are not always immune to fungal attack.

Deterioration of Underwater WCH by Biotic Agents

WCH submerged in seawater, lake, or mud for long periods can undergo structural and chemical alterations. Most water-saturated WCH with moisture content over 100–800%, are referred to as waterlogged wood. Some WAW were

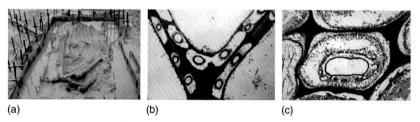

(a)   (b)   (c)

FIGURE 12.11 (a) Excavation of underwater ancient wooden ship (AD 11–14C) in Korea, (b) soft-rot cavities in the waterlogged archeological woods, excavated from Sweden, (c) erosion bacteria in the waterlogged archaeological woods, excavated Germany. *(Unpublished files from Kim, Y.S.)*

found to be heavily degraded within only a few 100 years, while others have survived thousands of years of exposure and in some cases appear to be relatively well preserved (Bjordal et al., 2012; Kim and Singh, 2000; Clausen, 1996; Qiu et al., 2013; Schmitt et al., 2005). This suggests that age alone does not determine the extent of deterioration in the WCH. Most wet-type WCH were found in a relatively good condition even after prolonged periods of exposure to water-saturated environments. This is because the activity of faster wood-degrading fungi (Basidiomycete) is restricted due to wood being water-saturated. Instead, bacteria and soft-rot fungi played an important role in the degradation process of such woods (Fig. 12.11).

Soft-rot fungi are tolerant to a wide range of temperature, humidity, and pH conditions and can attack a variety of wood substrates (Daniel and Nilsson, 1998). Bacteria are very tolerant to conditions considered extreme for most wood-decay fungi, for example high lignin and extractives content in wood, high preservative loading, and low levels of oxygen. Bacterial erosion has been found to be widespread in sediment where the archaeological woods had been buried (Fig. 12.11). Degradation of WAW by erosion bacteria is followed by scavenging bacteria in some situations (Fig. 12.12) (Kim and

FIGURE 12.12 Degradation of S2 layer by erosion bacteria (arrows) in *C. camphora* wood in an excavated wooden ship in Korea. Scavenging bacteria (arrowheads) are present within the degraded wall material (bar = 1 μm) (Kim et al., 1996).

Singh, 1994), some being sulfur-reducing bacteria, which produce hydrogen sulfide *in situ* (Fors et al., 2012). Co-occurrence of bacterial erosion, bacterial tunneling, and soft rot has also been noted in buried and waterlogged woods. Detailed information on the micromorphology of bacterial degradation can be found in a previous chapter in this book (Singh et al., 2014).

It is now widely accepted that attack by erosion bacteria is the main cause of destruction of WAW, as these bacteria are capable of degrading wood under conditions of limited oxygen availability (Huisman et al., 2008; Singh, 2012). Attack of WAW by erosion bacteria is mainly confined to the secondary cell wall (Nilsson et al., 2008), whereas attack of WAW walls by tunneling bacteria was found to extend to the middle lamella. This suggests that tunneling bacteria can degrade all areas of lignified cell walls, including the highly lignified middle lamella (Singh et al., 2003). Considering that tunneling bacteria have a relatively high requirement for oxygen (Daniel et al., 1989), the attack on WAW by tunneling bacteria is likely to have occurred before the objects were completely submerged in the ocean or mud. The micromorphology of waterlogged archaeological bamboo was similar to that for WAW. However, typical soft-rot cavities, which were frequently observed in waterlogged archaeological woods, were not found (Cha et al., 2014).

The degradation of WAW walls did not occur uniformly; relatively-sound tracheids and fibers were found among extensively degraded regions. Degradation occurred mainly in tracheids and fibers, whereas vessel walls were relatively intact. Separation of the secondary wall layers from middle lamella was frequently observed in all cell types. Degradation occurred preferentially in the S2 layer, whereas middle lamellae were mostly intact. The degraded tracheids and fibers showing a granular appearance were distinctly porous and lacked birefringence when observed under a polarized microscope, indicating a loss in the crystalline cellulose. In hardwoods, vessels in WAW were more intact compared to other cell types, such as fibers and parenchyma cells. In advanced stages of degradation, vessel walls were also degraded with only the middle lamella remaining. Ultrastructural examination on the WAW showed that cell outlines were distinguishable despite extensive losses in the integrity of the bulk of cell wall material. This is primarily because middle lamellae remained usually intact, although often highly convoluted. Resistance of middle lamella to erosion bacteria under waterlogged situations is likely to be due to its high lignin concentration. Compound middle lamella and vessel walls in woods are reported to exhibit higher lignin concentration compared to fiber secondary walls (Kim et al., 2008) and the composition of lignin is also different from the secondary cell walls (Faix, 1991; Fengel and Wegener, 1984).

Microscopic examinations showed that bacterial attack was the main cause of degradation of waterlogged woods and bamboos (Fig. 12.13) (Kim, 1989; Kim et al., 1996; Kim and Singh, 2000; Blanchette, 2000; Singh, 2012; Cha et al., 2014). Cell wall components, particularly lignin, were the main factors that influenced microbial degradation in underwater WCH. However, other

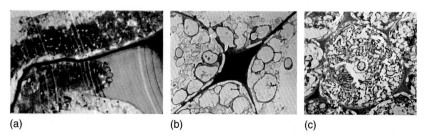

(a)   (b)   (c)

FIGURE 12.13   Degradation of waterlogged archeological woods and bamboos. *((a) Degradation of tracheids by erosion bacteria, (b) co-occurrence of tunneling bacteria, (c) soft-rot fungi and degradation of waterlogged archaeological bamboos by erosion bacteria.)*

factors, such as leaching of degraded cell wall materials and impregnation of cell walls by inorganic substances over time, may also affect the degree of deterioration of WAW. Nonbiological factors, such as salinity, temperature, nutrients, composition of sediment layers, and other factors, can also influence the rate of microbiological decay of WAW (Bjordal et al., 2012). Further studies are needed to determine whether abiotic reactions in seawater, such as hydrolysis, might also partly contribute to degradation of wood components of WAW, although it is now well documented that microbiological factors are the primary cause for the deterioration of woods subjected to waterlogging.

SOME REMARKS ON THE CONSERVATION OF WCH

Conservation includes maintenance, preservation, restoration, reconstruction, and adaptation; and commonly a combination of these, whereas preservation means maintaining the fabric of a place in its existing state and retarding deterioration. Conservation means all the processes of looking after a place so as to retain its cultural significance (Jokilehto, 2005). Consequently, conservation work is a very time-consuming, labor-intensive, and fairly expensive process. The conservation of WCH has been on empirical approaches rather than hard scientific data because each WCH is unique in terms of wood species, environment, deterioration rate, and manufacturing techniques. Individual assessment, evaluation, and solutions are required for the conservation and restoration of WCH. The application of general standard solutions into the conservation of WCH should be carefully exercised. In practice, different people, each with different sets of skills and paradigms, have carried out the selection and application of conservation treatments. The conservation professionals strive to select methods and materials that do not adversely affect cultural property. Techniques of conservation of WCH range from artistic forms to scientific technologies. Holistic approaches are required for WCH conservation; combination of curator's focus on context, archaeologist's deep awareness of the past, and wood scientist's comparative methods are needed for a rational integration of technical and scientific knowledge for effective conservation (Peterson, 1990). The mastery of an artist as well as the analytical abilities of a scientist is required. To avoid miscommunication and misunderstanding among the conservators, curators, the governmental authorities, and wood scientist, a coordinated approach for conservation should be developed. For the conservation of WCH, some recent books can be referred, for example, Bjordal et al. (2012), Christensen et al. (2012), Lucejko (2010), Unger and Schniewind (2001), and the proceedings from the International Council of Museums.

ACKNOWLEDGMENTS

The authors sincerely thank Dr Ick-Joo Kim, Dr Kwang Ho Lee, Dr Jong Sik Kim, and Dr Mi Young Cha for their generous help in the collection of valuable wooden samples and involvement in microscopic and chemical analyses of archaeological woods excavated from the Korean peninsula. The authors also thank Dr T. Itoh, Emeritus Professor of Kyoto University, for his critical reading of the manuscript and valuable comments.

REFERENCES

Bamford, M., 2010. Wood and Wooden Artefacts Among the Earliest Hominids in South and East Africa. Proc. Int. Wood Culture Symp., Shaanxi, China, Oct 23–24, 2010.

Bardet, M., Gerbaud, G., Giffard, M., Doan, C., Hediger, S., Pape, L.L., 2009. [13]C high-resolution solid-state NMR for structural elucidation of archaeological woods. Prog. Nucl. Magn. Reson. Spectrosc. 55, 199–214.

Bjoerdal, C.G., 2012. Evaluation of the microbial degradation of shipwreck in the Baltic Sea. Rev. Int. Biodet. Biodeg. 70, 126–140.

Bjordal, C.G., Gregory, D., Trakadas, A. (Eds.), 2012. Decay and protection of archaeological wooden shipwrecks. Archaeopress Ltd, Oxford, p. 155.

Blanchette, R.A., 2000. A review of microbial deterioration found in archaeological wood from different environments. Int. Biodet. Biodeg. 46, 189–204.

Blanchette, R.A., 2010. Assessment of wood deterioration in the furniture and coffin from Tumulus MM. The Furniture from Tumulus MM. In: Simpson, E. (Ed.), The Gordion Wooden Objects, vol. 1, Brill, Leiden, pp. 171–176 and plates 139–149.

Blanchette, R.A., Held, B.W., Jurgens, J.A., McNew, D.L., Harrington, T.C., Duncan, S.M., Farrell, R.L., 2004. Wood-destroying soft-rot fungi in the historic expedition huts of Antarctica. Appl. Environ. Microbiol. 70, 1328–1335.

Cha, M.Y., Lee, K.H., Kim, Y.S., 2014. Micromorphological and chemical aspects of archaeological bamboos under long-term waterlogged condition. Int. Biodet. Biodeg. 86, 115–121.

Christensen, M., Kutzke, H., Hansen, E.K., 2012. New materials used for the consolidation of archaeological wood – past attempts, present struggles, and future requirements. J. Cult. Herit. 13, 5183–5190.

Christiernin, M., Notley, S.M., Zhang, L., Nilsson, T., Henriksson, G., 2009. Comparison between 10,000-year-old and contemporary spruce lignin. Wood Sci. Technol. 43, 23–41.

Clausen, C.A., 1996. Bacterial association with decaying wood: a review. Int. Biodet. Biodeg. 37, 101–107.

Colombini, M.P., Lucejko, J.J., Modugno, F., Orlandi, M., Tolppa, E.-L., Zoia, L., 2009. A multi-analytical study of degradation of lignin in archaeological waterlogged wood. Talanta 80, 61–70.

Cymorek, S., 1985. Schadinsekten in Kunstwerken und AntiquitaetenausHolz in Europa. In: Cymorek, S. et al.(Ed.), Holzschutz–Forschung und Praxis. DRW-Verlag, Weinbrenner-KG, pp. 37–56.

Daniel, G.F., Nilsson, T., 1998. Developments in the study of soft rot and bacterial decay. In: Bruce, A., Palfreyman, J.W. (Eds.), Forest Products Biotechnology. Taylor and Francis Ltd, London, pp. 37–62.

Daniel, G.F., Nilsson, T., Singh, A.P., 1989. Degradation of lignocellulosics by unique tunnel-forming bacteria. Can. J. Microbiol. 33, 943–948.

Dedic, D., Sandberg, T., Iversen, T., Larsson, T., Monica, E.M., 2014. Analysis of lignin and extractives in the oak wood of the 17th century warship Vasa. Holzforschung 68, 419–425.

d'Errico, F., Backwell, L., Villa, P., Degano, I., Lucejko, J., Bamford, M.K., Highham, T.F.G., Colombini, M.P., Beaumont, P.B., 2012. Early evidence of San material culture represented by organic artifacts from Border Cave, South Africa. Proc. Natl. Acad. Sci. USA 109, 13214–13219.

Faix, O., 1991. Classification of lignins from different botanical origins by FT-IR spectroscopy. Holzforschung 45, 21–27.

Fazio, A.T., Papinutti, L., Gomez, B.A., Parera, S.D., Romero, A.R., Siracusano, G., Maier, M.S., 2010. Fungal deterioration of a Jesuit South American polychrome wood sculpture. Int. Biodet. Biodeg. 64, 694–701.

Feist, W.C., 1990. Outdoor wood weathering and protection. Archaeological Wood: Properties, Chemistry, and PreservationAmerican Chemical Society, Washington, DC, pp. 263–298.

Fengel, D., Wegener, G., 1984. Wood: Chemistry, Ultrastructure, Reactions. Walter De Gruyter, Berlin, pp. 132–181.

Fors, Y., Jalilehvand, F., Risberg, D.E., Bjordal, C., Phillips, E., Sandstrom, M., 2012. Sulfur and iron analysis of marine archaeological wood in shipwrecks from the Baltic Sea and Scandinavian waters. J. Arch. Sci. 39, 2521–2532.

Gelbrich, J., Mai, C., Militz, H., 2008. Chemical changes in wood degraded by bacteria. Int. Biodet. Biodeg. 61, 24–32.

Giachi, G., Bettazzi, F., Chimichi, S., Staccioli, G., 2003. Chemical characterization of degraded wood in ships discovered in a recent excavation of the Etruscan and Roman harbour of Pisa. J. Cult. Herit. 4, 75–83.

Grosser, D., 1985. Pflanzliche and tierischeBau- und Werkholzschaedlinge. DRW-Verlag, Leinfelden-Echterdingen.

Huisman, D.J., Manders, M.R., Kretschmar, E.I., Klaassen, R.K.W.M., Lamersdorf, N., 2008. Burial conditions and wood degradation at archaeological sites in the Netherlands. Int. Biodet. Biodeg. 61, 33–44.

Irbe, I., Karadelev, M., Andersone, I., Andersons, B., 2012. Biodeterioration of external wooden structures of the Latvian cultural heritage. J. Cult. Herit. 13, 579–584.

Itoh, T. (Ed.), 2008. Wood Culture and Science. Kaiseisha Press, Tokyo, Japan, (in Japanese).

Itoh, T., Yamada, M., 2012. Archaeological Wood in Japan. Kaiseisha Press, Tokyo, Japan, (in Japanese).

Jokilehto, J. 2005. Definition of cultural heritage: references to document in history. ICCROM Working Group 'Heritage and Society'.

Kim, J.S., Singh, A.P., Wi, S.G., Koch, G., Kim, Y.S., 2008. Ultrastructural characteristics of cell wall disintegration of *Pinus* spp. in the windows of an old Buddhist temple exposed to natural weathering. Int. Biodet. Biodeg. 61, 194–198.

Kim, Y.S., 1989. Micromorphology of degraded archaeological pine wood in waterlogged situations. Mat. u. Org. 24, 271–286.

Kim, Y.S., 1990. Chemical characteristics of waterlogged archaeological wood. Holzforschung 44, 169–172.

Kim, Y.S., Singh, A.P., 1994. Ultrastructural aspects of bacterial attacks on a submerged ancient wood. Mokuzai Gakkaishi 40, 554–562.

Kim, Y.S., Singh, A.P., 2000. Micromorphological characteristics of wood degradation in wet environments: a review. IAWA J. 21, 135–155.

Kim, Y.S., Singh, A.P., Nilsson, 1996. Bacteria as important degraders in waterlogged archaeological woods. Holzforschung 50, 389–392.

Kisternaya, M., Kozlov, V., 2012. Preservation of historic monuments in the "Kizhi" open-air museum. J. Cult. Herit. 13 (3), 74–78.

Kohara, J., 1958. Study on the old timber. Res. Rep. Faculty Eng. Chiba Univ. 9 (15), 1–55, (Japanese).

Koestler, R.J., Koestler, V.R., Charola, A.E., Nieto-Fernandez, F.E. (Eds.), 2010. Art, Biology, and Conservation: Biodeterioration of Works of Art. The Metropolitan Museum of Art, New York, NY.

Latour, G., Echard, J.-P., Didier, M., Schanne-Klein, M.-C., 2012. In situ 3D characterization of historical coatings and wood using multimodal nonlinear optical microscopy. Opt. Express 20, No. 22: 24623–24635.

Lee, K.H., Cha, M.Y., Chung, W.Y., Bae, H.J., Kim, Y.S., 2009. Chemical and morphological change and discoloration of cedar wood stored indoor. Mokchae Konghak 37 (6), 66–77, (in Korean).

Li, M.-Y., Fang, B.-S., Zhao, Y., Hou, X.-H., Tong, H., 2014. Investigation into the deterioration process of archaeological bamboo strips of China from four different periods by chemical and anatomical analysis. Polym. Degrad. Stability 109, 71–78.

Lucejko, J.J., 2010. Waterlogged Archaeological Wood: Chemical Study of Wood Degradation and Evaluation of Consolidation Treatments. VDM Verlag, Saarbrucken, Germany, pp. 157.

Maierhofer, C., Rollig, M., Krankenhagen, R., 2010. Integration of active thermography into the assessment of cultural heritage buildings. J. Mod. Opt. 57, 1790–1802.

Mannes, D.C. 2009. Non-destructive testing of wood by means of neutron imaging in comparison with similar methods. Diss. ETH Zurich.

Matsuo, M., Yokoyama, M., Umemura, K., Sugiyama, J., Kawai, S., Gril, J., Yano, K., Kubodera, S., Mitsutani, T., Ozaki, H., Sakamoto, M., Imamura, M., 2008. Evaluation of the aging wood from historical buildings as compared with the accelerated aging wood and cellulose-analysis of color properties. In: Gril, J. (Ed.), Wood Science for Conservation of Cultural Heritage – Bragen 2008. Proc. Int. Conf. by COST Action IE 0601. pp. 57–61.

Mizuno, S., Torizu, R., Sugiyama, J., 2010. Wood identification of wooden mask using a synchrotron X-ray microtomography. J. Arch. Sci. 37, 2842–2845.

Nakao, T., Tanaka, C., Takahashi, A., Okano, T., Nishimura, H., 1989. Long-term changes in degree of crystallinity of wood cellulose. Holzforschung 43, 419–420.

Nilsson, T., Daniel, G., 1990. Structure and the aging process of dry archaeological wood. Archaeological Wood: Properties, Chemistry, and Preservation. Am. Chem. Soc., Washing, DC, pp. 67–86.

Nilsson, T., Bjordal, C.G., Fallma, E., 2008. Culturing erosion bacteria: procedures for obtaining pure culture. Int. Biodet. Biodeg. 61, 17–23.

Noldt, U., 2009. Monitoring wood-destroying insects in wooden cultural heritage. In: Uzielli, L. (Ed.), Wood Science for Conservation of Cultural Heritage – Florence 2007. Firenze Univ. Press, pp. 70–79.

Nunes, L., 2010. Termite infestation in Portuguese historic buildings. In: Grill, J. (Ed.), Wood Science for Conservation of Cultural Heritage – Braga 2008. Firenze Univ. Press, pp. 117–122.

Ortiz, R., Parraga, M., Narvarrete, J., Charrasco, I., dela Vega, E., Ortiz, M., Herrera, P., Jurgens, J.A., Held, B.W., Blanchette, R.A., 2014. Investigations of biodeterioration by fungi in historic wooden churches of Chilo. Chile. Microb. Ecol. 67 (3), 568–575.

Palanti, S., Pecoraro, E., Scarpino, F., 2012. Wooden doors and windows in the church of the Nativity: evaluation of biotic and abiotic decay and proposals of interventions. J. Cult. Herit. 13, 82–92.

Park, S.J., Kang, A.K., Park, S.Y., 1996. Changes in cell structure in dry archaeological wooden plates Tripitaka Koreana. In: Donaldson, L.A. et al., (Ed.), Recent advances in wood anatomy. New Zealand Forest Research Institute, Rotorua, pp. 120–122.

Passialis, C.N., 1997. Physico-chemical characteristics of waterlogged archaeological wood. Holzforschung 51, 111–113.

Pedersen, N.B., Gierlinger, N., Thygesen, L.G., 2014. Bacterial and abiotic decay in waterlogged archaeological *Picea abies* (L.) Karst studied by confocal Raman imaging and ATR-FTIR spectroscopy. Holzforschung 68, 791–798.

Peterson, C.E., 1990. New directions in the conservation of archaeological wood. In: Archaeological Wood: Properties, Chemistry and Preservation. American Chemical Society, Washington, DC, pp. 433–449.

Pilt, K., Pau, K., Oja, J., 2009. The wood destroying fungi in buildings in Estonia. Structural Studies, Repairs and Maintenance of Heritage Architecture XI. WIT Press, UK, pp. 243–251.

Proietti, N., Presciutti, F., Tullio, V.D., Doherty, B., Marinelli, A.M., Provinciali, B., Macchioni, N., Capitani, D., Miliani, C., 2011. Unilateral NMR ^{13}C CPMAS NMR spectroscopy and microanalytical techniques for studying the materials and state of conservation of an ancient Egyptian wooden sarcophagus. Anal. Bioanal. Chem. 399, 3117–3131.

Qiu, J., Min, R., Kuo, M., 2013. Microscopic study of waterlogged archaeological wood found in Southwestern China and method of conservation. Wood Fiber Sci. 45, 396–404.

Querner, P., Simons, S., Morelli, M., Furenkranz, S., 2013. Insect pest management programmes and results from their application in two large museum collections in Berlin and Vienna. Int. Biodet. Biodeg. 84, 275–280.

Richardson, B.A., 1993. Wood Preservation, second ed E& FN Spoon, London.

Rowell, R.M., Barbour, R.J. (Eds.), 1990. Archaeological Wood: Properties, Chemistry and Preservation. American Chemical Society, Washington, DC.

Saito, Y., Shida, S., Ohita, M., Yamamoto, H., Tai, T., Ohmura, W., Mahihara, H., Nohiro, S., Goto, O., 2008. Deterioration character of aged timbers insect damage and material aging of rafters in a historic building of Fukushoji-temple. Mokuzai Gakkaishi 54, 255–262.

Salanti, A., Zoia, E.-L., Giachi, G., Orlandi, M., 2010. Characterization of waterlogged wood by NMR and GPC techniques. Microchemical J. 95, 345–352.

Schmitt, U., Singh, A.P., Thieme, H., Friedrich, P., Hoffman, P., 2005. Electron microscopic characterization of cell wall degradation of the 400,000-year-old wooden Shoeningenspears. Holz Roh- Werkst. 63, 118–122.

Schniwind, A.P., 1990. Physical and mechanical properties of archaeological wood. Archaeological Wood: Properties, Chemistry, and Preservation. American Chemical Society, Washington, DC, pp. 87–109.

Singh, A.P., Kim, Y.S., Wi, S.G., Lee, K.H., Kim, I.J., 2003. Evidence of the degradation of middle lamella in a waterlogged archaeological wood. Holzforschung 57, 115–119.

Singh, A.P., 2012. A review of microbial decay types found in wooden objects of cultural heritage recovered from buried and waterlogged environments. J. Cult. Herit. 13, 516–520.

Singh, A.P., Kim, Y.S., Singh, T., 2014. Bacterial degradation of wood. In: Kim, Y.S., Singh, A.P., Funada, R. (Eds.), Secondary Xylem Biology. Elsevier Publications.

Steppe, K., Cnuddle, V, Girard, C., Lemeur, R., Cnudde, J.P., Jacobs, P., 2004. Use of X-ray computed microtomography for non-invasive determination of wood anatomical characteristics. J. Struct. Biol. 148, 11–21.

Sterflinger, R., 2010. Fungi: their role in deterioration of cultural heritage. Fungal Biol. Rev. 24, 47–55.

Stienen, T., Schmidt, O., Huckfeldt, T., 2014. Wood decay by indoor basidiomycetes at different moisture and temperature. Holzforschung 68, 9–15.

Strzelczyk, A.B., 2004. Observations on aesthetic and structural changes induced in Polish historic objects by microorganisms. Int. Biodet. Biodeg. 53, 151–156.

Sutter, H.-P., 2002. HolzschaedlingeanKulturgueternerkennen und bekaempfen–Handbuchfuer Denkmalspflege, Restauratoren, Architekturen und Holzfachleute. Verlag Paul Haupt, Bern, 166 pp.

Takei, T., Hamajima, M., Kamba, N., 1997. FT-IR analysis of the degradation of structure lumber in Horyu temple. Mokuzai Gakkaishi 43 (3), 285–294, (in Japanese).

Tamburini, D., Lucejko, J.J., Modugno, F., Colombini, M.P., 2014. Characterization of archaeological waterlogged wood from Herculaneum by pyrolysis and mass spectrometry. Int. Biodet. Biodeg. 86, 142–149.

Tsuchikawa, S., Yonenobu, H., Siesler, H.W., 2005. Near-infrared spectroscopic observation of the aging process in archaeological wood using a deuterium exchange method. Analyst 130, 379–384.

Unger, A., Schniewind, A.P., 2001. Conservation of Wood Artifacts. A Handbook. Springer-Verlag, Berlin.

Uzielli, L., 2011. Assessing and monitoring Italian Renaissance panel paintings with a special reference to Mona Lisa. Proc. Wood Cult. Sci. Kyoto, 1–6.

van Bergen, P.F., Poole, I., Ogilvie, T.M., Caple, C., Evershed, R.P., 2000. Evidence for demethylation of syringyl moieties in archaeological wood using pyrolysis-gas chromatography/mass spectrometry. Rapid Commun. Mass Spectrom. 14 (2), 71–79.

Viel, S., Capitani, D., Proietti, N., Ziarelli, F., Segre, A., 2004. NMR spectroscopy applied to the cultural heritage: a preliminary study on ancient wood characterization. Appl. Phys. A 79, 357–361.

Wang, S.-Z., 2010. Study on the material of ancient tomb furniture in China. Proc. 2010 Int. Wood Culture Symp., Shaanxi, China, Oct 23–24, 2010.

Wilmering, A.M., 1996. The conservation treatment of the Gubbio Studiolo. vol. LIII No. 4Metropolitan Museum of Art Bull, NY.

Yokoyama, M., Gril, J., Matsuo, M., Yano, H., Sugiyama, J., Clair, B., Kubodera, S., Mitsutani, T., Sakamoto, M., Ozaki, H., Imamura, M., Kawai, S., 2009. Mechanical characteristics of aged Hinoki wood from Japanese historical buildings. C. R. Phys. 10, 601–611.

Chapter 13

Biomaterial Wood: Wood-Based and Bioinspired Materials

Ingo Burgert*,**, Tobias Keplinger*,**, Etienne Cabane*,**,
Vivian Merk*,**, Markus Rüggeberg*,**
*Swiss Federal Institute of Technology Zürich (ETH Zürich), Institute for Building Materials,
Zürich, Switzerland; **Swiss Federal Laboratories for Materials Science and
Technology (EMPA), Applied Wood Materials, Dübendorf, Switzerland*

INTRODUCTION

Bioinspired materials research has attracted increasing attention in recent years. A deeper understanding of the key mechanisms and principles that drive the specific functions of biological materials and systems allows for transferring them into the engineering world for the design of new functional materials. The superhydrophobicity and self-cleaning capabilities of lotus-leaf surfaces, mussel glues that function in salt water, or the exceptional mechanical property profile of bones, combining high stiffness and toughness, are peculiar highlights

Secondary Xylem Biology. http://dx.doi.org/10.1016/B978-0-12-802185-9.00013-9
259

of such sources of bioinspiration (Lee et al., 2006; Munch et al., 2008; Neinhuis and Barthlott, 1997). Hence, when seeking natural role models for bioinspired materials design one can access a multitude of organisms, organs, and materials in nature that can provide helpful information on the structural adaptation toward specific functionalities. In this chapter we would like to concentrate on wood and highlight its unique standing in being a source of bioinspiration and engineering materials at the same time. In consequence, the term "biomaterial wood" as reflected in this chapter considers two major aspects. On the one hand it addresses the potential of wood as it functions in the living tree to serve as an important source of bioinspiration for materials design (*bioinspired materials*). This approach is based on the fundamental assumption that trees invented an optimized wood material, which has to fulfill similar mechanical functions in the living tree compared to engineering applications. Consequently, various principles of its formation and functionality may be applied to technical problems striving to improve wood properties or opening new fields of utilization. The main features to be mentioned in this regard are the hierarchical structure of wood, cell wall assembly processes, the light-weight fiber composition, adaptive growth, or reaction wood formation (Emons and Mulder, 1998; Gibson, 2012; Speck and Burgert, 2011). On the other hand, the term also points to the fact that wood is per se a natural and renewable (bio)material with an intrinsic evolutionary optimization, which has been used from the beginning of humankind for engineering applications and can be further exploited to develop advanced wood materials, nowadays and in the future (*wood-based materials*). Consequently, in this definition the question is not about learning from wood but rather to directly make use of its exceptional material features. Obviously, both aspects can go hand in hand, and ideally research may combine bioinspiration and cutting-edge wood modification and functionalization to develop novel renewable wood-based materials with improved material properties that do not only advance the utilization in common fields of applications but also generate new fields of implementation.

An example that introduces and illustrates the potential of the approach, to better understand the basic principles of the biological system and to use it for a technical transfer, is the cell wall assembly process. The way cellulose fibrils, hemicelluloses, and lignin are assembled in the living tree is partly understood (Cosgrove, 2005; Somerville et al., 2004; Salmen and Burgert, 2009), but we still miss important basics for a (bio)technological transfer. Most is known about the secretion of the biomacromolecules in the forming cell wall but we lack information on the local spatial distributions as well as physical and chemical constraints of molecule interactions in the assembly process. Furthermore, it is unclear how this process can be adaptive, to optimize structural features based on mechanical stimuli. In particular the control and fine-tuning of the cellulose fibril orientation and their parallel formation in highly organized patterns (Mutwil et al., 2008; Paredez et al., 2006) could be used as a great source of bioinspiration to develop advanced fiber-composites. Brown (1996) already introduced the potential and the challenges by using the term "nature's assembly machines" for the cellulose synthesis, illustrating the analogy between cellulose fibril insertion in cell wall formation processes and fiber spinning technologies.

Research on wood-based or wood-derived materials, such as cellulose nanocrystals or nanofibrillated cellulose (Zimmermann et al., 2004; Eichhorn et al., 2010; Olsson et al., 2010), could definitely highly benefit from a better understanding of these principles, but also other fields of nanotechnology still lack adequate tools to sufficiently upscale material functionalization via assembly processes. For materials based on nanofibrillated cellulose, mostly only a random network of cellulose fibrils or crystals is obtained and its exceptional stiffness and strength is only marginally utilized in the composites. Procedures to orient fibers in a parallel fashion or to implement a certain directionality in the composites are demanding and not easily integrated in production processes (Bordel et al., 2006; Ebeling et al., 1999; Kvien and Oksman, 2007). Hence, one important breakthrough would be a successful transfer of biological principles of cellulose orientation control for the making of cellulose composites. A further important aspect of cell wall formation and assembly relates to the fiber–matrix interactions in the cell wall, which result in a stiff but still rather tough nanocomposite. In particular the basic principles of the mediating function of hemicelluloses by being matrix component and coupling agent to the cellulose fibrils, could be in the future exploited for technical composites in order to improve their toughness and damage tolerance without sacrificing the material stiffness and strength (Altaner and Jarvis, 2008; Keckes et al., 2003; Köhler and Spatz, 2002).

Consequently, to make better use of wood as a source of bioinspiration, we need a deeper understanding of the basic principles and mechanisms that drive the exceptional properties of the biomaterial wood. However, in terms of solid wood utilization it is also important to reflect which peculiar properties are lost or newly gained, when we transfer wood from its natural function in a tree to its engineering application purposes. This can be well illustrated taking wood–water interactions as an example, since a major shortcoming of wood is its hygroscopic nature, which results in dimensional instabilities. Swelling and shrinking under changing humidity conditions is clearly a problem of its use as an engineering material, because for the function in the living tree this problem is completely irrelevant as the tree keeps the wood well above fiber saturation. To overcome this problem in the engineering world various wood modification protocols have been developed that aim at reducing the water accessibility of the hydroxyl group of the cell wall components (Hill, 2011; Mai and Militz, 2004; Rowell and Dickerson, 2014). Interestingly, even though dimensional stability is not an immediate issue for the living tree one can still use wood as a source of bioinspiration in this regard, since living trees manage to insert hydrophobic substances in the wood cell wall during heartwood formation (Magel et al., 1991; Taylor et al., 2002). Hence, the way how trees achieve an insertion of phenolic substances in the hydrophilic environment of the already entirely formed cell wall under mild (natural) reaction conditions and guarantee their stable accumulation is not only relevant in understanding heartwood formation, but could also function as a biological role model to improve wood modification protocols aiming at increasing the dimensional stability of wood. However, for a full transfer, we do not only need to understand the chemical nature of the modifying agents and physiological aspects of the processes, but also the basic principles and mechanisms that drive the heartwood formation.

In the following, this chapter deals with aspects of wood modification and functionalization as well as their characterization, which reflect the mentioned strategy, with emphasis on the bioinspiration and/or the wood (bio)materials research. The subchapters review recent developments in the field and give examples mainly based on the work of the research group. Since modification and functionalization address the nano- and microscale of wood, it is highly important to make use of sophisticated characterization techniques that ideally allow for gaining structural, chemical, and mechanical (physical) information of cell walls with high spatial resolution. Hence, we start with a subchapter on characterization, which introduces the general concept and specifically highlights Raman microscopy and scanning near-field optical microscopy as powerful tools for the analysis of natural and modified cell walls. This is followed by a subchapter on cell wall modification protocols, with a focus on improving wood dimensional stability. Further concepts of modularity in polymerization will be highlighted, which do not aim at making the cell wall particularly inert, but allowing for polymerizing various monomers in the cell wall and thereby implementing new functions and properties. The aspect of new functionalities is further developed in the fourth subchapter with a focus on wood actuation based on two principles. The first part reports on the making of wood-hybrid composites with emphasis on the insertion of iron-oxide particles in the microstructure of wood, which results in a wood hybrid that can be actuated in a magnetic field, utilizing the anisotropic and hierarchical scaffold of the wood structure to achieve a directional property profile. The second part deals with a bioinspired approach to specifically make use of the swelling and shrinking of wood. As with the bending movements in pine cones or wheat awns, which are actuated in nature for seed dispersal (Dawson et al., 1997; Elbaum et al., 2007), wood layers can be combined in a way that they become active elements upon humidity changes. The resulting movements take place autonomously without further energy supply and can be utilized, for instance, for shading purposes at building facades or solar tracking.

WOOD CELL WALL ASSEMBLY CHARACTERIZATION

A deeper understanding of the basic principles of cell wall assembly or the impact of modifications at the nano- and microscale requires comprehensive cell wall characterization with high resolution. Already the native wood cell wall is a complex multicomposite structure that can become even more complicated

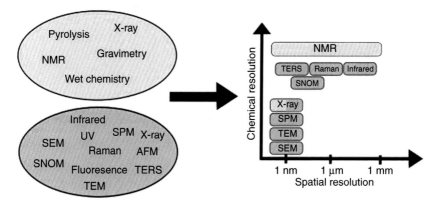

FIGURE 13.1 Classification of common analytical methods for wood cell wall analysis based on their sample preparation.

to analyze due to the insertion of further chemical substances in a modification or functionalization process. For a comprehensive characterization of wood cell walls it is important to be able to study the composition of the cell wall and the spatial arrangement of the components within the wood structure. Various methods have been applied to wood to gain further insight into the cell wall organization, for example, nuclear magnetic resonance (NMR) (Mansfield et al., 2012), thermogravimetry (Korosec et al., 2009), pyrolysis (Alves et al., 2006), vibrational spectroscopy (IR and Raman) (Gierlinger et al., 2012; Tsuchikawa and Schwanninger, 2013), UV spectroscopy (Koch and Kleist, 2001), fluorescence spectroscopy (Donaldson and Radotic, 2013), or scanning probe microscopy techniques (Fahlen and Salmen, 2005; Zimmermann et al., 2006). The methods differ substantially in terms of the chemical information that can be gained and in their spatial resolution. Some of these methods require a disintegration of the cell wall resulting in a loss of the information on the arrangement of the components. In Fig. 13.1 a rough classification of the main analytical methods based on their sample preparation (disintegrating (red) or not (blue)) and on their spatial/chemical resolution is shown. Here we intend to focus on vibrational spectroscopy (especially Raman spectroscopy) and new developments in scanning probe microscopy as two representatives of analytical methods that allow for analyzing the plant cell wall in its original anatomical state and can be used to visualize the distribution of cell wall modifying substances.

Scanning Probe Microscopy Techniques

In recent years, different techniques have been further developed and optimized to generate knowledge of the organization of the cell wall components at the nanoscale, for example, scanning probe microscopy (SPM) techniques such as atomic force microscopy (AFM) and scanning near field optical microscopy (SNOM). In all SPM techniques the interactions between a very sharp tip and

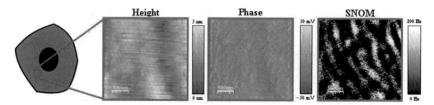

FIGURE 13.2   SNOM analysis of the secondary cell wall of beech wood. *(Adapted from Keplinger et al. (2014).)*

the surface are measured. AFM has been extensively used for the characterization of the architecture of plant cell walls, for example, studies on the pore size distribution or the organization of the microfibrils within the plant cell wall (Fahlen and Salmen, 2005). Another very promising area is the measurement of mechanical properties at a very small scale (compared to nanoindentation) with the potential of imaging these properties based on force-distance curves with AFM. There exist also alternative methods for determining mechanical properties, for example, resonant contact AFM or mode synthesizing AFM. Thereby a good sample preparation and a careful evaluation of the results to identify influences caused by indenter geometry or internal stresses are highly important (Burgert and Keplinger, 2013). AFM is an excellent tool in providing structural or mechanical information with high resolution, but the possibility to generate chemical information is limited. One way to overcome this limitation is the combination of AFM with other optical or spectroscopic methods to build so-called hybrid instruments, for example, SNOM (Flores and Toca-Herrera, 2009; Yarbrough et al., 2009). In SNOM, a gold- or silver-coated tip into which a laser is coupled in, with a subwavelength aperture at the end, is scanned in AFM feedback over the sample. As the light source is very close to the surface, the diffraction limit is not valid anymore and as a result a higher resolution can be achieved (for more information on SNOM see Lewis et al., 2003). Recently, SNOM was used for the first time to analyze wood secondary cell walls (Keplinger et al., 2014). In Fig. 13.2, the result of a SNOM scan of the secondary cell wall of beech wood is shown. Compared to the homogeneous surface of the height and phase image (classical AFM parameters) the SNOM image reveals a circumferential lamellar structure with a resolution of about 100 nm (limited by the size of the aperture), which was concluded to be a result of differences in lignin amount and/or structure (Keplinger et al., 2014).

Raman Spectroscopy of Wood Cell Walls

Raman spectroscopy is based on the inelastic scattering of a photon from a laser light source interacting with the molecule – leading to a polarization of the electron cloud and as a result a virtual energy state is formed before the reradiation takes place. For a more detailed description of the Raman technique and its theoretical background see Smith and Dent (2006). It should be mentioned that only a very small number of photons (one every 10^6–10^8 photons) is affected in Raman spectroscopy. This partly explains why the development of the technique, from the first experiments proving the Raman effect in 1927, until now where Raman spectroscopy is considered as one of the most promising analytical methods in plant cell wall characterization – took so long. The development of lasers (the Raman signal is proportional to the excitation power), powerful CCD detectors, and the increasing computational power brought together, contributed to crucial advances for the method. The main method used in wood cell wall characterization is the Raman-imaging approach – in short, spectra are recorded over a defined area with a predefined step size. The possible resolution is determined by the so-called diffraction limit depending on the numerical aperture of the objective and the wavelength of the laser (resolution down to 300 nm possible). Compared to traditional vibrational methods, such as IR spectroscopy, the resolution is increased by up to a factor of 10. The information gained in Raman spectroscopy is manifold, ranging from the visualization of wood components, for example, lignin, cellulose, hemicellulose, and pectin, the microfibril angle, or the effect of mechanical stress on the ultrastructure of cellulose fibrils (Agarwal, 2006; Gierlinger and Schwanninger, 2006; Gierlinger et al., 2006). In order to profit from the full potential of Raman spectroscopy, the data analysis is a crucial point and should be discussed here in more detail.

Raman Imaging Data Analysis – Univariate Versus Multivariate Methods

The easiest possibilities for the analysis of Raman images of wood cell walls based on thousands of spectra are so called univariate methods, for example, integration of specific wavelengths that are characteristic of certain components of the wood cell wall. Based on the results of the integration (area or height of the peak) the distribution of the components of interest can be visualized – in the case of wood cell walls, for example, the CC (cell corner), CML (compound middle lamella), S1, S2, S3 layers of the secondary cell wall. However, univariate methods suffer from some intrinsic drawbacks. On one hand, the intensity of the Raman peaks can be influenced by the focus on the sample and on the other hand, all components of the sample are probed at the same time resulting in overlapping bands. The different intensities due to the changing focal plane can be avoided by assuring a flat surface of the sample or by building ratios of two spectral regions (one has to keep in mind the different depth sensitivity of the peaks). Nevertheless, the problem of overlapping bands cannot be bypassed by univariate analytical methods (Gierlinger, 2014). In order to tackle this problem, multivariate methods are used since, in contrast to univariate procedures, a wider spectral range can be analyzed. There exist many different multivariate methods to deconvolute the hyperspectral dataset of Raman measurements (readers are referred to Geladi, 2003 and Gierlinger, 2014). One method used very recently for nonmodified and modified wood cell walls is vertex component analysis

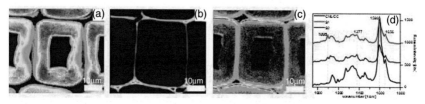

FIGURE 13.3 Vertex component analysis of spruce cell walls with the false-color images visualizing the secondary cell wall (a), CC and CML (b), and the S1 layer (c). (d) The corresponding EM spectra (d) are shown with the main peaks of the components marked.

(VCA). In VCA, so-called endmembers (EM) representing pure components are built, and each of the measured spectra is assigned to one of the EM. This assignment is visualized via false color images and as a result the distribution of each EM can be seen (Fig. 13.3).

The largest potential for progress in wood cell wall analysis can be foreseen in the field of combined SPM methods, such as tip-enhanced Raman spectroscopy, as they have the potential to generate full chemical information at a nanoscale without disintegration of the wood cell wall structure. Nevertheless, these new methods are still far away from being routine methods and remain highly challenging in terms of data acquisition, interpretation, and analysis. In addition, there is still room for improvement in terms of data analysis of conventional Raman measurements to gain further insights.

RECENT ADVANCES IN WOOD CELL AND CELL WALL MODIFICATION

The growing concerns for a more sustainable world are driving societies to look for alternatives to 100% oil-based solutions, especially in the development of materials to be used in everyday life. It appears that wood, as a renewable and sustainable material that has been used for thousands of years, could very well be a material of the future. Wood offers the advantage of being a light material in regard to its excellent mechanical properties, but it also has a few drawbacks that result from the transfer from the biological material function to the engineering application. To name a few, wood is vulnerable to fungi, degraded by UV light, and is sensitive to moisture fluctuations. These problems may be suppressed or reduced, according to a number of technologies that were developed in the past decades to obtain wood and wood-based materials with extended lifetime and wider range of utilization (Homan and Jorissen, 2004).

Driven by this demand, a large body of literature on the chemical modification of wood is now available, and research in this field remains an important topic nowadays (Hill, 2011). In particular, a majority of papers deal with dimensional stability issues resulting from the uptake of liquid water and moisture sorption by cell wall components. To reduce moisture fluctuations in the cell

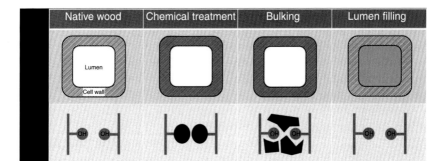

FIGURE 13.4 Overview of wood modification treatments. As depicted, chemical treatment and bulking treatment generate a new material, the new properties of which are a result of the cell wall alteration. In contrast, the lumen filling treatment is best described as a composite material with two distinct parts: wood and filling material.

wall, that is, to render wood more hydrophobic, there exist two main modes of action (see Fig. 13.4). In the bulking treatment, nanovoids in the cell wall are filled with a hydrophobic material to replace adsorbed water. In the modification treatment, the hydrophilic nature of wood is changed by reacting the hydroxyl groups in the cell wall with hydrophobic moieties. One prominent example of such a successful treatment is the acetylation process (Rowell and Dickerson, 2014). Filling up the lumen of wood cells may also be used as treatment, but in contrast to cell wall modification and bulking, the process is usually less efficient, since the properties of the wood cell wall itself are not altered while the density is substantially increased.

The treatment of wood cell wall requires the incorporation and stabilization of new material at the cell wall level. The abundant hydroxyl functionalities from cell wall components are therefore used to covalently attach small molecules and polymers, but new material may also be stabilized through physical interactions such as hydrogen bonding. Proper stabilization of the added material is crucial to obtain a long-lasting treatment.

Chemical Modification Toward Wood Materials with Extended Lifetime

Recently, further progress was made in the field of wood modification toward materials with improved properties. New achievements in this research field are based on novel methods using impregnation of organic and inorganic materials (or a combination of both) in the wood cell walls, addressing various issues such as dimensional stability and biodegradability in presence of fungi.

Recent reports have shown that the combination of organic materials with inexpensive inorganic fillers (such as clay) is suitable for wood modification. Work from Devi et al. (2012) shows that wood modified with nanoclay and styrene-acrylonitrile copolymer exhibits improved mechanical properties as well as a slightly higher resistance to fire. Wang et al. (2014) were able to insert clay into cell walls in an *in situ* synthesis process, which resulted in modified wood material with improved dimensional stability, compression strength, and surfaces hardness. As reported by Chen et al. (2014a), *in situ* gel polymerization of methylolurea and sodium silicate sol in the wood structure resulted in a reduced water uptake.

Silane-based formulations have attracted a lot of interest in wood research as well, and several systems were described recently. Organic–inorganic silanol products were used by Reinprecht et al. (2013) who showed that methyltripotassiumsilanol can be used to reduce water uptake and swelling, and also as a preservative treatment against fungal decay. By impregnating wood with amino-silicones of short chain lengths Ghosh et al. (2013) were able to render the wood more hydrophobic and improve the antiswelling efficiency by 60%. Similar hydrophobization results were achieved by Van Opdenbosch et al. (2013), using a mixture of phenyltrimethoxysilane (PTMOS) and phenyltriethoxysilane (PTEOS).

Using the formation of heartwood as a source of inspiration, Ermeydan et al. (2012) reported on the introduction of flavonoid molecules inside cell walls of spruce wood. Since the basic principles of the insertion of extractive molecules in the cell wall of the living tree are not yet fully understood or ready for a bioinspired technology transfer, in this work, a premodification of the cell wall via tosylation was necessary to render the wood more hydrophobic, and to provide accessible aromatic rings for stabilization of a flavonoid-like molecule (3-hydroxy flavone). Characterization with spectroscopic techniques (such as Raman) revealed the presence of the molecule stabilizing the cell walls, and the resulting wood showed reduced shrinking of the cell walls. The technique was further utilized to facilitate the penetration of styrene monomers in the cell wall, followed by *in situ* polymerization (see Fig. 13.5a) (Ermeydan et al., 2014a). The obtained wood-polystyrene material was shown to be more resistant to dimensional changes, due to a reduced water uptake upon immersion in water. Another approach, developed by Chen et al. (2014b), was designed to exploit natural wood features to improve impregnation of wood. The method, based on pulsating pressure cycles, allowed direct impregnation of green wood without the necessity for predrying, due to the efficient circulation of liquids (phenolic methylol urea) in the poplar sap wood.

In the field of hydrophobization, a further achievement was reported, with the production of a wood-based material using a biodegradable polyester (Ermeydan et al., 2014b). In this work, the hydroxyl functionalities from the cell wall components were used as coinitiator for the ring opening polymerization (ROP) of a cyclic ester, ε-caprolactone (see Fig. 13.5b). The approach reported yielded a completely renewable hydrophobic wood material, which represents a clear advantage in terms of recyclability. However, the technique certainly lacks modularity and hence, so far could only be utilized to produce hydrophobized wood.

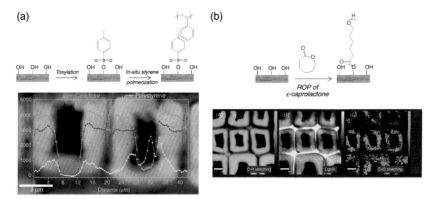

FIGURE 13.5 (a) Modification scheme according to a two-step procedure (tosylation and subsequent *in situ* polymerization) and Raman pictures and profiles showing the presence of polystyrene within the cell wall of spruce wood. (b) Synthesis route to fully biodegradable hydrophobized wood, and Raman images revealing the distribution of natural polymers in the cell wall: (A) cellulose, (B) lignin, and (C) poly(caprolactone). *(Adapted from (a) Ermeydan et al. (2014a) and (b) Ermeydan et al. (2014b).)*

Modification Toward Functional Wood-Based Materials

A key feature of many biological materials is their inherent hierarchical and anisotropic structure (Sanchez et al., 2005; Munch et al., 2008; Salmen and Burgert, 2009; Fratzl and Weinkamer, 2007), which can be used to implement new functionalities that go beyond their naturally given attributes. Such hierarchical biological tissues offer an ideal platform for common template-directed approaches, such as biomimetic mineralization, self-organization of nanostructures, and *in situ* polymerization (Weiner and Addadi, 1997; Mann, 1993; Aizenberg et al., 2005; Ikkala and ten Brinke, 2002; Cabane et al., 2014; Trey et al., 2012b; Liu et al., 2013). In wood, the rigid cellulose nanofibrils embedded in an amorphous organic matrix constitute a stable multilayered cell wall composite at the nanoscale. At the microscale, fibers with a high aspect ratio result in a high degree of unidirectional anisotropy of the wood structure. Following up on the aforementioned modification principles, wood has been equipped with several unprecedented functionalities, for example, antistatic (Trey et al., 2012b), magnetic (Merk et al., 2014; Oka et al., 2002), or superhydrophobic properties (Fu et al., 2012b). Hence, smart wood-based composites promise novel perspectives in materials science outside of conventional wood applications (Ansell, 2013).

Many applications require engineered composites with high specific surface areas and tailored grains, such as heat insulators, filters, absorbers, catalyst substrates, battery supports, or medical implants (Zollfrank et al., 2004; Zhu et al., 2013; Greil et al., 1998). Whereas chemically modified wood retains most of its natural attributes, such as high fracture strength and toughness, biomorphic ceramics replicate the artful interconnected pore network with inorganic matter while the organic matrix is carbonized (Paris et al., 2013).

Hence, besides aiming at making wood more inert for common engineering applications, one can look at wood also from a different perspective, and try to make use of the wood multiscale hierarchical structure, in order to develop a new class of renewable functional materials by applying *in situ* polymerization principles in a top–down approach. For this, it is important to open new perspectives in the field of wood modification and find chemical treatment alternatives that are more modular, which means that they can be used to render the wood more hydrophobic but also to insert other functionalities. Typically, polymerization techniques provide access to many functional materials in a bottom–up approach, and are also largely used to graft polymer chains on a variety of surfaces. As an example, nearly all products derived from wood disintegration have been used as macroinitiators for further polymerization (Carlsson et al., 2012; Hilburg et al., 2014; Kim and Kadla, 2010; Meng et al., 2009; Tastet et al., 2011). However, to date, there are only a handful of papers dealing with the introduction of functional materials in the solid wood structure.

Trey et al. (2012a) described the *in situ* polymerization of aniline in beech veneers, to obtain semiconductive wood materials. The veneers obtained show an uncommon conductivity brought by the polyaniline chains located in the cell walls. In addition, the modified veneers showed increased resistance against fire, and the group reported on the decay resistance of the polyaniline-veneers (Treu et al., 2014). A possible alternative reported recently, consists in the use of the wood structure as a scaffold to perform surface-initiated polymerization (Fig. 13.6) (Cabane et al., 2014). In the first step, an azo-based radical initiator modified with reactive acyl chloride groups was attached to the wood hydroxyl groups. The anchored initiator was then used to grow polymer chains from the cell wall itself. The reaction was proven successful with styrene monomers, yielding traditional hydrophobized wood materials. The authors also demonstrated that more functional monomers can be polymerized in wood. As a proof of concept, two pH-responsive acrylate monomers were used: acrylic acid and dimethylaminoethyl methacrylate (AA and DMAEMA). A simple experiment showed that cubes modified with those pH-responsive polymers exhibit opposite water absorption capacities, according to the pH of the water.

Finally, it is also worth mentioning the pioneering works on the implementation of atom transfer radical polymerization (ATRP) on wood materials. By grafting methyl methacrylate (MMA) to the wood surface, Fu et al. (2012a) developed wood products with highly hydrophobic surfaces. The technique was used by Yu et al. (2013) to grow poly(DMAEMA) brushes on wood surfaces. As reported, the obtained positively charged surface had a clear biocidal effect against *Escherichia coli* bacteria. Although in both studies the authors aimed at wood surface functionalization, the future may show that ATRP, a versatile polymerization technique, can be used to functionalize wood cell walls as well, and would therefore offer one more synthesis tool to perform chemical modification in wood.

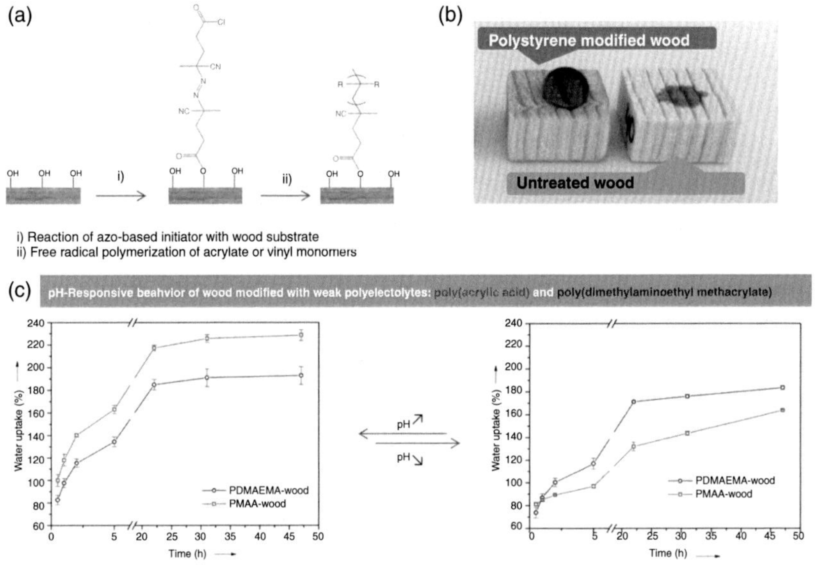

FIGURE 13.6 (a) New route to functional wood materials. (b) Water droplet on wood modified with polystyrene showing the hydrophobic character of the new material. (c) Water uptake of wood samples modified with different polyelectrolytes. As expected, the total water uptake shows an opposite trend depending on the nature of the added polymer in basic or acidic media. *(After Cabane et al. (2014).)*

WOOD FUNCTIONALIZATION TOWARD ACTUATION

In this last subchapter we intend to introduce two functionalization principles of the wood structure, which have in common that they both utilize the intrinsic anisotropy of wood to achieve an actuation and movement of wood by an external stimulus. In the first example, the insertion of iron oxide into the wood structure allows to manipulate the wood-hybrid samples in a magnetic field. In the second example, no prior functionalization but just a thoughtful combination of wooden elements to use ambient relative humidity changes to drive movements of wood elements, based on the anisotropic swelling and shrinking of cell walls, will be introduced.

Magnetic Hybrid-Wood Materials

Creating anisotropic and stimuli-responsive engineered materials by bottom–up approaches is highly challenging, whereas the inherent structural architecture of wood can be used as a scaffold for a three-dimensional assembly of colloidal iron oxide prepared *in situ* inside bulk wood (Merk et al., 2014). By classical co-precipitation of ferric and ferrous salts in aqueous media with an alkaline solution (Laurent et al., 2008; Massart, 1981), a simple and affordable modification route was applied to wood, in order to implement magnetic functionalities into

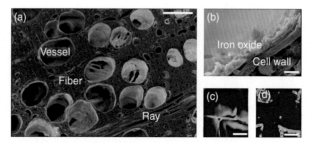

FIGURE 13.7   (a and b) Cryogenic FE-SEM images of beech cross-section collected in the secondary electron mode. (c and d) Raman mappings of 20 μm microtome cross-sections show the distribution of the aromatic cell wall polymer lignin (1707–1540 cm^{-1}) in spruce tracheids and colloidal iron oxide (825–566 cm^{-1}) attached to the inner lumina walls. Scale bars correspond to 500 μm (a), 500 nm (b), and 10 μm (c, d) *(After Merk et al. (2014).)*

the wood superstructure with an only slightly increased density. Microstructural imaging using cryogenic scanning electron microscopy and spectral Raman mapping (Fig. 13.7) was used to fathom the origins of magnetic anisotropy on various levels of hierarchy. Images of wood cross-sections show thin layers of ferrite particles firmly attached to the inner lumina walls of the axially oriented structural wood elements; thus, a simple lumen filling was largely avoided. Based on XRD and Raman phase analysis, the precipitated iron oxide mineral was identified as a mixture of the spinel phases magnetite and maghemite.

Direction-dependent magnetic hysteresis loops (Fig. 13.8a) reveal a predominant magnetization along the wood fibers (longitudinal). The shape of the hysteresis curves indicates the formation of small, multidomain ferrite particles carrying a net magnetization only in the presence of an external field, so that the wood blocks can be lifted by a permanent magnet (Fig. 13.8b). The slope of the hysteresis curve in the low-field regime (inset in Fig. 13.8a) is sensitive

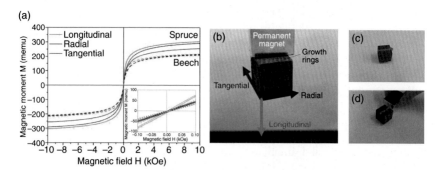

FIGURE 13.8   (a) Magnetic hysteresis loops of magnetic spruce (continuous line) and beech (dashed line) and a magnification of the low-field regime. (b) Magnetic spruce cube lifted by a household magnet. The arrows indicate the main wood directions. (c and d) Flipping of a magnetic wood block mounted on a rotary, orienting itself with the wood fiber (longitudinal) direction parallel to the magnetic flux. *(After Merk et al. (2014).)*

to the ferromagnetic fraction of the particles. Exposed to a magnetic flux, the magnetic wood specimens experience a torque (Fig. 13.8c and d) giving rise to complex motion such as flip, drag, alignment, (re)orientation, and actuation. Easy processing, handling and up-scaled production allows for production in arbitrary shapes and sizes up to macroscopic scales; thus, we envisage applications of these highly sophisticated biobased composites as magnetic switches and actuated elements.

Bioinspired Hydro-Actuated Wood

Although plants are sessile organisms, an astonishing variety of movements have evolved in the plant kingdom (Skotheim and Mahadevan, 2005). Several organs of the plant perform nastic (undirected) and tropic (directed) movements in response to external stimuli such as touch (Braam, 2005), temperature, and relative humidity. The movements can span several orders of magnitude in size and time, considering very fast movements like the snapping of the Venus flytrap (Forterre et al., 2005) or the slow opening of pine cones (Dawson et al., 1997). Leaving aside movements that depend on growth, such as phototropism, the tremendous diversity of movements is based on just two principles. Water is exchanged within living plant cells or between the hygroscopic plant cell walls and the ambient air (Skaar, 1988; Skotheim and Mahadevan, 2005). In the first case the turgor of plant cells is adapted and changed, which leads to dimensional changes of the affected cells. This principle requires living cells and an active metabolism. In contrast, the exchange of water between plant cell walls and the surrounding air is of a passive nature and the actuation, which is based on this principle, is independent of a living organism. These passive movements are solely driven by an alternating ambient relative humidity, which leads to a time-dependent change of the moisture content of the cell wall, which in turn results in dimensional changes of the cells and the tissues.

Hierarchical structuring and the anisotropy of cell walls and tissues transform these water driven dimensional changes of single cells into the drivers of complex actuation patterns in different plant organs and plant species (Elbaum et al., 2007; Harrington et al., 2011). The anisotropic structure of the plant cell wall with parallel orientation of the stiff and virtually inextensible cellulose microfibrils leads to a pronounced swelling anisotropy at the level of the plant cell wall. Swelling and shrinking almost exclusively occurs perpendicular to the cellulose microfibrils (Skaar, 1988). Due to the hierarchical structuring, the swelling anisotropy of the plant cell wall is up-scaled to the tissue level and enables directed dimensional changes. By smart structuring and layering of tissues, these directed dimensional changes are then transformed into complex movements. If two layers with perpendicular fiber orientation are connected, bending or twisting of this bilayer is generated in case of dimensional changes within the single layers. The well-known humidity-driven opening and closing of pine cones relies on this principle (Fig. 13.1a) as each scale consists of a

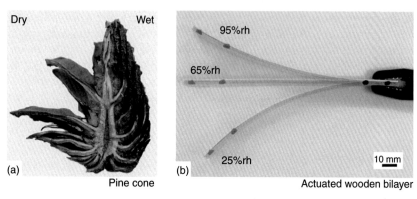

Dry   Wet   95%rh

                                             65%rh

                                             25%rh   10 mm

(a)   (b)
        Pine cone   Actuated wooden bilayer

FIGURE 13.9   Actuation of pine cone scales and of wooden bilayers. (a) Composite image of a wet and a dry pinecone. The scales represent bilayers with differential orientation of the cellulose microfibrils and bend upon wetting and drying. (b) Composite image of a wooden bilayer with perpendicular orientation of the fibers at different relative humidity. rh, relative humidity.

bilayer with almost perpendicular cellulose microfibril orientations in the single layers (Dawson et al., 1997).

Hence, plant movements can be used as a source of inspiration for developing smart, actuated materials (Burgert and Fratzl, 2009; Erb et al., 2013; Reyssat and Mahadevan, 2009). Especially those movements that are independent of a living organism, such as the actuation of the scales of the pine cones are suitable for a biomimetic transfer to technical application.

The bilayer principle has been frequently employed and adapted for technical application using various classes of materials such as metals, polymers, hydrogels, and ceramics. These bilayers may be responsive to temperature, relative humidity (Erb et al., 2013), pH (Bassik et al., 2010), light (Viola et al., 2010), or electric fields (Bi et al., 2013). Hereby actuators were successfully developed down to the micrometer size, which is of great interest for biomedical applications or for the semiconductor industry.

Following the bioinspired actuation principle, the dimensional instability of wood upon changes in relative humidity, which is regarded as one of the major drawbacks for the use of timber in construction, can be utilized to establish directional wood element movements (Menges and Reichert, 2012). Its anisotropy and high stiffness facilitates the fabrication of bilayers with differential orientation of the fibers. Thus, macroscopic, reversible actuation is generated upon changes of relative humidity (Fig. 13.9).

By using wood, the fabrication of anisotropic bilayers is greatly simplified and up-scaling is not limited by the manufacturing process or the lack of mechanically self-supporting capabilities. Amplitude and velocity of the actuation can be tuned by adapting the thicknesses of the single wood layers. The curvature and thus the amplitude of actuation can be calculated using the theory of bending of bimetals, which has been developed by Timoshenko

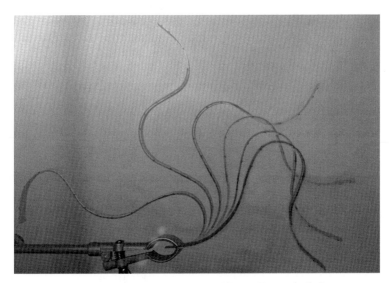

FIGURE 13.10 Actuation of a steam-bent wooden bilayer. The composite image.

(1925). By steam-bending, more complex shapes can be created and long elements that generate larger motion amplitudes can be packed into a small space. Figure 13.10 shows a rotation of the tip of a steam-bent bilayer close to 360° within 3 h upon a change in relative humidity of 50%.

Based on the fundamental understanding of these processes, such wooden bilayers can be integrated in adaptive architectural and constructional elements. In outdoor applications the actuation of such elements would be driven by the daily changes of relative humidity. Relative humidity is typically highest at sunrise going down until late afternoon and rising again at night. Long-term tests with wooden bilayers in the field in full weathering condition have revealed sufficient long-term stability of the actuation. Whereas the daily amplitude of actuation is variable as it depends on the respective change in relative humidity, the specific amplitude (related to the difference in relative humidity) decreased only slightly over time despite the visible impacts of weathering such as cracking and splitting. Ultimately, the actuation is driven and controlled by the sun. As the actuator and sensor are intrinsically incorporated in the wooden bilayer, it thus represents a very simple and efficient motor.

CONCLUSIONS AND OUTLOOK

We conclude that nature provides an inexhaustible reservoir for transferring mechanisms and principles for the development of advanced bioinspired materials. Combining this approach with wood science by using methodological tools from nanotechnology and polymer chemistry allows for designing advanced wood materials for given and novel applications. Wood is a peculiar biomaterial in this regard as it is a source of bioinspiration due to its function in the living tree as well as an engineering material that can be optimized by applying the mentioned principles. However, certainly, one would unnecessarily limit the potential of biomimetic approaches to improve wood materials in concentrating exclusively on using wood as a biological role model.

A great potential arises from the common principle of structural hierarchy in nature since it provides an inherent and tunable pattern for the up-scaling of nanotechnological treatments. In terms of wood, it seems promising to utilize the nanostructure of the wood cell wall to make use of it as a strong and rigid scaffold for further functionalization. Related approaches have been followed by using wood as a template for the fabrication of biomorphic ceramics to determine the nano- and microstructure based on the wood features. However, here the concept is to retain the cell wall or the cellulose scaffold, as with bone, which is formed by a mineralization of the collagen matrix. In analogy, we are currently inserting minerals in the cell wall to change the material properties and develop advanced wood-hybrid materials. Another option is the insertion of nanoparticles, which so far turns out to be difficult since the nanoporosity of the native cell wall would require very small particles. A useful solution could be the chemical degradation of matrix polymers to retain just the more porous cellulose scaffold for functionalization. However, in the future, chemical posttreatments may even become unnecessary. In case of a consolidation of nanotechnology and biotechnology, trees could directly produce wood that is more suitable for subsequent functionalization, for instance due to reduced lignin contents or other modifications that open the micro- and nanostructure of the cell wall scaffold.

ACKNOWLEDGMENTS

We thank the SNF for financial support, in the frameworks of the NFP66 and the R'Equip program. The financial support of the Wood Materials Science group by BAFU (Bundesamt für Umwelt) and Lignum, Switzerland is gratefully acknowledged.

REFERENCES

Agarwal, U.P., 2006. Raman imaging to investigate ultrastructure and composition of plant cell walls: distribution of lignin and cellulose in black spruce wood (*Picea mariana*). Planta 224, 1141–1153.

Aizenberg, J., Weaver, J.C., Thanawala, M.S., Sundar, V.C., Morse, D.E., Fratzl, P., 2005. Skeleton of *Euplectella* sp.: structural hierarchy from the nanoscale to the macroscale. Science 309, 275–278.

Altaner, C.M., Jarvis, M.C., 2008. Modelling polymer interactions of the 'molecular Velcro' type in wood under mechanical stress. J. Theor. Biol. 253, 434–445.

Alves, A., Schwanninger, M., Pereira, H., Rodrigues, J., 2006. Analytical pyrolysis as a direct method to determine the lignin content in wood – part 1: comparison of pyrolysis lignin with Klason lignin. J. Anal. Appl. Pyrolysis 76, 209–213.

Ansell, M.P., 2013. Multi-functional nano-materials for timber in construction. In: Proceedings of ICE - Construction Materials, 248–256.

Bassik, N., Abebe, B.T., Laflin, K.E., Gracias, D.H., 2010. Photolithographically patterned smart hydrogel based bilayer actuators. Polymer 51, 6093–6098.

Bi, H., Yin, K., Xie, X., Zhou, Y., Wan, S., Banhart, F., Sun, L., 2013. Microscopic bimetallic actuator based on a bilayer of graphene and graphene oxide. Nanoscale 5, 9123–9128.

Bordel, D., Putaux, J.L., Heux, L., 2006. Orientation of native cellulose in an electric field. Langmuir 22, 4899–4901.

Braam, J., 2005. In touch: plant responses to mechanical stimuli. New Phytol. 165, 373–389.

Brown, J.R.M., 1996. The biosynthesis of cellulose. J. M. S. - Pure Appl. Chem. A33, 1345–1373.

Burgert, I., Fratzl, P., 2009. Actuation systems in plants as prototypes for bioinspired devices. Philos. Trans. R. Soc. A 367, 1541–1557.

Burgert, I., Keplinger, T., 2013. Plant micro- and nanomechanics: experimental techniques for plant cell-wall analysis. J. Exp. Bot. 64, 4635–4649.

Cabane, E., Keplinger, T., Merk, V., Hass, P., Burgert, I., 2014. Renewable and functional wood materials by grafting polymerization within cell walls. ChemSusChem 7, 1020–1025.

Carlsson, L., Utsel, S., Wågberg, L., Malmström, E., Carlmark, A., 2012. Surface-initiated ring-opening polymerization from cellulose model surfaces monitored by a quartz crystal microbalance. Soft Matter 8, 512.

Chen, H., Miao, X., Feng, Z., Pu, J., 2014a. *In situ* polymerization of phenolic methylolurea in cell wall and induction of pulse–pressure impregnation on green wood. Ind. Eng. Chem. Res. 53, 9721–9727.

Chen, H., Lang, Q., Bi, Z., Miao, X., Li, Y., Pu, J., 2014b. Impregnation of poplar wood (*Populus euramericana*) with methylolurea and sodium silicate sol and induction of *in-situ* gel polymerization by heating. Holzforschung 68, 45–52.

Cosgrove, D.J., 2005. Growth of the plant cell wall. Nat. Rev. Mol. Cell Biol. 6, 850–861.

Dawson, J., Vincent, J.F.V., Rocca, A.M., 1997. How pine cones open. Nature 390, 668–1668.

Devi Rashmi, R., Mandal, M., Maji Tarun, K., 2012. Physical properties of simul (red-silk cotton) wood (*Bombax ceiba* L.) chemically modified with styrene acrylonitrile co-polymer and nanoclay. Holzforschung 66, 365.

Donaldson, L.A., Radotic, K., 2013. Fluorescence lifetime imaging of lignin autofluorescence in normal and compression wood. J. Microsc. 251, 178–187.

Ebeling, T., Paillet, M., Borsali, R., Diat, O., Dufresne, A., Cavaille, J.Y., Chanzy, H., 1999. Shear-induced orientation phenomena in suspensions of cellulose microcrystals, revealed by small angle X-ray scattering. Langmuir 15, 6123–6126.

Eichhorn, S.J., Dufresne, A., Aranguren, M., Marcovich, N.E., Capadona, J.R., Rowan, S.J., Weder, C., Thielemans, W., Roman, M., Renneckar, S., et al., 2010. Review: current international research into cellulose nanofibres and nanocomposites. J. Mater. Sci. 45, 1–33.

Elbaum, R., Zaltzman, L., Burgert, I., Fratzl, P., 2007. The role of wheat awns in the seed dispersal unit. Science 316, 884–886.

Emons, A.M.C., Mulder, B.M., 1998. The making of the architecture of the plant cell wall: how cells exploit geometry. Proc. Nat. Acad. Sci. USA 95, 7215–7219.

Erb, R.M., Sander, J.S., Grisch, R., Studart, A.R., 2013. Self-shaping composites with programmable bioinspired microstructures. Nat. Commun., 4.

Ermeydan, M.A., Cabane, E., Masic, A., Koetz, J., Burgert, I., 2012. Flavonoid insertion into cell walls improves wood properties. ACS Appl. Mater. Interfaces 4, 5782–5789.

Ermeydan, M.A., Cabane, E., Gierlinger, N., Koetz, J., Burgert, I., 2014a. Improvement of wood material properties via *in situ* polymerization of styrene into tosylated cell walls. RSC Adv. 4, 12981–12988.

Ermeydan, M.A., Cabane, E., Hass, P., Koetz, J., Burgert, I., 2014b. Fully biodegradable modification of wood for improvement of dimensional stability and water absorption properties by poly([ε]-caprolactone) grafting into the cell walls. Green Chem. 16, 3313–3321.

Fahlen, J., Salmen, L., 2005. Pore and matrix distribution in the fiber wall revealed by atomic force microscopy and image analysis. Biomacromolecules 6, 433–438.

Flores, S.M., Toca-Herrera, J.L., 2009. The new future of scanning probe microscopy: combining atomic force microscopy with other surface-sensitive techniques, optical and microscopy fluorescence techniques. Nanoscale 1, 40–49.

Forterre, Y., Skotheim, J.M., Dumais, J., Mahadevan, L., 2005. How the Venus flytrap snaps. Nature 433, 421–425.

Fratzl, P., Weinkamer, R., 2007. Nature's hierarchical materials. Prog. Mater. Sci. 52, 1263–1334.

Fu, Y., Li, G., Yu, H., Liu, Y., 2012a. Hydrophobic modification of wood via surface-initiated ARGET ATRP of MMA. Appl. Surf. Sci. 258, 2529–2533.

Fu, Y.C., Yu, H.P., Sun, Q.F., Li, G., Liu, Y.Z., 2012b. Testing of the superhydrophobicity of a zinc oxide nanorod array coating on wood surface prepared by hydrothermal treatment. Holzforschung 66, 739–744.

Geladi, P., 2003. Chemometrics in spectroscopy. Part 1. Classical chemometrics. Spectrochim. Acta Part B 58, 767–782.

Ghosh, S.C., Militz, H., Mai, C., 2013. Modification of *Pinus sylvestris* L. wood with quat- and amino-silicones of different chain lengths. Holzforschung 67, 421–427.

Gibson, L.J., 2012. The hierarchical structure and mechanics of plant materials. J. R. Soc. Interface 9, 2749–2766.

Gierlinger, N., 2014. Revealing changes in molecular composition of plant cell walls on the micron-level by Raman mapping and vertex component analysis (VCA). Front. Plant Sci. 5, 306.

Gierlinger, N., Schwanninger, M., 2006. Chemical imaging of poplar wood cell walls by confocal Raman microscopy. Plant Physiol. 140, 1246–1254.

Gierlinger, N., Schwanninger, M., Reinecke, A., Burgert, I., 2006. Molecular changes during tensile deformation of single wood fibers followed by Raman microscopy. Biomacromolecules 7, 2077–2081.

Gierlinger, N., Keplinger, T., Harrington, M., 2012. Imaging of plant cell walls by confocal Raman microscopy. Nat. Protoc. 7, 1694–1708.

Greil, P., Lifka, T., Kaindl, A., 1998. Biomorphic cellular silicon carbide ceramics from wood: I. Processing and microstructure. J. Eur. Ceram. Soc. 18, 1961–1973.

Harrington, M.J., Razghandi, K., Ditsch, F., Guiducci, L., Rüggeberg, M., Dunlop, J.W.C., Fratzl, P., Neinhuis, C., Burgert, I., 2011. Origami-like unfolding of hydro-actuated ice plant seed capsules. Nat. Commun. 2, 337.

Hilburg, S.L., Elder, A.N., Chung, H., Ferebee, R.L., Bockstaller, M.R., Washburn, N.R., 2014. A universal route towards thermoplastic lignin composites with improved mechanical properties. Polymer 55, 995–1003.

Hill, C.A.S., 2011. Wood modification. Bioresources 6, 918–919.

Homan, W.J., Jorissen, A.J.M., 2004. Wood modification developments. Heron 49, 361–386.

Ikkala, O., ten Brinke, G., 2002. Functional materials based on self-assembly of polymeric supramolecules. Science 295, 2407–2409.

Keckes, J., Burgert, I., Fruhmann, K., Muller, M., Kolln, K., Hamilton, M., Burghammer, M., Roth, S.V., Stanzl-Tschegg, S., Fratzl, P., 2003. Cell-wall recovery after irreversible deformation of wood. Nat. Mater. 2, 810–814.

Keplinger, T., Konnerth, J., Aguie-Beghin, V., Rueggeberg, M., Gierlinger, N., Burgert, I., 2014. A zoom into the nanoscale texture of secondary cell walls. Plant Methods 10, 1.

Kim, Y.S., Kadla, J.F., 2010. Preparation of a thermoresponsive lignin-based biomaterial through atom transfer radical polymerization. Biomacromolecules 11, 981–988.

Koch, G., Kleist, G., 2001. Application of scanning UV microspectrophotometry to localise lignins and phenolic extractives in plant cell walls. Holzforschung 55, 563–567.

Köhler, L., Spatz, H.C., 2002. Micromechanics of plant tissues beyond the linear-elastic range. Planta 215, 33–40.

Korosec, R.C., Lavric, B., Rep, G., Pohleven, F., Bukovec, P., 2009. Thermogravimetry as a possible tool for determining modification degree of thermally treated Norway spruce wood. J. Therm. Anal. Calorim. 98, 189–195.

Kvien, I., Oksman, K., 2007. Orientation of cellulose nanowhiskers in polyvinyl alcohol. Appl. Phys. A 87, 641–643.

Laurent, S., Forge, D., Port, M., Roch, A., Robic, C., Elst, L.V., Muller, R.N., 2008. Magnetic iron oxide nanoparticles: synthesis, stabilization, vectorization, physicochemical characterizations, and biological applications. Chem. Rev. 108, 2064–2110.

Lee, H., Scherer, N.F., Messersmith, P.B., 2006. Single-molecule mechanics of mussel adhesion. P Proc. Nat. Acad. Sci. USA 103, 12999–13003.

Lewis, A., Taha, H., Strinkovski, A., Manevitch, A., Khatchatouriants, A., Dekhter, R., Ammann, E., 2003. Near-field optics: from subwavelength illumination to nanometric shadowing. Nat. Biotechnol. 21, 1377–1386.

Liu, Y.D., Goebl, J., Yin, Y.D., 2013. Templated synthesis of nanostructured materials. Chem. Soc. Rev. 42, 2610–2653.

Magel, E.A., Drouet, A., Claudot, A.C., Ziegler, H., 1991. Formation of heartwood substances in the stemwood of *Robinia pseudoacacia* L. 1. Distribution of phenylalanine ammonium lyase and chalcone synthase across the trunk. Trees 5, 203–207.

Mai, C., Militz, H., 2004. Modification of wood with silicon compounds. Treatment systems based on organic silicon compounds – a review. Wood Sci. Technol. 37, 453–461.

Mann, S., 1993. Molecular tectonics in biomineralization and biomimetic materials chemistry. Nature 365, 499–505.

Mansfield, S.D., Kim, H., Lu, F., Ralph, J., 2012. Whole plant cell wall characterization using solution-state 2D NMR. Nat. Protoc. 7, 1579–1589.

Massart, R., 1981. Preparation of aqueous magnetic liquids in alkaline and acidic media. IEEE Trans. Magn. 17, 1247–1248.

Meng, T., Gao, X., Zhang, J., Yuan, J., Zhang, Y., He, J., 2009. Graft copolymers prepared by atom transfer radical polymerization (ATRP) from cellulose. Polymer 50, 447–454.

Menges, A., Reichert, S., 2012. Material capacity: embedded responsiveness. Architectural Design, 52–59.

Merk, V., Chanana, M., Gierlinger, N., Hirt, A.M., Burgert, I., 2014. Hybrid wood materials with magnetic anisotropy dictated by the hierarchical cell structure. ACS Appl. Mater. Interfaces.

Munch, E., Launey, M.E., Alsem, D.H., Saiz, E., Tomsia, A.P., Ritchie, R.O., 2008. Tough, bio-inspired hybrid materials. Science 322, 1516–1520.

Mutwil, M., Debolt, S., Persson, S., 2008. Cellulose synthesis: a complex complex. Curr. Opin. Plant Biol. 11, 252–257.

Neinhuis, C., Barthlott, W., 1997. Characterization and distribution of water-repellent, self-cleaning plant surfaces. Ann. Bot. 79, 667–677.

Oka, H., Hojo, A., Seki, K., Takashiba, T., 2002. Wood construction and magnetic characteristics of impregnated type magnetic wood. J. Magn. Magn. Mater. 239, 617–619.

Olsson, R.T., Samir, M.A.S.A., Salazar-Alvarez, G., Belova, L., Strom, V., Berglund, L.A., Ikkala, O., Nogues, J., Gedde, U.W., 2010. Making flexible magnetic aerogels and stiff magnetic nano-paper using cellulose nanofibrils as templates. Nat. Nanotechnol. 5, 584–588.

Paredez, A.R., Somerville, C.R., Ehrhardt, D.W., 2006. Visualization of cellulose synthase demonstrates functional association with microtubules. Science 312, 1491–1495.

Paris, O., Fritz-Popovski, G., Van Opdenbosch, D., Zollfrank, C., 2013. Recent progress in the replication of hierarchical biological tissues. Adv. Funct. Mater. 23, 4408–4422.

Reinprecht, L., Panek, M., Dankova, J., Murinova, T., Mec, P., Plevova, L., 2013. Performance of methyl-tripotassiumsilanol treated wood against swelling in water, decay fungi and moulds. Wood Res. (Bratislava, Slovakia) 58, 511–520.

Reyssat, E., Mahadevan, L., 2009. Hygromorphs: from pine cones to biomimetic bilayers. J. R. Soc. Interface 6, 951–957.

Rowell, R.M., Dickerson, J.P., 2014. Acetylation of wood. Deterioration and protection of sustainable biomaterials. Am. Chem. Soc., 301–327.

Salmen, L., Burgert, I., 2009. Cell wall features with regard to mechanical performance. A review COST Action E35 2004-2008: wood machining – micromechanics and fracture. Holzforschung 63, 121–129.

Sanchez, C., Arribart, H., Giraud Guille, M.M., 2005. Biomimetism and bioinspiration as tools for the design of innovative materials and systems. Nat. Mater. 4, 277–288.

Skaar, C., 1988. In: Timmel, T.E. (Ed.), Wood-Water Relation. Springer Verlag, Berlin Heidelberg.

Skotheim, J.M., Mahadevan, L., 2005. Physical limits and design principles for plant and fungal movements. Science 308, 1308–1310.

Smith, E., Dent, G., 2006. Modern Raman Spectroscopy – A Practical Approach. John Wiley & Sons, Chichester, England, UK.

Somerville, C., Bauer, S., Brininstool, G., Facette, M., Hamann, T., Milne, J., Osborne, E., Paredez, A., Persson, S., Raab, T., et al., 2004. Toward a systems approach to understanding plant cell walls. Science 306, 2206–2211.

Speck, T., Burgert, I., 2011. Plant stems: functional design and mechanics. Annu. Rev. Mater. Res. 41, 169–193.

Tastet, D., Save, M., Charrier, F., Charrier, B., Ledeuil, J.-B., Dupin, J.-C., Billon, L., 2011. Functional biohybrid materials synthesized via surface-initiated MADIX/RAFT polymerization from renewable natural wood fiber: grafting of polymer as non leaching preservative. Polymer 52, 606–616.

Taylor, A.M., Gartner, B.L., Morrell, J.J., 2002. Heartwood formation and natural durability – a review. Wood Fiber Sci. 34, 587–611.

Timoshenko, S., 1925. Analysis of bi-metal thermostats. J. Opt. Soc. Am. Rev. Sci. Instrum. 11, 233–255.

Treu, A., Bardage, S., Johansson, M., Trey, S., 2014. Fungal durability of polyaniline modified wood and the impact of a low pulsed electric field. Int. Biodeterior. Biodegrad. 87, 26–33.

Trey, S., Jafarzadeh, S., Johansson, M., 2012a. In situ polymerization of polyaniline in wood veneers. ACS Appl. Mater. Interfaces 4, 1760–1769.

Trey, S., Jafarzadeh, S., Johansson, M., 2012b. In situ polymerization of polyaniline in wood veneers. ACS Appl. Mater. Interfaces 4, 1760–1769.

Tsuchikawa, S., Schwanninger, M., 2013. A review of recent near-infrared research for wood and paper (Part 2). Appl. Spectrosc. Rev. 48, 560–587.

Van Opdenbosch, D., Doerrstein, J., Klaithong, S., Kornprobst, T., Plank, J., Hietala, S., Zollfrank, C., 2013. Chemistry and water-repelling properties of phenyl-incorporating wood composites. Holzforschung 67, 931–940.

Viola, E.A., Levitsky, I.A., Euler, W.B., 2010. Kinetics of photoactuation in single wall carbon nanotube-nafion bilayer composite. J. Phys. Chem. C 114, 20258–20266.

Wang, W., Zhu, Y., Cao, J., Liu, R., 2014. Improvement of dimensional stability of wood by *in situ* synthesis of organo-montmorillonite: preparation and properties of modified Southern pine wood. Holzforschung 68, 29–36.

Weiner, S., Addadi, L., 1997. Design strategies in mineralized biological materials. J. Mater. Chem. 7, 689–702.

Yarbrough, J.M., Himmel, M.E., Ding, S.Y., 2009. Plant cell wall characterization using scanning probe microscopy techniques. Biotechnol. Biofuels 2, 17.

Yu, H., Fu, Y., Li, G., Liu, Y., 2013. Antimicrobial surfaces of quaternized poly[(2-dimethyl amino) ethyl methacrylate] grafted on wood via ARGET ATRP. Holzforschung 67, 455–461.

Zhu, H., Jia, Z., Chen, Y., Weadock, N., Wan, J., Vaaland, O., Han, X., Li, T., Hu, L., 2013. Tin anode for sodium-ion batteries using natural wood fiber as a mechanical buffer and electrolyte reservoir. Nano Lett. 13, 3093–3100.

Zimmermann, T., Pohler, E., Geiger, T., 2004. Cellulose fibrils for polymer reinforcement. Adv. Eng. Mater. 6, 754–761.

Zimmermann, T., Thommen, V., Reimann, P., Hug, H.J., 2006. Ultrastructural appearance of embedded and polished wood cell walls as revealed by atomic force microscopy. J. Struct. Biol. 156, 363–369.

Zollfrank, C., Kladny, R., Sieber, H., Greil, P., 2004. Biomorphous SiOC/C-ceramic composites from chemically modified wood templates. J. Eur. Ceram. Soc. 24, 479–487.

Chapter 14

Biological, Anatomical, and Chemical Characteristics of Bamboo

Benhua Fei, Zhimin Gao, Jin Wang, Zhijia Liu
International Centre for Bamboo and Rattan, Beijing, China

Chapter Outline

Bamboo is a vital component of plant and forest resources in the world, including over 70 genera and 1200 species. There are 220,000 km^2 of bamboo resources in total and the annual production of bamboo is estimated to be 15–20 million tons (Jiang, 2002). Bamboo grows more rapidly than trees and starts to yield within 4 to 5 years of planting. They can be selectively harvested annually and nondestructively (Devi et al., 2007). Bamboo resources are mainly distributed in the tropical and subtropical zones, and a few in the temperate and frigid zones. Table 14.1 shows some of the major bamboo producers: China, India, Japan, Myanmar, Thailand, Bangladesh, Cambodia, Vietnam, Laos, Malaysia, Indonesia, and the Philippines.

Secondary Xylem Biology. http://dx.doi.org/10.1016/B978-0-12-802185-9.00014-0

TABLE 14.1 Species, Genera, and Area of Bamboo Forest in Some Countries

Country	Species	Genera	Area (km^2)
China	500	40	61,586
India	136	23	108,630
Myanmar	100	17	8,950
Thailand	60	13	8,100
Bangladesh	30	13	863
Cambodia	10	4	2,870
Vietnam	101	15	10,000
Malaysia	44	7	5,920
The Philippines	55	12	1,560
Japan	230	13	1,413

As a mega-diverse country in bamboo resources, China enjoys a better utilization of the bamboo industry with about 601,000 km^2 of bamboo resources, among which over 443,000 km^2 is of moso bamboo (*Phyllostachys edulis*). Bamboo resources are intensively distributed in the mountainous regions of Zhejiang, Jiangxi, Anhui, Hunan, Hubei, Fujian, Guangdong, Guangxi, Guizhou, Sichuan, Chongqing, and Yunnan provinces, municipalities, and autonomous regions, and so on. With the outstanding characteristics of short rotation, high economic value, and advantage for sustainable management, bamboo has become one of the most important nontimber forest products. About half of the world population is associated with bamboo products, such as bamboo house, bamboo panels or composites, bamboo mat, bamboo chopsticks, bamboo sticker, bamboo charcoal or active carbon, bamboo pulp and papermaking, and bamboo shoot. The manufacturing process of these products is affected by bamboo's characteristics. The research on the biological, anatomical, and chemical characteristics of bamboo is therefore reviewed in this section.

BIOLOGICAL CHARACTERISTICS OF BAMBOO

Vegetative Growth

Bamboo is one of the fastest-growing plants in the world, which can increase to 10–30 m in 40–60 days, and reach complete height growth and diameter growth within one growing season. According to the phenotype, there are two main types of bamboo. The first one is sympodial bamboo, which tends to be found in tropical or subtropical zones, with underground stems that sprout vertical shoots much closer to their parental plants and grow slowly outward. The other type of bamboo is monopodial bamboo mainly found in temperate regions,

TABLE 14.2 Comparison Between Growth Characteristics of Bamboo and Trees (Jiang, 2007)

Growth	Bamboo	Trees
Height growth	Bamboo reaches the final height during 2–4 months, which is relatively short; height growth mainly depends on the intercalary meristem kept during the whole height growth phase.	Trees keep growing higher in their whole lifetime; Height growth of trees mainly depends on the primordial meristem on the tip of the trees.
Diameter growth	During the stage of sprouting and height growth, the culm diameter and wall thickness increases a little, and both will stop growing absolutely after reaching the final height.	Diameter growth of trees is a result from the activity of the cambium, and occurs during the whole lifetime.

which often sends out underground stems to various distances and sends up vertical shoots (Stapleton, 1997). The rhizomes of a monopodial bamboo contain dormant buds at each rhizome nodal ring, which often grow into large thickets or groves if they grow alone. Furthermore, there are some bamboo species that have both types of sympodial and monopodial rhizome systems in a genet. Bamboo can sustain itself for several decades by means of a rhizomatous growth habit. Bamboo has a good vegetative propagation ability and strong antibarren ability, which is suitable for mountainous afforestation, especially with the dwindling of afforestation land. Bamboo usually can reach optimum material properties at 4 years old, and bamboo timber has special properties of better split, easy preparation, high strength, moderate rigidity, and good toughness, which is much higher than that of general timber. Therefore, as an important biomass material, bamboo could be sustainable for utilization once planted (Table 14.2).

Reproductive Growth

Study on bamboo reproductive growth is in low progress, which is limited due to the long flowering cycle of up to 120 years, infrequent and unpredictable flowering events, coupled with peculiar monocarpic behavior, seed sterility, and short seed viability. The characteristics of flowering and pollinating show that the pistil and stamen are vigorous and mature at different times in *Dendrocalamus latiflorus*, so the pollination is easy; while in the *Dendrocalamopsis* genus the pistils are degenerated and the pollens are rare, comparatively difficult for pollination. In the *Bumbusa* genus the pistil and stamen are strong and mature at the same time, so the pollination is simple (Yuan et al., 2005). The structure of the reproductive organs and their development in *Dendrocalamus sinicus* indicate that the cytokinesis of the microspore mother cells in meiosis is successive and the microspore tetrads are tetrahedral; the mature pollen grains are two-celled and monoaperturate, dichogamous, and protogynous in the same flower (Wang et al., 2006). The observation of blooming progress in five bamboo species showed that they bloomed fast in the morning; the percentage of pollen germination was low and different among five bamboo species with the highest at 76.5% and the lowest at 5.3% (Lin et al., 2008). The reproductive biology of bamboo needs to be further characterized.

Molecular Biology

Molecular Marker

With the fast development of molecular biology, a number of PCR-based molecular markers, including RAPD, SCAR, AFLP, and SSR, have been developed and used for bamboo genetic diversity and phylogenetic analysis, germplasm identification, and assisted breeding. Further joint analysis of molecular markers effectively improved the identification rate; the best one was AFLP joint ISSR and SRAP, followed by AFLP joint SRAP, SSR joint SRAP, and AFLP joint ISSR.

Functional Genes

The wood properties of bamboo mainly depend on the components and structure of the cell wall. The genes involved in the pathway of cell wall biosynthesis were the main objects of our study. Lignin biosynthesis involves many genes such as *PAL, C4H, 4CL, COMT, CCoAOMT,* and *F5H. PePAL* isolated from *Phyllostachys edulis* was well-studied, which had similar biochemical properties and K_m value with PALs reported in other plants (Hsieh et al., 2011). The expression patterns of *C3H, C4H,* and *PheCYP-1* in different parts of *P. edulis* indicated that they are involved in the thickness of the cell wall during the development of vascular tissues (Yang and Peng, 2010). The transgenic tobacco of antisense *BoCOMT1* displayed a dwarf and slow growth phenotype, of which lignin content declined (Li et al., 2012).

Omics

The advent of high-throughput sequencing technologies provides a turning point for bamboo research; *de novo* transcriptome sequencing for bamboo developed rapidly recently. To reveal the reason for the fast growth of bamboo shoots through the rapid elongation of internodes, 13 ESTs were preferentially expressed in elongating internodes and seven in static internodes (Zhou et al., 2011). A total of 51 out of 307 functional ESTs were found fiber-specific in *Bambusa balcooa* (Rai et al., 2011). To better understand the anatomical structure of bamboo culms, according to the interaction of 64 genes and correlation

FIGURE 14.1   Web page of bamboo genome database.

between miRNA–mRNA and mRNA–protein, a complex network was built and the 64 genes were in the core position (He et al., 2013), which played an important role in the development of bamboo culm.

Recent genomic studies in bamboo have been implemented, including genome-wide full-length cDNA sequencing, chloroplast and mitochondrial genome sequencing, identification of syntenic genes between bamboo and other grasses, and phylogenetic analysis of the Bambusoideae subspecies. The first draft genome of bamboo (*Phyllostachys heterocycla*) was published in 2013; 19 cellulose synthase (CesA) and 38 cellulose synthase-like (Csl) genes were detected in the genome, as well as 35 genes involved in the phenylpropanoid and lignin biosynthetic pathways (Peng et al., 2013). In order to facilitate the use of these bamboo genome data, a bamboo genome database (www.bamboogdb.org) had been developed (Fig. 14.1), which provided the majority of researchers with a comprehensive, accurate, convenient, and timely- updated bamboo database with multiple omics data on genome. The achievement has played a significant role in the study of the evolution, improvement, and genetic information of the grass family including the bamboo subfamily, as well as the oriented cultivation of bamboo resources, development of bamboo industry, and prosperity of bamboo culture.

ANATOMICAL CHARACTERISTICS OF BAMBOO CULM

Macrostructure

On the cross-section, the macrostructure of bamboo culm comprises the epidermis, cortex, ground tissue, and lacuna, as shown in Fig. 14.2. The epidermis is

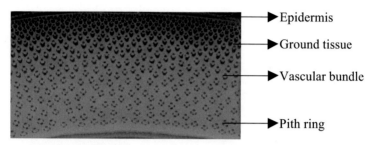

→ Epidermis

→ Ground tissue

→ Vascular bundle

→ Pith ring

FIGURE 14.2   Cross-section of a culm wall with vascular bundles.

the outer layer of the culm, including elongated cells, short cork and silica cells, and the stomata. The outer cell wall usually is covered by tiny protuberances, a cutinized layer, and a wax coating. Size, shape, distribution, and arrangement of protuberances vary among different bamboo species. These protuberances mainly diffuse in aggregates. There is also reticulate, grid shape, or undulation for protuberances. Some special bamboo have unicellular hair on the epidermis. The hypodermis is close to the epidermis, containing two to three layers of thick-walled sclerenchymatous cells. The thickness of the cell wall varies with its location. The cortex is located between the hypodermis and ground tissue, including several layers of parenchyma cells. The ground tissue is also composed of parenchyma cells and vascular system. Parenchyma tissue has two types of vertically positioned cells, such as the short and long parenchyma cells. Generally, the culm consists of about 52% parenchyma, 40% fiber, and 8% conducting tissue. But they vary among different bamboo species. There are vascular bundles in parenchyma tissues.

Liese (1985) found different types of vascular bundles, such as *P. edulis, Cephalostachyum pergracile, Oxytenanthera albociliata,* and *Thyrsostachys oliveri.* The outside of the vascular bundle is known as the phloem and the inside is known as the xylem. The vascular bundles of bamboo internodes are composed of two metaxylem vessels, phloem, protoxylem, and attached fiber sheaths. Some bamboo species have additional fiber bundles (Jiang, 2007). Figure 14.3 shows the shape of the vascular bundles in the cross-section of a culm wall, as shown in Fig. 14.2. Bamboo has many similarities and differences with woody plants. The formation of a cell wall in bamboo proceeds in a different manner with that in woody plants. Thickening growth of the culm does not occur. Fibers and parenchyma cells are characterized by a polylamellate structure (Parameswaran and Liese, 1975, 1980; Liese, 1998 cited by Cha et al., 2014). Table 14.3 shows the structure differences between bamboo and woody plants.

From outer to inner

FIGURE 14.3   The shape of vascular bundles in cross-section of a culm wall.

TABLE 14.3 Comparison of Structure Between Bamboo and Trees (Yin, 1996)

	Bamboo	Woody plants
Origin	Bamboo is the primary product, and without secondary growth. The height growth is happened only during short time. Bamboo has no diameter growth.	The proportion of the primary product is the minimum and the most part of wood is the secondary product. A tree has height growth and diameter growth during the whole lifetime.
Structural elements	Structural elements of internodes arrange in the longitudinal direction. Bamboo has no rays	Structural elements arrange in the longitudinal and radial directions. Wood has rays.
Distribution of vascular bundles	The vascular tissues are embedded in the ground tissue.	Vascular tissues are around the pith at the position of treetop with primary growth. Vascular tissues connect each other and present a sole tube-shape.
Vascular tissue	Vascular tissue has no cambium. Fibers distribute around the vascular tissue together. Xylem contains vessels.	Vascular tissue has cambium. Fibers are dispersed under microscopic scale. Xylem of most hardwood trees and a few softwood trees contains vessels.
Nodes	With nodes	Without nodes

Microstructure

It was confirmed that the anatomical and physical properties of bamboo culms had significant effects on their durability and strength (Liese, 1985). To better use bamboo resources, the anatomical characteristics of bamboo need to be investigated. The anatomical characteristics of four types of Chinese bamboo species are shown in Tables 14.4–14.6. Mustafa et al. (2011) also determined the anatomical and microstructural features of the tropical bamboo *Gigantochloa brang, Gigantochloa levis, Gigantochloa scotechinii,* and *Gigantochloa wrayi*. It was confirmed that the fibers for each species had different lengths, diameters, thickness of cell wall, and lumen sizes. The size of the vascular bundle was smaller at the outer position and became bigger at the inner position. He et al. (2011) found that the fiber length of bamboo was shorter than that of soft wood, but longer than that of hard wood. Ding and Liese (1995) found that the fibers in the nodes were forked and had obtuse ends, which differed from fibers in the internodes. Node fibers were shorter than those in internodes and sometimes the fibers in the nodes even had cell inclusions.

Variation in anatomical properties of bamboo culm has a very strong correlation with their age, location where samples are taken, moisture content and basic density, and so on. Murphy and Alvin (1992) found that the extent

TABLE 14.4 Fiber, Parenchyma, Vessel, and Sieve Tube Proportion of Different Bamboo Types in China

Bamboo types	Parenchyma proportion	Fiber proportion	Vessel proportion	Sieve tube proportion
Dendrocalamus giganteus	60.9	34.5	4.0	2.1
Bambusa sinospinosa	65.7	29.3	4.6	1.4
Dendrocalamus farinosus	51.0	39.8	6.2	3.0
Bambusa rigida	60.1	32.3	7.0	2.2
Bambusa pervariadilis × Grandis nin	55.8	37.8	5.6	3.6

of polylamellation was influenced by the position of the vascular bundle in the culm wall, in certain positions by age of the culm and, most strikingly, with position of the vascular bundle. The number of wall lamellae was variable but tended to be greatest in fibers adjacent to either vascular elements or ground tissue at the periphery of the fiber bundles. Wahab et al. (2009) found that the frequency of vascular bundles was greater at the bottom and top portion than in the middle portion. The cell wall thickness of parenchyma and fiber were greater in the 4-year-old than in the 2-year-old culms. The authors also confirmed that the main anatomical properties of *Bambusa pervariabilis × Dendrocalamopsis daii* including the parenchyma cell, fiber, vessel, and sieve tube varied from outer to inner in the radial direction. Figure 14.4 shows that the 0.5- and 6.5-year-old bamboo fibers had circular cross-sections and cell cavities. Li and Shen (2011)

TABLE 14.5 Shape Characteristics of Vascular Bundles for Different Bamboo Types in China

Bamboo types	Radial width (μm)	Tangential width (μm)	Frequency (Individual/mm^{-2})
D. giganteus	862.803	702.743	1.133
B. sinospinosa	710.170	742.441	1.056
D. farinosus	664.177	553.427	3.077
B. rigida	762.978	640.210	1.686
B. pervariadilis × Grandis pin	778.445	700.162	1.512

TABLE 14.6 Fiber Shape Characteristics of Different Bamboo Types in China

Bamboo types	Diameter (μm)	Lumen diameter (μm)	Double-wall thickness (μm)	Length (μm)	Length-width ratio	Double-wall thickness–lumen diameter ratio	Lumen diameter–width ratio
D. giganteus	19.043	9.525	9.518	2546.235	133.707	0.999	0.500
B. sinospinosa	20.999	14.868	6.186	1686.100	80.296	0.416	0.708
D. farinosus	13.963	4.816	9.173	2258.791	161.766	1.905	0.345
B. rigida	20.271	4.092	16.178	1901.775	93.820	3.953	0.202
B. pervariadilis × Crandis nin	15.364	5.811	9.552	1671.993	108.828	1.644	0.378

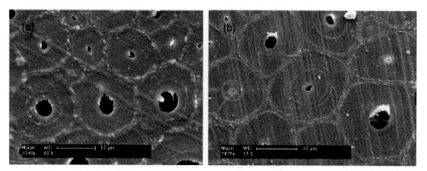

FIGURE 14.4 **Bamboo cross-sections.** (a) 0.5 years old, (b) 6.5 years old. SEM Bar = 10 μm.

extracted the vascular bundles from different height locations of a moso bamboo with an alkali treatment method and found that the longitudinal Young's modulus and strength of the vascular bundles linearly increased from the inner to the outer side. Wang et al. (2011) found that the radial length/tangential diameter ratio of the vascular bundle, tangential diameter of metaxylem vessel, hemicelluloses, and fiber characteristics varied with age and portion. Cha et al. (2014) investigated the micromorphological and chemical aspects of archaeological bamboos under long-term waterlogged condition (Fig. 14.5). They confirmed

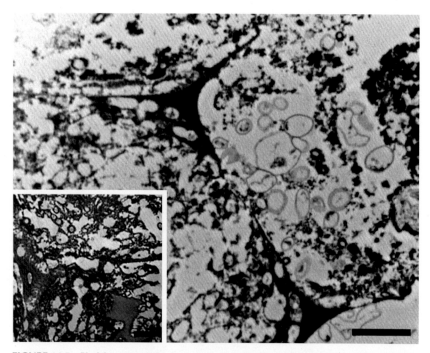

FIGURE 14.5 **Vesicles around the degraded cell walls.** Note: the degradation of the middle lamella. TEM Bar = 2 μm. Inset: numerous vesicles around the degraded cell wall (Cha et al., 2014).

TABLE 14.7 Comparison of Microstructure Between Bamboo and Trees

Bamboo	Woody plants
Vascular bundles are composed of vessels, fiber, sieve tube, and companion cell. The basic tissues mainly consist of elongated parenchyma cells, and short square-shaped parenchyma cells, which disperse among elongated parenchyma cells. The number of pits is small, and all of which are simple. Transverse ray tissue is absent.	Wood is mainly composed of tracheid for softwood, or vessels and fiber for hardwood, also containing axial parenchyma and horizontal ray tissues. Some species also contain intercellular canal. The number of pits is rich, and the kinds of which are single pit and bordered pit.

that parenchyma cells were more degraded than bamboo fibers. Degradation in fibers was confined mainly to secondary walls except the middle lamella.

It is well known that the anatomical characteristics affect bamboo utilization or the manufacturing process of bamboo products. Wang et al. (2012) found that the mechanical stability of the culms of monocotyledonous bamboos is highly attributed to the proper embedding of the stiff fiber caps of the vascular bundles into the soft parenchymatous matrix. Xie et al. (2014) confirmed that higher vessel and parenchyma percentages and lower cellulose and lignin contents in the inner layer contributed to lower residue content in the process of microwave-assisted liquefaction of bamboo wastes, while higher fiber percentage and cellulose or lignin contents in the epidermis layer resulted in higher residue content. Liese and Schmitt (2006) confirmed that the anatomical characteristics of the inner culm wall affected the diffusion treatment with preservatives (Table 14.7).

Ultrastructure

The bamboo's ultrastructure helps it to withstand the extreme natural environment, properties, and utilization. Authors have investigated the multilayer structure of the bamboo cell wall with FE-SEM, TEM, nanoindentation, and AFM. We determined the arrangement of cellulose microfibrils aggregated with AFM and the effect of FT-IR and XRD was also studied. Considering that the thickness of some layers in the bamboo cell wall is less than 100 nm, AFM with atomic level resolution was used to observe the multiple layers by measuring the modulus of the cross-section of the cell wall, as shown in Fig. 14.6. With the difference among the modulus of each layer in the bamboo cell wall, the multilayer structure was identified. The multilayer structure and the thickness of each layer in different bamboo cell walls varied. The thickness of the layer in parenchyma cell wall was smaller than in the fiber cell wall. There were also more layers in the parenchyma cell wall. Cellulose microfibril aggregates arrange randomly in the cross-section of the bamboo fiber cell wall and bamboo parenchyma cell wall.

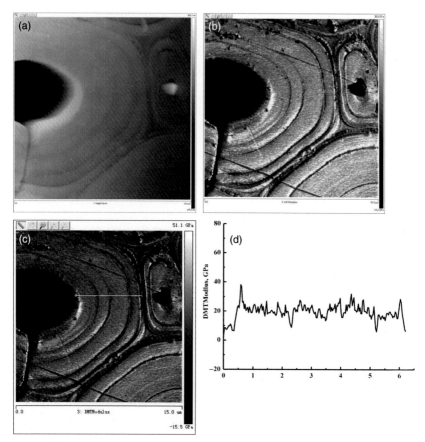

FIGURE 14.6 **The modulus mapping in the cross-section of bamboo cell wall.** (a) Height image, (b) modulus image, (c) modulus curve on chosen line, and (d) modulus curve.

Ray et al. (2004) found that the cross-section of bamboo fibrils was either pentagonal or hexagonal, arranged in a honeycomb pattern, separated by thin walls of matrix. The wall thickness between two sclerenchyma cells ranged from 1 to 5 μm and the width of the same varied from 10 to 20 μm. Again a fibril contains many continuous elongated cellulose, staggered in a twisted nature, like a metal cable formed by many twisted wires. The better mechanical properties of bamboo are mainly attributed to the structure and performance of the bamboo cell wall. Gritsch et al. (2004) found that the multilayered structure of fiber cell walls was formed mainly during the first year of growth by the deposition of new wall layers of variable thickness, resulting in a high degree of heterogeneity in the layering patterns among individual fibers.

Yu et al. (2007) confirmed the different deformation mechanisms of the cell walls when indented in the longitudinal and transverse direction of bamboo fibers, respectively. Yang et al. (2009) found that the radial variation of the

TABLE 14.8 Comparison of Ultrastructure Between Bamboo and Trees

Bamboo	Woody plants
Secondary wall of bamboo fiber is composed of more than one sublayer of the secondary wall whose thickness alternates from thin to thick. The number of layers can be up to 18, which relates to the age and the location of the fibers in the bamboo, but generally significantly greater than that of wood fibers. The orientation of microfibril is different, similarly parallel to the fiber axial direction in the thick layer, and its microfibril angle increases from outside to inside. In the thin layer, the orientation of microfibril is perpendicular to the fiber axial direction, and microfibril angle is bigger; the thickness is less than 100 nm. In addition, the level of lignification is different among the thick and thin layers, and the latter has high lignin content.	The cell walls of tracheids or fibers are composed of primary wall and secondary wall. The secondary wall is divided into the S1, S2, and S3 layers from outside to inside. The S1 layer is adjacent to the primary wall, whose thickness is 0.1–0.2 μm and occupies 10–22% thickness of the average cell wall. The deposition of microfibril is almost horizontal in the S1 layer, and the microfibril angle is 50–70°. The S2 middle layer has a thickness of 1–5 μm, and occupies 70–90% thickness of the average cell wall; and the microfibril is close to the fiber axis, Z-shaped with an angle of 10–30°. The S3 layer is adjacent to the cell lumen, and its thickness of commonly 0.03–0.3 μm varies based on species, and accounts for 2–8% of the average cell wall thickness. Microfibril deposition in this layer approximates to the S1 layer, almost perpendicular to the fiber direction, arranges approximately circular with an angle of 60–90°.

microfibril angle of the secondary cell wall within ages took an irregular tendency as the distance ranged from bark to pith. The distribution of microfibril aggregates varied in different layers of the cell wall. Gritsch and Murphy (2005) found that the structure of the primary cell wall comprised two layers. The secondary cell wall of the fiber began to be laid down while the cells were still undergoing some elongation, suggesting that it may act to cause the slow-down and eventual cessation of cell elongation. Crow (2000) defined three stages of fiber and parenchyma development that occurred during internode elongation in the temperate bamboo *Phyllostachys viridiglaucescens*. The first stage involved cell divisions in both cell types. During the second stage, as the elongation of the internode proceeded, parenchyma cells continued to divide whereas fibers began to elongate. Finally, in the third stage, both cell types elongated (Gritsch and Murphy, 2005). Parameswaran and Liese (1976) found that the thick-walled bamboo fibers exhibited a polylamellate structure with alternating broad and narrow lamellae. Cellulose fibrils in the broad lamellae were oriented almost parallel to the longitudinal axis of the fiber (2–20°), whereby there was a gradual but only slight increase in the angle from the middle lamella to the lumen. The narrow lamellae consisted of fibrils oriented almost perpendicular to the cell axis (85–90°). This angle remained constant in all the successive narrow lamellae (Table 14.8).

TABLE 14.9 Chemical Composition of Bamboo, Wood, and Agriculture Residue

Biomass type	Species	Cellulose (%)	Hemicelluloses (%)	Lignin (%)
Wood	Pine	40–45	25–30	26–34
	Maple	45–50	21–36	22–30
Bamboo	Moso bamboo	42–50	24–28	24–26
Agriculture residues	Rice straw	41–57	33	8–19
	Rice husk	35–45	19–25	20
	Bagasse	40–46	25–29	12.5–20
	Cotton stalk	43–44	27	27

CHEMICAL CHARACTERISTICS OF BAMBOO

Chemical Composition

The chemical constituents of a bamboo culm are very complicated. Bamboo, like wood and agricultural residue, is mainly composed of cellulose, hemicelluloses, and lignin, even though the contents of these compositions are different. Table 14.9 shows the difference of wood, bamboo, and agricultural residue chemical composition. The minor constituents consist of various soluble polysaccharides, and protein materials, resins, tannins, waxes, a little amount of ashes, and so on. Table 14.10 shows the chemical compositions of 10 bamboo species with different ages. It exhibits different chemical compositions in moisture, ash, extractives, lignin, pentose, holocellulose, and α-cellulose among the bamboo species with different ages. Li et al. (2007) investigated the chemical changes with maturation of the bamboo species *P. pubescens* and confirmed small but significant increases in holocellulose and α-cellulose contents from the base to the top of the culm for 1-, 3- and 5-year-old bamboo materials. Wahab et al. (2013) confirmed that the chemical constituents of *G. brang, G. levis, G. scortechinii,* and *G. wrayi* were different including holocellulose, α-cellulose, lignin, extractives, and ash contents. The ash of biomass materials depended on the composition of mineral constituents. It was confirmed that the main ash-forming elements of bamboo included Na, Mg, Si, Al, Fe, Ca, Na, K, Zn, and Ti (Liu et al., 2013a). The amount and types of heavy metals found in ash of bamboo and pine are shown in Table 14.11. It was confirmed that these elements varied with different positions of bamboo.

The structure of the chemical constituents of bamboo was also investigated. Lou et al. (2012) found that the structure of hemicelluloses was determined to be mainly arabinoxylans linked via (1-4)-β-glycosidic bonds with branches

TABLE 14.10 Chemical Composition of 10 Bamboo Species with Different Ages (Jiang, 2002)

Species	Ages	Moisture (%)	Ash (%)	Solvent of cold water (%)	Solvent of hot water (%)	Solvent of benzene–alcohol (%)	Solvent of 1% NaOH (%)	Lignin (%)	Pentose (%)	Holocellulose (%)	α-Cellulose (%)
P. pubescens	6 months	9.00	1.77	5.41	3.26	1.60	27.34	26.36	22.19	76.62	61.97
	1 year	9.79	1.13	8.13	6.31	3.67	29.34	34.77	22.97	75.07	59.82
	3 years	8.55	0.69	7.10	5.11	3.88	26.91	26.20	22.11	75.09	60.55
	7 years	8.51	0.52	7.14	5.17	4.78	26.83	26.75	22.01	74.98	59.09
Bambusa textilis	6 months	9.09	2.39	6.64	8.03	4.59	32.27	18.67	22.22	77.71	51.96
	1 year	10.58	2.08	6.30	7.55	3.72	30.57	19.39	20.83	79.39	50.40
	3 years	10.33	1.58	6.84	8.75	5.43	28.01	23.81	18.87	73.37	45.50
Lingnania chungii	6 months	9.21	2.73	8.10	9.70	4.16	35.17	17.58	23.91	79.00	47.63
	1 year	10.33	2.10	8.07	9.46	4.35	29.97	21.41	18.72	73.72	47.76
	3 years	10.26	1.50	6.34	9.21	3.98	30.57	22.70	18.88	71.70	43.65
B. pervariabilis	6 months	8.38	2.16	4.93	6.35	2.14	27.71	20.92	21.47	79.41	52.63
	1 year	11.66	2.29	7.64	7.71	2.15	29.99	21.43	20.26	73.34	48.15
	3 years	11.04	2.65	9.51	9.25	6.42	30.63	22.02	19.22	69.14	45.33

(Continued)

TABLE 14.10 Chemical Composition of 10 Bamboo Species with Different Ages (Jiang, 2002) (Cont.)

Species	Ages	Moisture (%)	Ash (%)	Solvent of cold water (%)	Solvent of hot water (%)	Solvent of benzene–alcohol (%)	Solvent of 1% NaOH (%)	Lignin (%)	Pentose (%)	Holocellulose (%)	α-Cellulose (%)
B. sinospinosa	6 months	9.17	2.69	7.29	8.23	4.23	29.98	19.90	21.81	78.29	52.58
	1 year	11.49	1.92	8.08	9.91	5.49	30.25	20.51	20.27	71.16	49.45
	3 years	11.13	1.84	9.07	9.27	5.88	26.92	24.17	19.72	72.77	17.10
Phyllostachys heteroclada	1 year	8.38	1.24	13.57	9.60	5.38	30.89	22.42	20.43	71.98	58.15
	3 years	10.87	1.27	9.68	15.94	9.11	34.84	22.75	21.83	59.92	38.96
P. nigra	6 months	10.31	1.98	6.72	8.30	4.12	31.83	28.49	22.21	70.77	45.38
	1 year	7.70	1.81	10.69	8.53	5.29	33.24	23.90	22.08	73.61	58.85
	3 years	11.61	1.71	6.50	8.36	5.58	33.65	25.00	22.30	68.64	13.79
Phyllostachys bambysoides	6 months	10.69	2.22	4.62	5.93	1.81	27.60	24.51	22.69	76.41	48.92
	1 year	9.44	1.25	10.49	8.97	7.34	29.93	23.39	22.46	72.65	56.74
	3 years	9.90	0.98	6.11	7.32	5.86	31.33	25.15	22.65	65.39	12.02
Phyllostachys meyeri	6 months	10.70	1.68	3.60	5.15	1.81	27.27	23.58	21.95	78.17	49.97
	1 year	8.29	1.29	10.70	8.91	7.04	34.28	23.62	22.35	72.84	57.88
	3 years	9.33	1.85	8.81	12.71	7.52	35.32	23.35	22.49	62.40	39.95
Phyllostachys praccox	6 months	10.64	3.24	6.72	8.57	2.25	33.36	26.74	21.98	72.83	42.23
	1 year	8.19	1.96	11.21	7.68	3.80	32.84	21.68	22.21	73.81	56.03
	3 years	11.29	2.26	7.18	9.09	5.64	23.26	25.65	22.39	65.77	40.81

TABLE 14.11 Amount and Types of Heavy Metals Found in Ash of Bamboo (mg/kg)

Type	Na	Mg	Al	Si	K	Ca	Ti	Fe
OB	233.64	264.58	29.56	450.54	1224.02	579.65	3.09	111.80
MB	195.75	195.65	22.70	82.83	3103.03	421.69	2.17	73.54
IB	234.26	208.81	33.24	93.99	2539.90	974.56	2.08	121.23
BL	429.71	923.83	324.24	784.49	9761.55	2525.11	27.07	681.61
Moso bamboo	242.83	229.46	35.09	358.14	4032.58	965.02	18.06	111.93
Pine	827.15	234.98	48.89	250.75	1261.39	1160.94	6.74	192.60

Note: OB is outer layer of moso bamboo; MB is middle layer of moso bamboo; IB is inner layer of moso bamboo; and BL is leaf of moso bamboo.

of arabinose and 4-O-methyl-D-glucuronic acid. The molecular weight was 6387 Da and 4067 Da, corresponding to the hemicelluloses HA and HB, respectively. Sun et al. (2012) found that the fractions isolated with 0.5 and 1.0 M NaOH were composed of a linear (1→4)-β-D-xylopyranosyl main chain substituted by a small amount of L-arabinofuranosyl at C-2 and/or C-3 together with a minor quantity of 4-O-methylglucuronic acid at C-2, representing a typical polysaccharide structure in bamboo. Wen et al. (2012) found that the heterogeneous lignin polymers in bamboo samples were HGS-type and partially acylated at the γ-carbon of the side chain by p-coumarate and acetate groups. Liang et al. (2011) analyzed the compositions of alkaline extractives in bamboo sulfonated chemomechanical pulps (SCMP) by UV spectroscopy and gas chromatography–mass spectrometry, and found that the extractives were composed of short-chain aliphatic compounds, phenols, and fatty acids.

Surface Chemical Characteristics of Bamboo

The surface chemical behavior of bamboo plays an important role in its utilization. Bamboo is a complex polymer consisting primarily of carbon (C), hydrogen (H), oxygen (O), nitrogen (N), and sulfur (S). Authors have found that C, H, O, N, and S contents were different for OB, MB, IB, moso bamboo, pine, and rice straw as shown in Table 14.12. Jiang et al. (2006) also confirmed that the proportion of oxygen atoms on the outer surface of bamboo was higher than that on the inner surface. The content of cellulose and hemicelluloses on the surface of bamboo were obviously lower than those of lignin and abstracts. Hemicelluloses content on the outer layer of bamboo was lower than that on the inner layer. There is a waxy or siliceous layer on the bamboo surface, affecting the property of interfacial conglutination between bamboo particle and adhesive in the manufacturing process of bamboo composite.

TABLE 14.12 Ultimate Analysis of Moso Bamboo

Types	OB	MB	IB	Moso bamboo	Pine	Rice straw
C (%)	51.58	49.02	49.42	49.40	49.71	41.38
H (%)	5.74	5.75	5.70	4.60	6.36	5.39
O (diff.) (%)	42.52	45.12	44.79	45.82	42.50	52.52
N (%)	0.16	0.11	0.09	0.18	1.43	0.71
S (%)	0.02	0.04	0.03	0.04	0.02	0.07

Note: OB is outer layer of bamboo; MB is middle layer of bamboo; and IB is inner layer of bamboo.

The waxy or siliceous layer on the bamboo surface prevents adhesive from permeating into bamboo, which results in the poor bonding strength of bamboo composites. Liu et al. (2012) also confirmed that the glass transition of wax on the surface of bamboo likely overlapped with the higher temperature of hemicelluloses. The softening and the subsequent flow of the waxes at lower temperature results in the formation of weak boundary layers that are responsible for the low physical properties of bamboo pellets. Liu et al. (2013b) investigated surface characteristics of bamboo treated with sodium hydrate and acetic acid solution. It was found that the main polar chemical groups of OB (outer layer of bamboo) included hydroxyl group (O—H) and ester carbonyl ($C=O$). Some new polar chemical groups appear on the IB (inner layer of bamboo) surface, such as aromatic ethers (C—O—C) and phenolic hydroxyl (C—O). The chemical group difference of OB and MB confirms that there is a waxy layer on the OB surface. Figure 14.7 shows that nonpolar chemical groups decrease and polar chemical groups increase on the OB surface when it is treated with an acetic acid solution. The waxy layer of the OB surface is removed and the lignin structure is also destroyed by sodium hydrate solution.

Utilization of Bamboo Extractives

The chemical components in bamboo extracts include flavonoids, phenolic acids, polysaccharides, essential oil, and so on.

Flavonoids

Flavonoids are a large group of plant secondary metabolites. Many flavonoids are active principles of medicinal plants and exhibit pharmacological effects (Zhang et al., 2005).

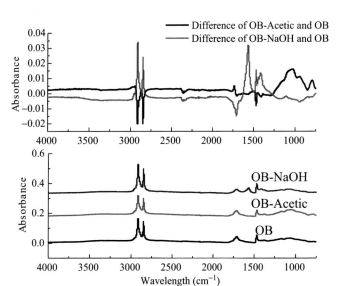

FIGURE 14.7   **The chemical groups of OB treated by different chemical agents from FT-IR.**

Bamboo leaves are rich in flavonoids. Bamboo-leaf flavonoids (also called antioxidant of bamboo leaves), a brown–yellow powder extracted from bamboo leaves (*P. pubescens* or *Phyllostachys nigra* var. *henonis*), is effective as anti-free-radical, antioxidation, antiaging, antibacteria, and can be used as a pharmaceutical intermediate, cosmetic ingredient, and food additive (Wang et al., 2010). Table 14.13 shows the content of the main ingredients of the antioxidant of bamboo leaves from *P. nigra* var. *henonis* (Lu et al., 2005).

As for the flavonoids in bamboo leaves, flavone *C*-glycosides, such as isoorientin, isovitexin, orientin, and vitexin (Fig. 14.8) are a group of representative flavonoids.

TABLE 14.13 Specifications of Antioxidant of Bamboo Leaves

Parameters	Specifications
Total flavonoids	32.4%
Total lactone	15.6%
Phenolic acids	7.9%
Ash	1.24%
Protein	2.3%
Total heavy metals (Plumbum, Pb)	<0.001
Arsenicum (As)	<0.0003
Moisture	4.9%

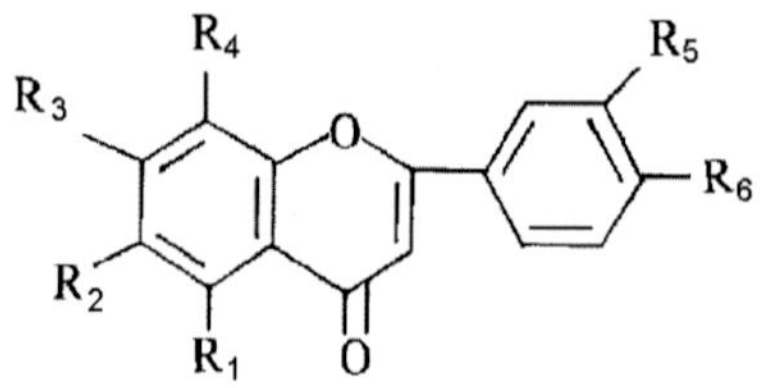

(1): $R_1 = R_3 = R_5 = R_6 = OH$, $R_4 = H$, $R_2 =$ Glucose
(2): $R_1 = R_3 = R_6 = OH$, $R_4 = R_5 = H$, $R_2 =$ Glucose
(3): $R_1 = R_3 = R_5 = R_6 = OH$, $R_2 = H$, $R_4 =$ Glucose
(4): $R_1 = R_3 = R_6 = OH$, $R_2 = R_5 = H$, $R_4 =$ Glucose

FIGURE 14.8   Chemical structures of isoorientin (1), isovitexin (2), orientin (3), and vitexin (4).

Flavonoids occur widely in bamboo species. Several flavonoids, such as tricin and its derivatives, were found in the bamboo leaves of the genus *Sasa* (Sultana and Lee, 2009). Lee et al. (2010) isolated and identified eight compounds from the leaves of *P. nigra*, namely, orientin, vitexin, coumaric acid, tricin, isoorientin, etc.

Phenolic Acids

Two phenolic acid compounds (*cis*-coumaric acid and *p*-coumaric acid) were extracted from the leaves of *P. nigra* (Jung et al., 2007). Three phenolic acid compounds have been isolated from bamboo (*P. edulis*): (1) 3-O-(3'-methylcaffeoyl) quinic acid, (2) 5-O-caffeoyl-4-methylquinic acid, and (3) 3-O-caffeoyl-1-methylquinic acid (Kweon et al., 2001). Besides bamboo leaves, phenolic acid compounds, such as vanillic acid and salicylic acid, have been found in bamboo roots and bamboo shoots (Zhang et al., 2011).

Antioxidant and Anticancer Activity of Bamboo Extracts

Antioxidant Activity

In China, the antioxidant of bamboo leaves (AOB) has been listed in the national standards (standard No.GB2760) in 2004. AOB as a kind of food antioxidant was capable of blocking the chain reactions of lipid auto-oxidation, chelating metal ions of transient state, and scavenging nitrite compounds (Zhang et al., 2005). Several studies showed that the antioxidant properties of bamboo leaves could be mainly attributed to the presence of the flavonoids and phenolic acids (Park et al., 2007).

Anticancer Activity

The *Sasa* species of bamboo has been described to have anticancer activity (Kuboyama et al., 1981). The active polysaccharides in bamboo leaves could be the effective component (Seki and Maeda, 2010). Ando et al. (2004) found that the xylo-oligosaccharides in bamboo extract induced apoptosis in the leukemia cell line derived from acute lymphoblastic leukemia. Furthermore, a methanol extract of bamboo leaves induced rapid apoptosis in the human leukemia CMK-7 cell line. The active compounds in the extract were identified as 201-hydroxypurpurin-7 delta-lactone esters, which also could be promising lead compounds as photosensitizers for the photodynamic therapy of cancer (Kim et al., 2003).

Safety

For safely using AOB in food systems, safety evaluation studies were conducted on animal models. A 90-day oral toxicity study showed that the maximum tolerated dose of AOB was > 10 g/kg body weight in rats and in mice and no mutagenicity evidence was detected. Moreover, no mortality occurred in the test for teratogenicity (Lu et al., 2005). The safety studies showed a no-observed-adverse-effect level of 4.30 g/kg per day indicating safe use as a food additive (Lu et al., 2006).

THE KEY SCIENTIFIC RESEARCH IN THE FUTURE

With the rapid development of modern biotechnology and its related technologies, bamboo molecular biology has been investigated including the development and application of molecular markers, the identification of gene function, and the omics of bamboo. Molecular assistant breeding and genetic engineering of bamboo will be one of the key areas of biological research. For the anatomical and chemical characteristics of bamboo, the multilamellate structure of thick-walled bamboo fibers will be investigated to explore the natural laws of the cell wall forms, investigating the layers and the thicknesses of the cell wall at nano levels, determining the orientation of microfibrils in each layer and how the microfibril angle changes in the cell wall, analyzing the chemical composition of the cell wall, and building a three-dimensional multilamellate structure model of thick-walled bamboo fibers. Quantitatively characterizing the nanostructure of each layer in the cell wall, orientation of microfibrils, and distribution of chemical compositions is conducive to further studying the features of the multilamellate structure of thick-walled bamboo fibers.

REFERENCES

Ando, H., Ohba, H., Sakaki, T., Takamine, K., Kamino, Y., Moriwaki, S., Bakalova, R., Uemura, Y., Hatate, Y., 2004. Hot-compressed-water decomposed products from bamboo manifest a selective cytotoxicity against acute lymphoblastic leukemia cells. Toxicol. In Vitro 18, 765–771.

Cha, M.Y., Lee, K.H., Kim, Y.S., 2014. Micromorphological and chemical aspects of archaeological bamboos under long-term waterlogged condition. Int. Biodeterior. Biodegrad. 86, 115–121.

Crow, E., 2000. Development of fibre and parenchyma cells in the bamboo *Phyllostachys viridiglaucescens* (Carr.) Riv. & Riv. PhD Thesis. Imperial College London, Univ. London, UK.

Devi, M.R., Poornima, N., Guptan, P.S., 2007. Bamboo-the natural, green and eco-friendly new-type textile material of the 21st century. J. Textile Assoc..

Ding, Y.L., Liese, W., 1995. On the nodal structure of bamboo. J. Bamboo Res. 14, 24–32.

Gritsch, C.S., Murphy, R.J., 2005. Ultrastructure of fibre and parenchyma cell walls during early stages of culm development in *Dendrocalamus asper*. Ann. Bot. 95, 619–629.

Gritsch, C.S., Kleist, G., Murphy, R.J., 2004. Developmental changes in cell wall structure of phloem fibers of the bamboo *Dendrocalamus asper*. Ann. Bot. 94, 497–505.

He, L., Zhou, G.Y., Zhang, H.Y., 2011. Research and utilization status of natural bamboo fiber. Adv. Mater. Res. 159, 236–241.

He, C.Y., Cui, K., Zhang, J.G., Duan, A.G., Zeng, Y.F., 2013. Next-generation sequencing-based mRNA and microRNA expression profiling analysis revealed pathways involved in the rapid growth of developing culms in Moso bamboo. BMC Plant Boil. 13, 119.

Hsieh, L.S., Hsieh, Y.L., Yeh, C.S., Cheng, C.Y., Yang, C.C., Lee, P.D., 2011. Molecular characterization of a phenylalanine ammonia-lyase gene (*BoPAL1*) from *Bambusa oldhamii*. Mol. Biol. Rep. 38, 283–290.

Jiang, Z.H., 2002. Bamboo and Rattan in the World (In Chinese). Liaoning Sci. Technol. Press, Shen Yang.

Jiang, Z.H., 2007. Bamboo and Rattan in the World (In English). China Forestry Publishing House, Beijing.

Jiang, Z.H., Yu, W.J., Yu, Y.L., 2006. Analysis of chemical components of bamboo wood and characteristics of surface performance. J. Northeast For. Univ. 34, 1–2, 6.

Jung, S.H., Lee, J.M., Lee, H.J., Kim, C.Y., Lee, E.H., Um, B.H., 2007. Aldose reductase and advanced glycation endproducts inhibitory effect of *Phyllostachys nigra*. Biol. Pharm. Bull. 30, 1569–1572.

Kim, K.K., Kawano, Y., Yamazaki, Y., 2003. A novel porphyrin photosensitizer from bamboo leaves that induces apoptosis in cancer cell lines. Anticancer Res. 223, 2355–2361.

Kuboyama, N., Fujii, A., Tamura, T., 1981. Antitumor activities of bamboo leaf extracts (BLE) and its lignin (BLL). Nihon Yakurigaku Zasshi 77, 579–596.

Kweon, M.H., Hwang, H.J., Sung, H.C., 2001. Identification and antioxidant activity of novel chlorogenic acid derivatives from bamboo (*Phyllostachys edulis*). J. Agri. Food Chem. 49, 4646–4655.

Lee, H.J., Kim, K., Kang, K.D., Lee, E.H., Kim, C.Y., Um, B.H., Jung, S.H., 2010. The compound isolated from the leaves of *Phyllostachys nigra* protects oxidative stress-induced retinal ganglion cells death. Food Chem. Toxicol. 48, 1721–1727.

Li, H.B., Shen, S.P., 2011. The mechanical properties of bamboo and vascular bundles. J. Mater. Res. 26, 2749–2756.

Li, X.B., Shupe, T.F., Peter, G.F., Hse, C.Y., Eberhardt, T.L., 2007. Chemical changes with maturation of the bamboo species *Phyllostachys pubescens*. J. Trop. For. Sci. 19 (1), 6–12.

Li, X.P., Peng, Z.H., Gao, Z.M., Hu, T., 2012. The effects of depressing expression of *COMT* on lignin synthesis of transgenic tobacco. Mol. Plant Breed. 10, 689–692.

Liang, C., Zhan, H.Y., Li, B.Y., Fu, S.Y., 2011. Characterization of bamboo SCMP alkaline extractives and the effects on peroxide bleaching. Bioresoures 6, 1484–1494.

Liese, W., 1985a. Anatomy and properties of bamboo. Recent research on bamboos. Proceedings of International Bamboo Workshop. October 6–14, 1985, Hangzhou, China.

Liese, W. 1985b. Bamboo biology, silvics properties, utilization. Gesellschaft fur technische Zusammenarbeit Schriftenreihe, Eschborn.

Liese, W., 1998. Anatomy and Properties of Bamboo. Technical Report 18. Int. Network for Bamboo and Rattan, Beijing.

Liese, W., Schmitt, U., 2006. Development and structure of the terminal layer in bamboo culms. Wood Sci. Technol. 40, 4–15.

Lin, S.Y., Ding, Y.L., Zhang, H., 2008. Pollen germination percentage and the floral character of five bamboo species. Scientia Silvae Sinicae 44, 159–163.

Liu, Z.J., Jiang, Z.H., Cai, Z.Y., Fei, B.H., Yu, Y., Liu, X.E., 2012. Dynamic mechanical thermal analysis of moso bamboo (*Phyllostachys heterocycla*) at different moisture content. Bioresources 7, 1548–1557.

Liu, X.M., Fei, B.H., Liu, Z.J., Jiang, Z.H., Cai, Z.Y., Liu, X.E., Yu, Y., 2013a. Mechanical properties of moso bamboo treated with chemical agents. Wood Fiber Sci. 45, 1–8.

Liu, Z.J., Liu, X.E., Fei, B.H., Jiang, Z.H., Cai, Z.Y., Yu, Y., 2013b. The properties of pellets from mixing bamboo and rice straw. Renew. Energ. 55, 1–5.

Lou, Q., Peng, H., Zhou, M.Y., Lin, D., Ruan, R., Wan, Q.Y., Zhang, Y.S., Liu, Y.H., 2012. Alkali extraction and physicochemical characterization of hemicelluloses from young bamboo (*Phyllostachys pubescens mazel*). Bioresources 7, 5817–5828.

Lu, B., Wu, X., Tie, X., Zhang, Y., Zhang, Y., 2005. Toxicology and safety of anti-oxidant of bamboo leaves. Part 1: acute and subchronic toxity studies on anti-oxidant of bamboo leaves. Food Chem. Toxicol. 43, 783–792.

Lu, B., Wu, X., Shi, J., Dong, Y., Zhang, Y., 2006. Toxicology and safety of antioxidant of bamboo leaves. Part 2: developmental toxicity test in rats with antioxidant of bamboo leaves. Food Chem. Toxicol. 44, 1739–1743.

Murphy, R.J., Alvin, K.L., 1992. Variation in fiber wall structure in bamboo. IAWA Bull. n.s. 13, 403–410.

Mustafa, M.T., Wahab, R., Sudin, M., Sulaiman, O., Kamal, N.A.M., Khalid, I., 2011. Anatomical and microstructures features of tropical bamboo *Gigantochloa brang*, *G. levis*, *G. scotechinii* and *G. wrayi*. Int. J. Forest, Soil Erosion 1, 25–35.

Parameswaran, N., Liese, W., 1975. On the polylamellate structure of parenchyma wall in *Plyllostachus edulis* Riv. IAWA Bull. n.s. 4, 57–58.

Parameswaran, N., Liese, W., 1976. On the fine structure of bamboo fibres. Wood Sci. Technol. 10, 231–246.

Parameswaran, N., Liese, W., 1980. Ultrastructural aspects of bamboo cells. Cellulose Chem. Technol. 14, 587–609.

Park, H.S., Lim, J.H., Kim, H.J., Choi, H.J., Lee, I.S., 2007. Antioxidant flavone glycosides from the leaves of *Sasa borealis*. Arch. Pharm. Res. 30, 161–166.

Peng, Z., Lu, Y., Li, L., Zhao, Q., Feng, Q., Gao, Z., Lu, H., Hu, T., Yao, N., Liu, K., Li, Y., Fan, D., Guo, Y., Li, W., Lu, Y., Weng, Q., Zhou, C., Zhang, L., Huang, T., Zhao, Y., Zhu, C., Liu, X., Yang, X., Wang, T., Miao, K., Zhuang, C., Cao, X., Tang, W., Liu, G., Liu, Y., Chen, J., Liu, Z., Yuan, L., Liu, Z., Huang, X., Lu, T., Fei, B., Ning, Z., Han, B., Jiang, Z., 2013. The draft genome of the fast growing non-timber forest species moso Bamboo (*Phyllostachys heterocycla*). Nat. Genet. 45, 456–461.

Rai, V., Ghosh, J.S., Pal, A., Dey, N., 2011. Identification of genes involved in bamboo fiber development. Gene 478, 19–27.

Ray, A.K., Das, S.K., Mondal, S., Ramachandrarao, P., 2004. Microstructural characterization of bamboo. J. Mater. Sci. 39, 1055–1060.

Seki, T., Maeda, H., 2010. Cancer preventive effect of Kumaizasa bamboo leaf extracts administered prior to carcinogenesis or cancer inoculation. Anticancer Res. 30, 111–118.

Stapleton, C.M.A., 1997. Morphology of woody bamboos. In: Chapman, G.P. (Ed.), The Bamboos. Academic Press, pp. 251–267.

Sultana, N., Lee, N.H., 2009. New phenylpropanoids from *Sasa quelpaertensis* Nakai with tyrosinase inhibition activities. Bull. Korean Chem. Soc. 30, 1729–1732.

Sun, S.N., Yuan, T.Q., Li, M.F., Cao, X.F., Xu, F., Liu, Q.Y., 2012. Structure characterization of hemicelluloses from bamboo culm (*Neosinocalamus affinis*). Cellulose Chem. Technol. 46, 165–176.

Wahab, R., Mohamed, A., Mustafa, M.T., Hassan, A., 2009. Physical characteristics and anatomical properties of cultivated bamboo (*Bambusa vulgaris* Schrad.) culms. J. Bio. Sci. 9, 753–759.

Wahab, R., Mustafa, M.T., Salam, M.A., Sudin, M., Samsi, H.W., Rasat, M.S.M., 2013. Chemical composition of four cultivated tropical bamboo in genus *Gigantochloa*. J. Agri. Sci. 5, 66–75.

Wang, S.G., Pu, X.L., Ding, Y.L., 2006. The structures of reproductive organs and development of the female and male gametophyte of *Dendrocalamus sinicus*. Bull. Bot. Res. 26, 270–274.

Wang, J., Tang, F., Yue, Y., Guo, X., Yao, X., 2010. Development and validation of an HPTLC method for simultaneous quantitation of isoorientin, isovitexin, orientin, and vitexin in bamboo-leaf flavonoids. J. AOAC Int. 93, 1376–1383.

Wang, S.G., Pu, X.L., Ding, Y.L., Wan, X.C., Lin, S.Y., 2011. Anatomical and chemical properties of *Fargesia yunnanesis*. J. Trop. For. Sci 23, 73–81.

Wang, X.Q., Ren, H.Q., Zhang, B., Fei, B.H., Burgert, I., 2012. Cell wall structure and formation of maturing fibers of moso bamboo (*Phyllostachys pubescens*) increase buckling resistance. J. Royal Soc. Interface 9, 988–996.

Wen, J.L., Xue, B.L., Xu, F., Sun, R.C., 2012. Unveiling the structural heterogeneity of bamboo lignin by in situ HSQC NMR technique. Bioenergy Res. 5, 886–903.

Xie, J.L., Huang, X.Y., Qi, J.Q., Hse, H.Y., Shupe, T.F., 2014. Effect of anatomical characteristics and chemical components on microwave-assisted liquefaction of bamboo wastes. Bioresources 9, 231–240.

Yang, X.W., Peng, Z.H., 2010. Cloning and expression of a cytochrome *P450* gene from moso bamboo. J. Anhui Agri. Univ. 37, 116.

Yang, S.M., Jiang, Z.H., Ren, H.Q., Fei, B.H., 2009. Variations of the microfibril angle in developmental moso bamboo culms. J. Nanjing Forestry Univ. (Natural Sci. Ed.) 33, 73–76.

Yin, S.C., 1996. Wood Science. China Forestry Publishing House, Beijing.

Yu, Y., Fei, B.H., Zhang, B., Yu, X., 2007. Cell-wall mechanical properties of bamboo investigated by *in-situ* imaging nanoindentation. Wood Fiber Sci. 39, 527–535.

Yuan, J.L., Fu, M.Y., Zhuang, J.K., Xiao, X.T., Wang, P.H., 2005. The characteristics of flowering and pollinating of several sympodial bamboos and elementary selection of *Dendrocalamus latiflorus* seedlings. J. Bamboo Res. 24, 9–13.

Zhang, Y., Bao, B., Lu, B., Ren, Y., Tie, X., Zhang, Y., 2005. Determination of flavone *C*-glucosides in antioxidant of bamboo leaves (AOB) fortified foods by reversed-phase high-performance liquid chromatography with ultraviolet diode array detection. J. Chromatogr. A 1065, 177–185.

Zhang, R., Xiong, S., Liu, J., Xue, D., Xue, P., 2011. Chemical constituents of bamboo root. Lishizhen Medicine Materia Medica Res. 22, 793–794.

Zhou, M., Yang, P., Gao, P., Tang, D., 2011. Identification of differentially expressed sequence tags in rapidly elongating Phyllostachys pubescens internodes by suppressive subtractive hybridization. Plant Mol. Biol. Rep. 29, 224–231.

Part IV

Advanced Techniques for Studying Secondary Xylem

Chapter 15

Microscope Techniques for Understanding Wood Cell Structure and Biodegradation

Geoffrey Daniel

Department of Forest Products/Wood Science, Swedish University of Agricultural Sciences, Uppsala, Sweden

Chapter Outline

Secondary Xylem Biology. http://dx.doi.org/10.1016/B978-0-12-802185-9.00015-2

309

GENERAL BACKGROUND: MICROSCOPE ANALYSIS OF WOOD STRUCTURE AND BIODEGRADATION

Since the first realization of the compound microscope in the late seventeenth century by Leeuwenhoek, microscopy in its very forms has helped to revolutionize our understanding of structures for life and physical sciences. In the fields of wood science and wood biology, microscopy has played revolutionary roles for our understanding of the fundamental aspects of wood cell formation, the hierarchical and ultrastructure of wood cell walls, and biomineralization of wood by fungi and bacteria. In its strictest sense, microscopy allows for the observation of objects that cannot be seen with the naked eye (i.e., less than 0.2 mm at 25 cm) and fundamentally may be divided into optical (i.e., light microscopy), electron (scanning and transmission electron microscopy (SEM/TEM)), and scanning probe microscopy. Scanning probe microscopy has seldom been applied for studies on wood structure and thus will not be considered further. Possibly, microscopy's greatest advantage is that it can provide complementary and detailed information on the structure of objects at low and high spatial resolutions not possible using other techniques. For example, while gross chemical analysis can give a global overview of the components in a biological object, microscopy techniques can often give precise information on how these components are distributed at tissue, cellular, and subcellular levels and even organelle level at low and high spatial resolutions. This is of considerable importance for understanding the basic wood anatomy and the variations in wood decay that can be developed particularly in relation to native tissue and cell types, native components (e.g., lignin type/concentration), or any form of artificial wood modification/protection. Microscopy in broad terms has developed into a multiarray of approaches that includes light, ultraviolet (UV), fluorescence, electron (SEM and TEM), X-rays, spectroscopy (Fourier transform infrared spectroscopy (FTIR), Raman, X-ray photospectroscopy (XPS), and time-of-flight secondary ion mass spectroscopy (TOF-SIMS)), and even atomic force signals (i.e., via atomic force microscopy, AFM) to image samples at different magnifications and resolutions providing morphological or morphological/chemical information (Fig. 15.1). Each of the approaches is also multifaceted and when combined with ancillary techniques allows for analysis of specific molecules in samples (e.g., localization of proteins and polysaccharides by fluorescence microscopy), the microdistribution of components (e.g., lignin via UV-microscopy or after binding metals that can be detected using SEM-X-ray or TEM-X-ray microanalysis), or the orientation of structural components (e.g., cellulose orientation using polarized light). Wood can also be viewed using complementary approaches, such as light microscopy and TEM, which can be carried out on the same sample areas thus combining global overviews with more detailed information. The choice of microscopy technique is generally dependent on the type of information required since morphological and chemical information on wood structure can be obtained at different levels (Fig. 15.1).

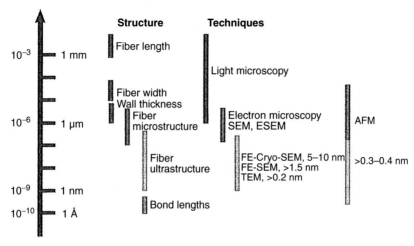

FIGURE 15.1 Simplified overview of techniques available and resolution with regard to wood structure.

In the following outline on the major microscopy techniques applied for studies on wood and wood decay, information is also given on sample preparation since few biological samples can be observed without some form of prior selection and preparation before observation. Since light and electron microscopy (SEM and TEM) and their ancillary techniques have been the predominant approaches used for the study of wood structure and wood decay, these approaches will be emphasized with lesser details given on the application of spectroscopy (FTIR, TOF-SIMS, XPS, Raman) approaches.

SAMPLE PREPARATION FOR MICROSCOPY

Importance of Sample Selection and Preparation – General Aspects

All forms of microscopy require some type of selection and preparation before objects can be observed. Prior selection is needed to make sure the area removed from a sample is representative of the material to be observed. This becomes all the more important where restrictions on sample size is governed by the microscopy approach selected, with TEM and high resolution techniques requiring greatest selection due to the maximum size that can be observed at any one time through physical restrictions of the specimen chamber and sample thickness used and often because of time and cost considerations. Good sample preparation is paramount for obtaining optimal and representative results especially where biological samples are prepared for high-resolution work or where the microscopy approach demands prior removal of water (i.e., conventional SEM, TEM).

Optical Microscopy: General Background and Applications

Light microscopy is the most frequent method used to study wooden objects and wood fibers of the order of magnitude 1 μm (0.001 mm) to 5 mm. Other light microscopy techniques that have frequently been used include fluorescence (Fig. 15.2a), ultraviolet (UV) (Fig. 15.2b), phase contrast, differential interference contrast (DIC), and dark field microscopy. Optical (bright field) microscopy is, however, the most common form of imaging carried out using conventional light approaches, with or without staining (Fig. 15.2c and d).

Optical (Light) Microscopy and Preparation Techniques

Bright field microscopy for the analysis of wood and degraded wood samples has been used over many decades for providing diagnostic information on basic wood anatomy (e.g., distribution of tissues and cell types) and for the identification of wood as well as for documenting the major patterns of wood decay produced by fungi and bacteria (Wheeler et al., 1989; Anagnost, 1998; Daniel, 2003; Richter et al., 2004; Schwarze, 2007). It has also been the given too used for a plethora of other fundamental aspects on the biology of wood (e.g., cell wall formation) and fungi/bacteria (taxonomy, etc.). Because of its physical nature, mature wood (as opposed to developing) is comparatively easy to prepare for light microscopy. Samples can simply be hand-sectioned using a razor blade to produce semithin sections of small areas (~1.0–4.0 μm thick) or large areas can be sectioned using a sliding microtome (e.g., 10–30 μm thick). Sections are then mounted in water on object glasses, covered with a coverslip and observed directly using bright field microscopy for taxonomic analyses. Problems arise if the wood material is very hard (e.g., dense tropical hardwoods), very soft (e.g., waterlogged archaeological wood), or partially/almost fully degraded by fungi/bacteria. In the case of hard samples, they can be softened using a mixture of alcohol/water/glycerin (ratio 1:1:3) over several hours/days, washed in water and then sectioned. In the case of very soft or degraded samples, the wood structure needs support (i.e., embedded in harder material to retain shape/structure) to allow semithin sectioning. This may be done by embedding samples in paraffin wax (suitable for large object areas), various methacrylate resins (e.g., Technovit®)/hydrophilic (e.g., London Resin) or hydrophobic resins (e.g., Agar and Spurr's resin (Spurr, 1969)). This is a lengthy process and it is therefore very important to decide the aim of the study beforehand. For example, if the aim is to conduct specific staining of wood cellular components (e.g., lignin via Mäule/Wiesner reactions, shown later) or polysaccharides (e.g., periodic acid–Schiff reaction for sugar vicinal diol groups) or use immunocytochemistry (e.g., detection of hemicelluloses, lignin) or visualize proteins or polysaccharides secreted and involved in fungal/bacterial wood decay, then the choice of resin type can be paramount for good detection. The primary advantage of resin embedding for light microscopy is that it allows thin sections to be cut (down to 0.1 μm) providing better focus and thereby detail compared with hand/microtome sections of nonembedded material. Use of hydrophilic/hydrophobic resin embedding and serial sectioning allows for the same sample area to be observed by light/fluorescence/confocal microscopy and by TEM. For example, observations of the same section area/ cellular region by complementing conventional fluorescence microscopy and TEM-immunocytochemistry using the same specific antibodies can provide correlative information (Kim and Daniel, 2012). The main disadvantage of embedding is that frequently, specific stains for light microscopy function poorly or not at all on sections after resin infiltration and polymerization (especially with epoxy resins) due to cross-linking of the wood material with the resin and/or because of hydrophobicity. This can sometimes be overcome through "etching" of the resin with sodium hydroxide, but there is always a risk that the components of interest are also removed or chemically modified. With samples embedded in paraffin or Technovit® staining is usually not a problem, but ultrathin sections (e.g., 70–120 nm) are normally difficult to cut from this material and observed using TEM because of instability (i.e., drift) under the electron beam. It is also possible to prestain wood samples before embedding. However, here problems can arise through loss of staining (i.e., bleaching) and component removal during subsequent ethanolic/acetone dehydration stages.

An alternative approach to resin embedding of soft material (e.g., waterlogged archaeological samples) is to freeze samples in LN_2 with or without 2.5M sucrose (protects against freezing artefacts) infusion and then use a cooled sliding microtome, cryostat or cryo-ultramicrotome for semithin sectioning. Advantages include the ease of staining and more rapid processing.

Examples Providing Information on Wood Structure and Wood Degradation

Wood sections are commonly observed without differentiation (i.e., staining/ contrasting) for taxonomic purposes (Wheeler et al., 1989; Richter et al., 2004). General staining with safranin (wood structure) (Fig. 15.3a), Alcian blue (Fig. 15.2d), Astra blue (Fig. 15.3c), and lactophenol blue (fungal hyphae/ bacteria) (Fig. 15.3d–f) may be used to emphasize general wood structure and presence of fungal and bacteria attack while more specific stains and "markers" (shown later) are applied to selectively localize wood components (e.g., polysaccharides/lignins) or fungal secretions (e.g., enzymes).

Specific staining and visualization of lignin can be carried out using Wiesner (phloroglucinol-HCL) or Mäule reagents (Nakagawa et al., 2012). Staining with Mäule's reagent allows the distribution of syringyl (i.e., S-lignin turns red) and guaiacyl units (G-lignin turn yellow) to be determined at tissue and cellular levels (Kim et al., 2012). The Mäule reaction is particularly useful for studies on the distribution of lignins in hardwoods, which usually possess both lignin types that may be distributed unevenly in different cell wall layers and cell types (e.g., vessel secondary walls are commonly G-lignified while

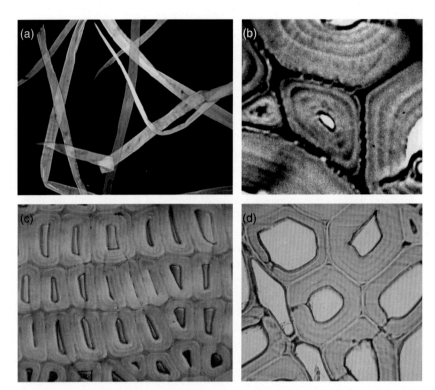

FIGURE 15.2   Comparison of light microscopy techniques for viewing aspects of wood structure. (a) Fluorescence of spruce wood fibers after staining with acridine orange; (b) cross-section of *Homalium foetium* viewed with UV-light at 280 nm; (c) *Elaterospermun tapos*; and (d) *Laurelia nova-zelandiae*.

fibers are S-lignified) and even across growth rings (early vs. latewood) (Kim et al., 2012). Total lignin can be visualized using the Wiesner reaction, which reacts with lignin hydroxycinnamaldehyde groups staining them pinkish red (Nakagawa et al., 2012). Such reactions are easily visualized using conventional bright field microscopy. Pectins in the primary cell walls of middle lamella regions and pit membranes can also be visualized as a violet color using 1% w/v toluidine blue or ruthenium red as can the presence of the gelatinous layer in tension wood (Daniel et al., 1996; Daniel, 2003).

The very close interaction and chemical bonding of individual wood chemical components presents difficulties for the specific staining of cellulose and major (i.e., hemicelluloses: xylan, glucomannans, galactans) and minor noncellulosic components (e.g., pectins; methyl-esterified and unesterified homogalactans). Hemicelluloses and pectins are best visualized using specific antibodies against whole or part (i.e., branched) of the homo/hetero-polysaccharide chain (Donaldson and Knox, 2012; Kim and Daniel, 2012; Kim et al., 2012). The simplest labeling approach consists of the following: (1) prior incubation of

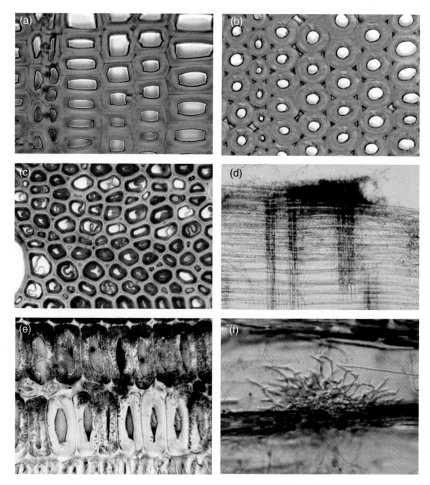

FIGURE 15.3 (a, b) Transverse sections of spruce mature and compression wood stained with safranin. (c) Birch tension wood stained with Astra blue. (d) Fungal hyphae stained with lactophenol penetrating rays in pine wood. (e, f) Tunneling bacteria in degraded spruce wood stained with lactophenol in transverse and longitudinal sections.

the semithin sections of wood tissue of interest in blocking agent (e.g., 3% w/v bovine serum albumin/chicken albumen) to cover nonspecific sites; (2) incubation with the specific antibody (requires time optimization); (3) washing in buffer and blocking agent; and (4) incubating sections with a secondary antibody bearing a fluorescent marker. Historically, the most common markers used are fluorescein isothiocyanate (FITC) and tetramethyl rhodamine (TRITC), although now a wide range of markers is available to enhance and reduce loss of signal through bleaching when examined in the microscope. This method is called "indirect labeling" and can be upgraded for the detection of several epitopes (antigens) on the same tissue sections using different secondary antibodies carrying markers excited at different wavelengths. An important aspect with double and multiple labeling approaches is that the primary antibodies (or manufactured peptides) should not show cross-reactions (i.e., usually facilitated by the antibodies derived from different immunized animal species). Cellulose and hemicellulose distributions in wood cells and tissues can also now be detected using "carbohydrate-binding domains (CBDs)" (i.e., part of cellulase/hemicellulase enzyme structure that normally binds (i.e., binding domain) the enzyme to their substrate) (Hildén et al., 2003; Filonova et al., 2007; Ruel et al., 2012). In principle, the microdistribution of "binding domains" can be visualized for light microscopy in much the same indirect way as for antibodies using secondary antibodies bearing fluorescent markers that specifically bind to the binding domains attached to sections. Observations on the microdistribution of the CBM markers can be carried out using conventional fluorescence microscopy or confocal laser microscopy, the latter allowing for better depth of field, observations of thick sections, and the combination of stacks of images (Donaldson and Knox, 2012).

With degraded wood samples the simplest approach is to stain transverse sections with 1% w/v safranin to visualize changes in wood cell walls and to highlight decay patterns (Daniel, 2003; Schwarze, 2007) and longitudinal sections with 1% w/v lactophenol blue, which stains fungal hyphae and bacteria blue without contrasting the wood material (Fig. 15.3d–f). With safranin, areas of fungal wood decay are normally easily detected by enhanced red staining (e.g., early stages of brown rot attack in the S_2 layer; as rings developing across wood cell walls with simultaneous white rot attack, or around soft rot cavities in wood fibers). Lactophenol blue staining is often used together with polarized light microscopy for examination of soft rotted wood for highlighting cavities and loss in cellulose birefringence with brown rotted wood (shown later). For visualizing preferential white rot decay (i.e., removal of lignin/hemicelluloses leaving cellulose), double staining with 1% w/v Astra blue followed by 1% w/v safranin can be used; the former staining areas blue where the lignin and hemicellulose have been removed/reduced and nondegraded regions (i.e., regions with remaining lignin/cellulose/hemicelluloses) red (Srebotnik and Messner, 1994). The same double staining approach can also be used for distinguishing tension wood in hardwoods (Fig. 15.3c) (Ruzin, 1999; Vazquez-Cooz and Meyer, 2002).

Use of polarized light with optical microscopy can also be used for providing information on the orientation of cellulose microfibrils (i.e., microfibril angle (MFA)) and loss of cellulose crystallinity in wood cell wall layers. The approach is useful for visualizing changes in MFAs across wood cell walls and for detecting changes in individual cell wall layers (e.g., presence of compression, tension wood) in fibers (Donaldson and Xu, 2005; Donaldson, 2008; Donaldson et al., 2010) and for detecting decay by soft rot and brown rot fungi (Fig. 15.4d, e and f). Using thin transverse sections (<0.5 μm), polarized light allows the transverse orientation (i.e., microfibril angles of 10–30°) of cellulose microfibrils in the S_1 and S_3 to shine brightly (i.e., show birefringence) against a dark, opaque S_2 and middle lamella regions. In radial longitudinal

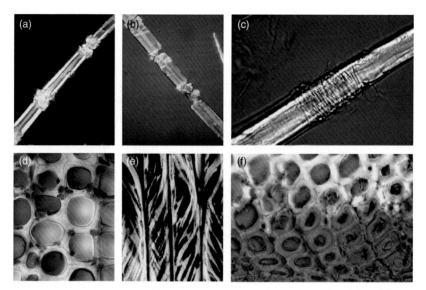

FIGURE 15.4   (a–c) Polarized light microscopy of delignified spruce fibers showing kinks, kinks degraded by enzymes (b) and perpendicular orientation of S1 secondary cell wall layer (c). (d, e) Half polarization of a cross-section of pine with soft rot cavities and loss of birefringence through cavity formation in longitudinal sections. (f) Use of polarized light microscopy for showing cellulose depolymerization and loss of birefringence in pine due to attack by a brown rot fungus.

(RLS) and tangential longitudinal (TLS) wood sections and delignified fibers (Fig. 15.4a–c), polarized light can provide information on disruptions in microfibrils in the fiber axis such as "dislocations" and "slip-planes." Polarized microscopy normally works best with thin hydrated (1.0–3.0 μm thick) nonembedded hand/microtome transverse sections. Polarized light is also useful for revealing the occurrence of different pit types, changes in MFA within and between pits, and for detecting the presence of crystalline deposits in wood sections for taxonomic purposes (Wheeler et al., 1989; Richter et al., 2004).

Because polarized light reveals the crystalline nature of the cellulose microfibrils in wood cell walls, it is particularly useful for demonstrating features of wood decay where cellulose has been preferentially attacked and depolymerized (Daniel, 2003). In this respect it can be used to detect incipient wood attack and development of brown rot decay where loss of cellulose birefringence (i.e., crystallinity) can be observed in large areas of the wood tissue and in single fibers in transverse wood sections (Fig. 15.4f). The same approach is used for showing decay in archaeological wood samples from waterlogged situations (e.g., shipwrecks) or from foundation poles and shipwrecks that have been degraded by erosion bacteria (Blanchette et al., 1990; Daniel and Nilsson, 1998). Polarized light is also useful for detecting soft rot attack (i.e., cavities/holes) produced by *Ascomycetes* and Fungi imperfecti in wood cell walls. Here, the cavities of various shapes/forms are shown best in longitudinal sections (TLS/RLS) by a loss in birefringence in the cell wall (see Fig. 15.4e). Since soft rot cavities are always orientated along the cellulose microfibrils, the cavities can be used to measure the MFA of secondary cell wall layers allowing comparison of MFAs between different xylem cell types (i.e., fiber, parenchyma cells) within and between wood species as well as differences in MFA orientation even along a single fiber, for example, between bordered pits (Khalili et al., 2001; Brändström et al., 2002, 2003). Comparisons made between MFAs measured using soft rot cavities and polarized confocal microscopy show quantitative agreement (Bergander et al., 2002).

SCANNING ELECTRON MICROSCOPY OF WOOD

Wood was found a perfect material for studies using SEM after the commercial availability of scanning electron microscopes in the late 1960s. SEM allows for three-dimensional images to be produced of wood structures after limited cleaning and planing and opens a whole new field of applications. Thus, a number of important reviews on wood structure and atlases on the fine structure of wood were produced during the 1970–1980s (e.g., Butterfield and Meylan, 1980).

The SEM approach has been used enormously over the years for looking at nearly all aspects of wood structure, initially simply from an anatomical point of view of the global structure, later for individual cell types and innate structure and then as a tool for relating structure to function, for example, pit structure and relation to closure (Choat et al., 2008) and adsorption of water. The great advantage of SEM is that it can be readily applied to all types and parts of wood samples if large enough to be accommodated for SEM processing. In addition, modern SEMs with large specimen chambers allow real-time dynamic studies of wood samples and fibers where the effect of tension and compression forces can be measured quantitatively on small wood blocks and isolated fibers. Initially, dynamic experiments were done on dried samples under high vacuum conditions but over the last decade, studies have been carried out under moist conditions using environmental SEMs (E-SEM) (discussed later) where the sample moisture content can be maintained under specified conditions (Mott et al., 1996; Turkulin et al., 2005a, 2005b). Another area where SEM has played a significant role is for demonstrating the microdistribution of metal actives in wood preservatives in impregnated wood (discussed later).

Electron microscopy (SEM and TEM) is commonly used to study wood structures that are either too small to be observed by the naked eye or too small/irregular to be viewed using optical microscopy. Fundamentally, an SEM allows for three-dimensional imaging by scanning samples maintained at either high/low vacuum, under moist conditions (E-SEM) or at elevated or cryogenic temperatures using a focused beam of electrons in contrast to the photons used in light microscopy. Electrons are generated and emitted from an electron gun (fitted with tungsten filament, lanthanum hexaboride (LaB_6) crystal or tungsten needle) and accelerated under the influence of a strong electrical voltage gradient down a high vacuum column. Using electron magnetic coils, an electron beam is formed and used to scan objects.

Electrons interact with atoms in the samples producing a range of signals (e.g., secondary/backscatter electrons, X-rays, cathodoluminescence, specimen current, transmitted electrons, auger electrons) that can be subsequently detected thereby providing information on the microtopography of the sample's surface structure as well as elemental composition (SEM-EDX, shown later). An electron beam is "scanned" across samples in a raster fashion with its position correlated with the signal to produce images. Decreasing the area scanned increases magnification. While interaction of the electron beam can produce a number of signals, the most fundamental mode of imaging is by using secondary electrons emitted from atoms excited by primary beam irradiation. By scanning across samples in zig–zag formation, each point on the surface hit by a primary electron emits secondary electrons, which provide information on the microtopography of the surface structure and thus images can be created. Some electrons, known as backscatter electrons, are reflected back from samples and provide information on the atomic number and hardness of samples. Depending on the source and density of electrons available, the resolution of the SEM can be dramatically improved. For example, the electron beam produced by modern Field Emission (FE)-SEMs is $\sim 1000\times$ better than in a conventional-(C)SEM and resolutions may be in the order <1.0 nm (i.e., 10^{-9} m) (Fig. 15.1). Although maximum magnification and resolutions are often quoted for SEMs, it is should be emphasized that most modern instruments function over a very wide range of magnifications (e.g., $20–100,000\times$) and thus low magnifications are equally important for giving three-dimensional observations of objects (e.g., irregular samples) not possible with optical microscopy. For further details, readers are referred to the many texts available in the literature concerning the operation and comparison of SEMs (e.g., Bozzola and Russell, 1999; Goldstein et al., 2003; Echlin, 2009). Here, only some examples in which SEM and FE-SEM have been employed to give better insights into wood structure/ultrastructure and wood decay together with a brief overview of the importance of sample preparation will be given.

Preparation Approaches for SEM and Ancillary Techniques

Numerous schemes can be used for the preparation of biological samples for SEM that have been adapted for studies on wood and degraded wood samples. Sample selection is again very important so that representative material is taken, although sample size may be less critical.

Most SEMs, apart from the high-resolution FE-SEMs with specimen side entry, can take quite large samples (i.e., several centimeters in cross-section), although 1 cm^{-3} planed wood cubes are frequently a common size. On the other hand there is no minimal size that can be processed so long as the wood samples can be handled during preparation.

SEM preparation approaches are dependent on whether the samples studied are taken from living wood tissue or dead wood. In its simplest form, pieces of mature wood can be shaped (e.g., via prior planing with a microtome) to reveal the three faces of wood (TS, RLS, TLS), dried (e.g., in an oven at 60°C for a few hours/or left in a desiccator), and observed in the SEM after prior coating or without coating (E-SEM). Various SEMs instrument types can be used (shown later) and therefore sample preparation is normally governed by the SEM equipment available.

Living wood tissues or biologically degraded wood samples can be processed after selection and fixation in 3% v/v glutaraldehyde + 2% v/v paraformaldehyde in buffer (e.g., phosphate/sodium cacodylate; 0.1 M, pH 7.2) followed by ethanolic (20–100% in 20% steps; 10 min each) and acetone (or amyl acetate) dehydration (60:30; 30:60; 100%, 15 min steps) and finally critical point dried (CPD) using liquid CO_2 as the substitution fluid. For CPD, wood samples in absolute acetone are inserted into precooled ($\sim 10°C$; cooled by circulating water or a cooling electrode) CPD chamber. Liquid CO_2 is then allowed to enter and mix with the acetone in the sample. Depending on sample size, impregnation with liquid CO_2 continues up to 1 h with intermittent "flushing" of any remaining acetone. Subsequently, the chamber is warmed slowly to the critical point of CO_2 (i.e., 31.5°C, 1100 psi (75 bar)) when the liquid CO_2 changes into gas. The gas is then released with the samples dried and ready for the next step.

An alternative approach is to freeze-dry (FD) samples directly without prefixation. Here, samples are simply plunged into LN_2 ice ($-210°C$), retained frozen (e.g., on a copper block), and freeze-dried overnight. During this process, water in the samples is evacuated (sublimated) at reduced pressure. Samples can also be dehydrated using t-butyl alcohol and freeze-dried (Inoué and Osatake, 1988). The purpose of CPD and FD is to dry biological samples optimally in order to reflect native conditions as much as possible; the procedures however are quite time-consuming. Alternatively, more rapid processing includes dehydration to ether via ethanol and drying in a dessicator or use of hexamethyldisilazane (Nation, 1983). This approach is an improvement on simply air-drying as it reduces surface tension forces that can cause severe structural damage and changes to the ultrastructure of the sample.

After drying for C-SEM, the wood samples are mounted on stubs (i.e., using carbon tags and double-sided cellotape) and normally coated with a thin layer (2–5 nm) of metal (e.g., either Au, C, Pd, Ag, Ti) to make the wood samples conductive under the electron beam. With C-SEMs this is necessary to dissipate the electron charge that can build up in samples (termed charging) under electron irradiation that causes damage (e.g., Fig. 15.5a). Wood samples and fibers can also be observed without coating and in these cases E-SEM using dedicated gaseous secondary detectors (GSE, Figs 15.7a and c, and 15.8a–d) or low vacuum-SEM with backscatter detectors imaging are used (Fig. 15.6c, d and f). With modern SEMs, samples are often observed using low acceleration voltage (e.g., 0.5–10.0 kV) and thus the penetration of the electron beam is reduced in the sample and build-up of charge is less (Fig. 15.6e). Preosmication (discussed later under TEM) of samples after prefixation in aldehydes is a further way to introduce a heavy metal into wood samples and thereby reduce charging and also increase the secondary electron signal but even here, samples are normally coated.

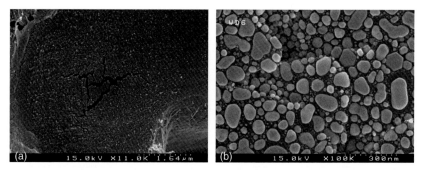

FIGURE 15.5   (a) Typical damage of a wood cell wall through electron irradiation without sufficient metal coating. (b) Exaggerated gold coating to show gold islands on the surface of wood at ×100,000.

The most common coating employed with conventional SEM is by a layer of Au (~2–5 nm), the thickness dependent on the type of SEM microscopy being performed and magnification required. Gold is the most common noble metal used because it is readily available, cheap, and easily evaporated in "sputter coating" devices at low atmospheric pressure. A disadvantage is that gold forms small islands on samples that are easily resolved at high magnifications (e.g., 100,000×) with modern FE-SEMs (Fig. 15.5b) and thereby can disguise the true surface topography of samples. Other metals (e.g., Cr, Ti) form much thinner and more homogeneous coatings, but require high vacuums (i.e., use of turbo/oil diffusion pumps) and energy to "sputter" the heavier atoms of the high atomic number elements. A very important aspect concerning the coating of samples concerns whether SEM-X-ray microanalysis will be performed (shown later).

In this case it is very important to choose the optimal metal for coating so the metal X-ray peaks do not interfere (i.e., overlap) with the characteristic peaks from the sample being studied. In this respect carbon was frequently used in the past because of its low atomic number ($Z = 12$) as a coating although its secondary electron signal is poor for morphological observations compared with the majority of noble metals. However, modern windowless detectors are able to detect down to boron ($Z = 5$) so the importance these days lies on selecting a metal outside the characteristic peaks of interest.

Conventional SEM and Field-Emission SEM

Conventional SEM (i.e., observation of samples under high-vacuum conditions) remains the most widely used approach of SEM for studying wood samples. Figure 15.6a and b shows typical examples of the structure of wood samples that are easily obtained using low-resolution C-SEM at medium high acceleration voltage under high-vacuum conditions. SEM gives three-dimensional morphological information on the surface topography of wood samples and at the cell wall level details of the structure of fibers, vessels, parenchyma secondary walls, and middle lamella regions (Fig. 15.6a and b). However, due to the nature

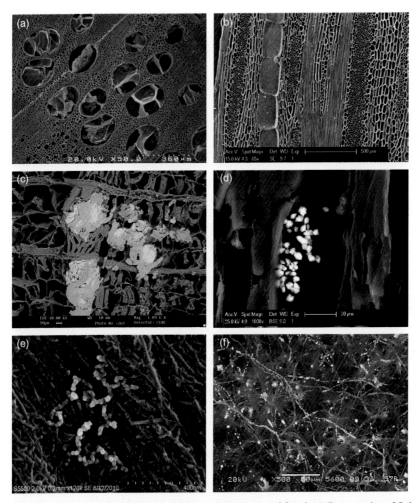

FIGURE 15.6 Use of C-SEM for studying wood structure and fungi. (a) Cross-section of Oak wood with vessels showing tyloses. (b) Radial longitudinal section of *Mitella* sp. with thick rays of parenchyma cells. (c, d) Cross- and longitudinal sections of deposits in waterlogged wood and silicon in preservative treated wood. (e) Macrofibrils of a fiber surface with deposits of xylan observed using 2.0 kV. (f) Fungal hyphae with oxalate deposits imaged using a BSE detector.

of preparation (sectioning) and resolution, the true details and native structure (e.g., relation between wood components cellulose, hemicelluloses, and lignin) of the secondary and primary cell walls are not easily revealed by C-SEM. With the advent of FE-SEM and improved resolution it was possible for the first time in the late 1990s to reveal aspects of the native structure of wood secondary cell walls previously only possible using freeze-fracture techniques (i.e., rapid freeze deep-etching, RFDE) in conjunction with TEM. The FE-SEM approach has been used with great success for revealing changes in wood fibers during different

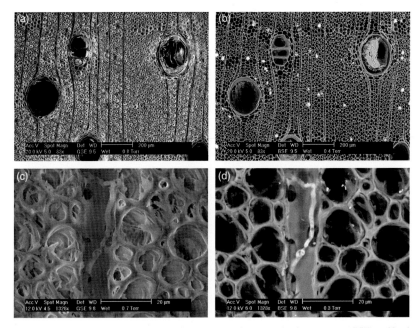

FIGURE 15.7 (a, b) Complementary E-SEM images of wet wood using gaseous (GSE) and back-scatter (BSE) detectors. Crystals are evident in parenchyma cells in the BSE but not SEI image. (c, d) White rotted wood observed by SEM with complementary GSE and BSE detectors showing fungal hyphae in a ray canal.

pulping (e.g., delignification) processes (Daniel and Duchesne, 1998; Duchesne and Daniel, 1999, 2000; Daniel et al., 2004) as well during wood decay.

Using FE-SEM, it has been possible to show that the cellulose microfibrils composing wood cell walls are coated by hemicelluloses and then lignin. Furthermore, the cellulose microfibrils have a tendency to aggregate into "macrofibrils" (*syn* = cellulose aggregates) on lignin/hemicellulose removal to form larger structures (e.g., 10–60 nm) (Figs 15.6e, 15.9d, and 15.12b–d) that are known to have a very important effect on fiber strength (Duchesne et al., 2001). How the microfibrils may group and regroup into aggregates is a very fundamental aspect for understanding the physical/mechanical aspects of wood cell walls and the wood structure itself. Initially, these FE-SEM observations were controversial but after they were shown in fiber samples after rapid freezing and Cryo-FE-SEM, it was confirmed that they were not artefacts of specimen preparation (Daniel et al., 2004). The fact that these observations provided a totally different view of wood cell walls previously accepted from TEM added to the controversy. In recent years, the existence of cellulose aggregates has been shown in variety of different wood cell types including the gelatinous layers of tension wood fibers (Daniel et al., 2006) and developing fibers during cell wall biosynthesis.

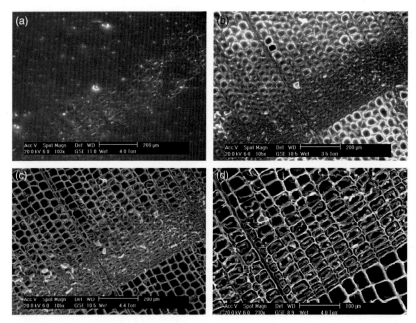

FIGURE 15.8 (a–c) E-SEM images using a GSE detector for visualizing bacterial (erosion bacteria) decayed waterlogged wood. (a–d) shows the effect of progressive water removal by changing the surrounding pressure in the specimen chamber. When water is present (a) and the sample waterlogged very little detail of the sample is shown.

C-SEM, FE-SEM, and Cryo-FE-SEM have helped greatly in our understanding of wood decay processes by fungi and bacteria. The SEM approach has confirmed many aspects of decay implied by light microscopy and added greater detail and additional information not possible previously. Typical examples include the concentric slime deposits produced in tunnels after tunneling bacterial attack of wood and ensheathment of bacteria in slime during the penetration of wood cells (Daniel et al., 1987), the "opening-up" of wood cell walls (Fig. 15.10c and d) and selective removal of lignin and hemicelluloses during white rot decay, the secretion of various types of extracellular polysaccharides during decay giving evidence for their involvement in biodegradation processes (Figs 15.9a, b, 15.10a), and the ability of germinating fungal spores (e.g., blue stain fungi) to grow through very small holes during the penetration of wood cells and coatings on wood (Daniel, 1994; Bardage and Daniel, 1997).

FE-SEM has not in any way replaced conventional SEM but provides possibilities of giving greater detailed observations because of the electron source and density of electrons possible.

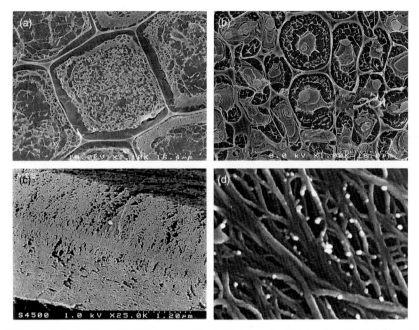

FIGURE 15.9 (a) Cryo-FE-SEM images of bacterial decayed waterlogged wood. (b) White rot decayed birch wood. (c) Cross-section of brown rotted birch wood. (d) Cellulose macrofibril structure of the surface of delignified spruce fiber showing porosity.

E-SEM, FE-E-SEM, FE-SEM, and Cryo-FE-SEM

For E-SEM and FE-E-SEM selected moist unfixed and never dried wood samples are simply placed on SEM stubs and introduced into the E-SEM chamber on a cooled specimen stage. The temperature of the specimen stage is normally kept around $-20°C$ and the stubs/sample at $1-5°C$. Observations in the wet (moist) mode are possible by retaining the pressure in the specimen chamber between 5.5 torr and 6.0 torr. The speed and lack of a lengthy preparation process is therefore very attractive. Images are produced in a dedicated E-SEM by the secondary electrons emanating from samples interacting with gas molecules in the chamber (held at atmospheric conditions) that are subsequently detected by a special gaseous detector (GSE). In contrast, other low-vacuum instruments use a variety of backscatter detectors (BSE) to produce the images (Fig. 15.7b, d). Figures 15.8a–d show typical E-SEM images of waterlogged wood degraded by erosion bacteria at low vacuum using a GSE detector and Figures 15.7a–d wood degraded by fungi comparing both BSE and GSE modes. E-SEM has been widely used for analyses of wood samples but difficulties can arise when examining samples to achieve the same quality of images possible with high vacuum or Cryo-FE-SEM because of the presence of water (Fig. 15.8a).

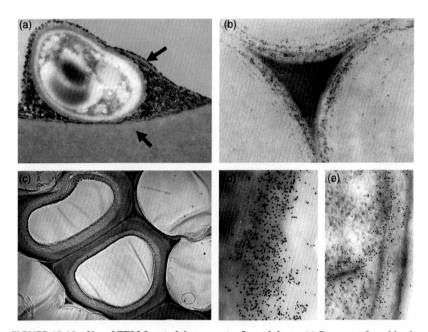

FIGURE 15.10   Use of TEM for studying aspects of wood decay. (a) Brown rot fungal hypha attached to the pine cell lumen wall. (b) Cross-section of remaining middle lamella cell corner after white rot decay of spruce fibers with immunogold labeling of manganese (Mn-)peroxidase. (c) Cross-section of spruce wood with zones of decay in the inner secondary cell wall. (d) Immunogold labeling of Mn-peroxidase (15 nm particles) and laccase (5 nm particles) in zones of white rot decay. (e) Part of a fungal hyphae with immunogold labeling of cellobiohydrolase in the periplasmic space.

Cryo-SEM and Cryo-FE-SEM

The advantage of the cryo-method is that samples can be rapidly frozen and processed without the use of solvents and any drying procedure, thus retaining samples in a more native state. Also, the use of cryo-fracturing allows samples to crack along weak points in the wood/fiber structure. The method has therefore potential application for viewing the morphology of organic materials in wood surfaces at high resolution not possible using conventional SEM preparation processes involving solvents. The difference between Cryo-SEM and Cryo-FE-SEM is simply related to the resolution possible rather than any differences in handling procedures.

Preparation of samples for Cryo-SEM and Cryo-FE-SEM is done by attaching samples to brass stubs using minimal amounts of Tissue-Tek® – a water soluble glycol/resin type formulation that provides a matrix for attaching samples at cryo-temperatures of $-10°C$ and below. Samples are then plunged into a LN_2 slush ($-210°C$) using a cryo-station (e.g., Oxford CT 1500 HF). After rapid freezing the samples are rapidly transferred into a cryo-preparation chamber (e.g., Oxford CT 1500 Cryo-chamber) held at −160°C. After raising the temperature to about −110°C, the samples can be fractured to expose the inner surfaces using a steel knife operated from the outside of the chamber. Following fracturing, the samples are "deep-etched" whereby ice on the surface is sublimed by raising the temperature between −90°C and −95°C (e.g., ~1–5 min for wood samples). This is a very important step for revealing the sample structure, otherwise only an ice profile will be visualized. Samples are then coated with Au/Pd using a "sputter coater" (e.g., Denton coater) and transferred into the Cryo-FE-SEM chamber and maintained at −140°C. Observations on wood structure can then be conducted at low acceleration voltage (0.5−10 kV) at varying working distance.

Despite its potential, surprisingly few studies have been performed using Cryo-SEM on wood or biologically degraded wood samples, a notable exception being that by Daniel et al. (2004). In particular, the Cryo-FE-SEM approach is very useful for examining fragile waterlogged wood samples degraded by bacteria, which retain their encapsulation in extracellular slime and breakdown materials (Fig. 15.9a; Daniel, unpublished observations). Using correlated Cryo-FE-SEM and TEM approaches, it was possible to show how extracellular polysaccharides/glycoprotein secretions are uniquely involved in fiber cell wall thinning of wood cells during simultaneous white rot attack and the different morphological forms present (Fig. 15.9b). The TEM-immunogold approach further showed the polysaccharides to carry diverse extracellular enzymes capable of lignin and polysaccharide degradation. The Cryo-FE-SEM approach has also been used to examine the ultrastructure of chemical pulped wood fiber cell walls revealing the true fiber wall structure and porosity after lignin and part hemicellulose removal allowing for better understanding of wood cell wall ultrastructure (Fig. 15.9d).

The advent of FE-SEM with improved resolution resulted in a somewhat recline in the numbers of studies previously carried out by TEM in the wood research field. As a technique, TEM is much more time-consuming, technically more challenging and requires a greater array of facilities, thus it was not surprising that an easier approach with improved resolution that could solve problems would be more attractive. As indicated earlier the challenge with FE-SEM concerns optimizing specimen preparation for biological specimens with Cryo-FE-SEM the best possible alternative offering possibilities for rapid freezing and retaining samples in optimal conditions as well as high resolution.

TRANSMISSION ELECTRON MICROSCOPY (CONVENTIONAL TEM)

TEM can be used to study the internal structure (ultrastructure) of biological samples like wood (e.g., during biosynthesis) and wood undergoing biodegradation. TEM is a technique whereby a beam of electrons accelerated (typically

40–400 kV) from an electron source (similar as for SEM) are transmitted through ultrathin specimens (typically <100 nm) providing information that can be visualized. In sample areas where the electrons encounter atoms of high atomic number (e.g., nuclei of heavy metals) they rebound while in areas where the sample is condense or the section thick, the electrons are repulsed or absorbed. However, in sample areas consisting of light atoms or which are thin and less concentrated, the electrons pass through the sample. The final image is formed from a greatly magnified (i.e., through electromagnetic lenses) projection of the interaction pattern of electrons transmitted through the specimen and focused onto a fluorescent screen, film or nowadays CCD camera. As an example, Fig. 15.10a shows a transverse section of a fungal hyphae from brown rotted wood. The lighter areas show where the electrons have passed through the sample (e.g., fungal cytoplasm) and the darker regions where greater absorbance has occurred. The contrasted image formed corresponds with the selective patterns of reflection or transmitted electrons, which in turn depend on the local properties of the sample.

When samples are sufficiently thin and small enough (e.g., fungal/bacteria proteins, cellulose microfibrils (Fig. 15.12a), isolated hemicelluloses, lignins) they can be viewed directly using TEM without embedding and sectioning. Modern high-resolution TEMs can image samples down to their various atoms and typically >0.5 Å is possible for most FE-TEMs. Like SEMs, TEMs can also function over a wide range of magnifications (e.g., 35–700,000×) depending on instrument type and main dedication.

Since its first application for studies on wood in the 1960s, TEM has contributed greatly to our understanding of the ultrastructure of wood (e.g., during biosynthesis) and wood decay processes. Wood samples, like most other biological materials studied by TEM, normally require extensive sample preparation to produce sections thin enough to allow passage of the electron beam during observations in the microscope. TEM analyses tend therefore to be time-consuming with low throughput of samples with great emphasis on optimal selection during the early stages of studies as well as specimen preparation. Despite these drawbacks, TEM and its host of ancillary approaches are unprecedented with regard to the information that can be obtained. Today TEM, like SEM, is a multifaceted technique with an array of ancillary techniques that can be used. Some of the most frequent approaches applied for studies on wood structure and wood decay include conventional (C-)TEM for morphological observations of ultrathin sections after general contrasting (Fig. 15.10a) or specific staining (Fig. 15.11a and b) or after immunocytochemistry/immunogold approaches (e.g., labeling of enzymes and wood components; Fig. 15.10b, d, and e), elemental analysis using TEM-EDX, TEM diffraction of cellulose, and TEM analysis of metal replicas from rapidly deep-etched freeze fractured (RFDE) samples (Fig. 15.12c). A brief overview of some of these approaches and sample preparation is given in the subsequent section.

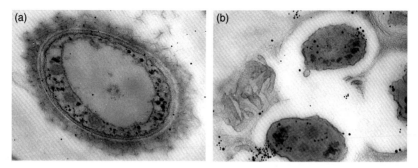

FIGURE 15.11   (a, b) TEM-PA-TSC-SP approach showing positive staining of extracellular poly-saccharides associated with a fungal hyphae and secretions from tunneling bacteria.

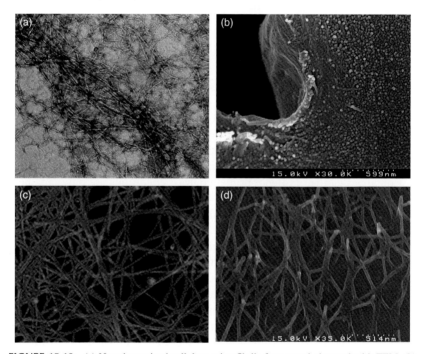

FIGURE 15.12   (a) Negative stained cellulose microfibrils from wood observed with TEM. (b) Cellulose macrofibrils in a cross-section of the secondary S_2 cell wall of spruce using FE-SEM. (c) Primary cell wall cellulose macrofibrils produced via RFDE and observed using TEM. (d) RFDE replica of cellulose macrofibrils observed using FE-SEM.

TEM Sample Preparation (Chemical Fixation, Dehydration, Resin Embedding, etc.)

Assuming the wood samples have living contents (e.g., undergoing bio-synthesis or decay), C-TEM normally involves the prefixation of samples in glutaraldehyde (e.g., 2.5% v/v) or a glutaraldehyde–formaldehyde (e.g., 2.5% + 2.0% v/v) mixture in 0.05–0.1 M buffer (pH 7.2) (e.g., sodium cacodylate, phosphate, Pipes buffer). After washing and postfixation in OsO_4 (e.g., 1% w/v overnight) and further washing in water, samples are dehydrated in alcohol or alcohol/acetone (for hydrophobic resins) series and slowly infiltrated with a hydrophilic (e.g., London resin) or epoxy resin (e.g., Spurr's resin) with the resin finally polymerized overnight at 60–70°C (i.e., as described earlier for light microscopy). Following ultrathin sectioning using an ultramicrotome, the resin-embedded wood sections are normally stained using lead citrate (1% w/v) and/or uranyl acetate (4% w/v) or aq. 1% w/v $KMnO_4$. Many variations on the general scheme exist, particularly the use of $KMnO_4$ as a prefixative or en-bloc prestaining with uranyl acetate shortly after fixation or during alcohol dehydration. $KMnO_4$ has been adopted by many wood researchers because of its specificity and general contrasting of lignin (Bland et al., 1971).

Historically, epoxy embedding has generally been employed for TEM studies because of the need for a hard supporting substrate and here Spurr's (1969) resin has been the choice because of its uniformity, low viscosity, and rapid penetration into wood materials. Major technical difficulties still exist in the preparation of wood samples and ultrathin sectioning (i.e., 70–100 nm) of undegraded, highly lignified (e.g., pine), or high-density wood is difficult, despite improvements in resins and equipment over the last few decades. The type of epoxy resin used also has an important effect on the degree of contrast obtained and very often, optimal contrast conditions are only obtained after changes in acceleration voltage (40–100 kV) and aperture selection in the TEM. Epoxy resins are less suitable for immunolabeling studies than hydrophilic resins.

A number of TEM-cytochemical approaches are available for use on ultrathin sections for detecting wood components. The limitation with cytochemical stains is that the contrasting agent must carry a heavy metal to be visible in the TEM during observations. One of the most useful stains for carbohydrates is periodic acid-thiosemicarbazide-silver proteinate (PA-TSC-SP).

Periodic acid oxidizes 1,2-glycol linkages of some carbohydrates to aldehydes, thiosemicarbazide then reacts with the aldehydes whose sites are revealed by thiosemicarbazide reducing silver proteinate and releasing silver at the same site. The approach is equivalent to the periodic acid–Schiff test at the light microscope level. The PA-TSC-SP approach has been very useful with contrasting isolated wood polymers and demonstrating extracellular polysaccharides produced by wood decay fungi and bacteria (Fig. 15.11a and b).

Conventional-TEM has been used to confirm the general patterns of wood decay by all the major groups of microbes colonizing wood including bacteria, actinomycetes, ascomycetes, higher ascomycetes, and basidiomycetes; the latter including white (i.e., simultaneous and preferential lignin degraders) and brown rot fungi (Chapter 8; Daniel, 1994, 2003).

TEM-Immunocytochemical/Immunogold Techniques for Labeling Extracellular Proteins and Wood Components

One of the major TEM approaches that have given important insights into our understanding of wood cell wall biodegradation processes, cell wall biosynthesis, and spatial distribution of cellular components in wood cell walls has been the application of immunocytochemical/immunogold labeling techniques. The approach is similar to the immunocytochemical/fluorescence labeling procedure described earlier during which specific primary antibodies (monoclonal/polyclonal) used to label epitopes (antigens) exposed in tissue sections are visualized by specific secondary antibodies conjugated in the TEM approach to gold-labeled (e.g., 1–15 nm gold particles) or gold-labeled protein A or G. This indirect method is carried out on resin embedded or cryo-sections after processing samples or by prelabeling whereby wood samples are exposed to the primary and secondary antibody steps before resin embedding thus revealing surface antigens. The difference between the light microscopy and TEM approach is that ultrathin sections carried on grids are used for the various labeling and washing processes.

By using immunogold cytochemistry, the occurrence and spatial distribution of many of the major fungal wood-degrading enzymes (i.e., hydrolytic, oxidative) identified and isolated by biochemical means (e.g., lignin peroxidase, Mn peroxidase, laccase, cellobiohydrolases) and sugar oxidizing/H_2O_2 producing enzymes (pyranose 2-oxidase, glucose 1-oxidase, alcohol oxidase; (Daniel, 1994; Daniel, 2003) from important white rot and brown rot fungi have been demonstrated during wood degradation providing strong evidence for their involvement during lignocellulose decay. The advantage of the immunogold technique lies in the discrete nature of the labeling process whereby gold particles can show the indirect location of epitopes within tissues. By varying the procedure, it is possible to label single, double, and multiple enzyme systems and thereby build up an understanding of the spatial distribution of several enzymes during specific phases of wood decay. Figure 15.10b, d, and e show examples of immunogold labeling of Mn-peroxidase and laccase associated with degraded wood cell walls and cellobiohydrolase in the periplasmic space of the white rot fungus *Phlebia radiata*. The main disadvantage of the approach when using lignocellulose as a substrate is that nonspecific labeling and false positive results can easily be obtained.

Using the same approach, wood cell wall components (e.g., xylans, glucomannans, pectins, xyloglucan) have also been detected using the indirect immunogold labeling approach thereby giving information on their spatial microdistribution and chemistry across wood cell walls (Donaldson and Knox, 2012; Kim and Daniel, 2012). This has been made easier over recent years through the commercial availability of numerous carbohydrate monoclonal antibodies (e.g., Plant Probes, Leeds UK; Carbon Source, Georgia, USA). Like immunofluorescence approaches, visualization of cellulose *in situ* using specific antibodies is still not possible and in recent years these problems have been overcome by the use of gold-labeled carbohydrate (cellulose)-binding modules (CBM) with indirect labeling employed in the same way as for antibodies (Filonova et al., 2007; Ruel et al., 2012). A similar approach has also been used for xylans (Filonova et al., 2007). The spatial distribution of lignins in wood cell walls has also been demonstrated using a variety of antibodies made to whole lignins (e.g., mill wood/sulfate lignins; Daniel et al., 1991) or S- and G-lignins (Yoshinaga et al., 2012). Here, the immunogold labeling process has also been indirect. Problems have however been experienced with characterization of the epitopes to which the original antisera were made. Immunolabeling of sections can also be applied as a direct method for visualization of epitopes. In this case the primary antibody needs to be conjugated directly to the label (i.e., gold).

Like C-TEM, with TEM immunocytochemistry, the wood samples are normally fixed, dehydrated, and resin-embedded prior to thin-sectioning, primary labeling with antibodies and post-immunogold labeling. Unfortunately, good ultrastructure and retention of antigenicity are often mutually exclusive and consequently samples are often weakly fixed in formaldehyde (e.g., 3% v/v) and low concentrations of glutaraldehyde (e.g., 0.1% v/v), dehydrated directly in ethanol and embedded in hydrophilic resin (e.g., London resin). OsO_4 is generally omitted during fixation because of its negative effects on antigenicity (especially proteins) and because it can cause premature cross-linking of some hydrophilic resins. Weak fixation and omission of osmication followed by direct dehydration, while providing good antigenicity often results in poor fungal ultrastructure and very often severe osmotic changes and shrinkage of fine structure. Like immunofluorescence, the type of resin used for embedding can also have an important effect on labeling, and hydrophilic resins (e.g., London resin) in contrast to hydrophobic resins (e.g., epoxy) are preferred. A negative feature of the weak cross-linking obtained with hydrophilic resins together with weak aldehyde fixation can be the loss of epitopes during EM processing and ultrathin sectioning. Poor embedding in hydrophilic resins can also result in poor polymerization, which in turn can cause nonspecific binding of primary antibodies and difficulties with sectioning of wood materials and instability in the TEM.

Originally, the drawback with immunolabeling procedures was the need for specific antibodies.

For the enzymes of wood-degrading fungi this remains true in most cases although modern molecular methods allow for antibodies to be produced to protein sequences determined by genomic studies thereby simplifying the process. Specific antibodies to wood carbohydrates are now available commercially and can be used in TEM-immunolabeling procedures.

TEM-Viewing of Cellulose Microfibrils and Macrofibrils

When wood samples are thin enough, they can be observed directly in the TEM without major preparation. This includes proteins (enzymes) isolated from fungi involved in wood cell wall decay and isolated cellulose microfibrils and carbohydrate polymers. Isolated cellulose microfibrils or aggregates can simply be pipetted onto formvar-coated or holey grids and negative stained using heavy metals like 1% w/v uranyl acetate or 1% w/v phosphotungstic acid. Studies on wood cellulose microfibrils can be quite challenging because of the size of fibrils (~3–4 nm in cross-section) and the rapid damage they experience directly on beam irradiation.

Figure 15.12a shows typical cellulose microfibrils from wood after negatively staining with uranyl acetate. Isolated polysaccharide chains (e.g., xylan, glucomannan) on TEM grids can also be negatively stained or labeled using antibody and immunogold procedures to characterize their homo/heterogeneous nature (e.g., degree of branching) (Reis et al., 1991; Reis and Vian, 2004).

Contrasting of isolated cellulose and hemicellulose chains on grids for TEM analysis can also be performed using shadowing procedures with heavy metals like Pd/Au – a technique that has largely been abandoned in favor of the negative staining approaches.

Some Recent Developments in TEM Analysis of Wood Cell Structure: TEM-Tomography

In the conventional mode TEM provides two-dimensional images of objects with no depth. Over the last decade together with the rapid development of computer software, it is now possible to conduct TEM-tomography (i.e., electron tomography, ET) of objects and construct three-dimensional tomograms. The technique involves progressively rotating the specimen holder with sample perpendicular to the axis of the beam through different angles (e.g., 1° increments) in the TEM and at the same time taking multiple stacks ("tilt" series) of images of a single sample area. The stacks of images are then used to construct three-dimensional representations (electron tomograms) of samples. The approach suffers from the fact that samples cannot be viewed with full 180° rotation in a TEM (termed "missing wedge") but can be improved using mechanical refinements like "multiaxis tilting" and "conical tomography" and by applying numerical approaches. The tilt series (stacks of images) can then be used for reconstruction purposes using a number of dedicated software programs (e.g., Amira). The reconstructions can thereby provide intricate details on dynamic processes (e.g., signaling) as well as morphological fine structure at high resolution. When the tilt series of images are taken, the sample area irradiated is subjected to considerable and repeated electron irradiation. To reduce this, samples are normally viewed using the technique of minimal dose ("low dose") whereby prefocusing and centering of images after tilting of specimens is conducted in an area adjacent to the area of interest. Thereafter, this area of interest is then switched back for image recording. To date the TEM-tomography has been mostly used in the medical and some plant research fields but has tremendous potential for understanding wood cell wall ultrastructure, biosynthesis, and decay mechanisms.

RAPID FREEZING APPROACHES

As outlined earlier, it can be very difficult to prepare satisfactory ultrathin sections (70–100 nm) from wood samples, thus the use of resins for embedding samples in order to give support. During the early days of TEM, neither epoxy resins nor the advanced ultramicrotomes available today were developed. Samples were embedded in early methacrylate resins, which suffered greatly from uneven polymerization and severe shrinkage. To circumvent these problems, wood samples were often rapidly frozen using LN_2, inserted in a freeze-fracture device, fractured, the surface ice sublimed, and the samples then coated with a layer of metal (Au/Pd) and then carbon. The process is known as "rapid freeze deep etching" (RFDE). Thereafter, the samples are removed from the RFDE apparatus, the organic materials dissolved by acid, and the final metal replicas of the original wood structure surface, washed and viewed in the TEM. The same approach is used today although the equipment is much more advanced. Typically, samples are viewed at 20,000×. Figure 15.12c shows an example of a metal replica of the primary cell wall from a spruce fiber after lignin and hemicellulose removal, the cellulose aggregates clearly visible. Such replicas can also be viewed with FE-SEM (Fig. 15.12d) and appear morphologically quite similar.

Development of Rapid Freezing Techniques for Wood Samples

Over the last two decades, great advances in the development of rapid freezing and tissue processing techniques has been developed in SEM and TEM for the life sciences. The approaches available are numerous but surprisingly relatively few have been used for studying aspects of wood decay and wood structure. Apart from the Cryo-FE-SEM example given earlier in which samples are rapidly frozen in LN_2 ice using "plunging," samples can be frozen using high-pressure freezing (HPF, high pressure under LN_2 cooling gives the greatest thickness of optimally frozen material known; ~200 μm). Thereafter, samples can be transferred for freeze substitution (FS) during which the sample can be fixed (e.g., glutaraldehyde/OsO_4 at −90°C in acetone/alcohol) and the water in the tissue subsequently removed progressively under low-temperature conditions (starting from −90°C). Samples can then be embedded directly at low temperature(s) using specially developed low-temperature embedding resins (e.g., Lowicryl) and the blocks polymerized using UV-light at −20°C. Samples can then be sectioned at room temperature for morphological/immunocytochemical studies. The advantage of this approach is the combination of rapid freezing and progressive water removal and embedding at low temperature, which provides for an improved ultrastructure of samples and retention of components compared with traditional chemical fixed and processed samples. It can also provide for improved retention of antigenicity of sensitive epitopes.

Other variations include the HPF freezing of samples followed by transfer to a cryo-ultramicrotome, cryo-sectioning (sections 70–100 nm cut between

−90°C and −110°C), transfer of sections under cryo-conditions (i.e., cryo-transfer device) to a cryo-TEM, and subsequent observations of frozen hydrated sections. This approach avoids the fixation, dehydration, and resin embedding phases, thus reducing the time needed as well as preserving samples in the best possible way. Although, the process is technically difficult to apply to wood (wood is an insulator and difficult to freeze) from freezing and sectioning points of view, the approach represents the future. In the same way, samples can also be HPF-frozen and transferred to the cryo-FE-SEM for direct viewing (or cryo-fracture) after surface ice sublimation or to a dedicated cryo-freeze fracture apparatus.

A discussion on the plethora of the different cryo-techniques currently available is outside the scope of this chapter and readers are referred to the several books now available on the topic (Cavalier et al., 2009; Hurbain and Sachse, 2011).

APPLICATION OF ANALYTICAL TECHNIQUES (SEM-EDX, TEM-EDX) FOR UNDERSTANDING WOOD STRUCTURE AND WOOD DEGRADATION

SEM and TEM energy dispersive X-ray (EDX) microanalysis or energy dispersive analysis of X-rays (EDXA) is an X-ray technique that allows identification and characterization of the elemental composition of materials. In wood research, EDX has been used for analysis of wood chemical structure (e.g., lignin distributions, presence of crystals) and modification (e.g., after impregnation with heavy metals, binding of cofunctional agents), and determination of the elemental composition of inorganic substances associated with wood during fungal decay (Daniel and Nilsson, 1985, 1989; Daniel et al., 1987). The EDX system is attached to the SEM/TEM column with the imaging capability of the microscope used for identification of areas of interest. EDX functions by the focused primary beam of electrons stimulating X-ray emissions from atoms in samples. In the unexcited ground state, atoms have electrons in discrete energy levels positioned in electron shells surrounding the nucleus. During excitation with the electron beam, electrons can be ejected from shells creating holes. When this occurs, an electron from an outer high-energy shell will fill the hole and in doing so the difference in energy between the higher- and lower-energy shells can be released in the form of an X-ray. Since the energy of the X-rays is always characteristic of the difference in energy between the two shells and the atomic structure of the element they are derived, this allows the elemental composition of samples to be measured. The number and characteristic energies emitted are detected using an energy dispersive detector attached to the microscope. Using this approach it is possible to capture information and correlate the three-dimensional morphological view of samples with global elemental "mapping" or at specific points using "spot" and "linescan" analyses of the diversity of elements present (Fig. 15.13).

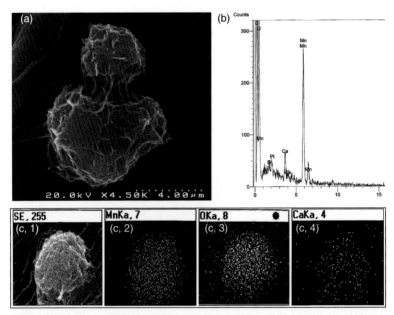

FIGURE 15.13   Use of SEM-EDXA for elemental analysis of precipitates in white rotted wood. (a) Mn precipitate on the fiber lumen surface. (b) X-ray spectra of the precipitate showing major peaks (i.e., Kα) for O, Ca, Mn, and P. (c) Complementary dot maps of the same precipitate showing signals for (1) SEI, (2) Mn, (3) O, and (4) Ca.

For wood analysis, SEM-EDX has been the method of choice because of preparation simplicity even though the resolution is inferior to TEM-EDXA. With the resolution now possible with FE-SEM-EDXA together with the high intensity beam and window-less detectors (detection down to B: $Z = 5$), SEM-EDXA approaches still dominate. With SEM-EDX, the characteristic elements emitted from surface regions and the depth (e.g., usually in micrometer range) from which X-rays are detected will vary with sample type (i.e., density) and operating conditions (i.e., accelerating voltage). As a general rule, an acceleration voltage of 2–3$\times$ the atomic number of the elements of interest in the sample is often used with heavier atomic elements needing greater excitement depending on electron shell used for analysis. TEM-EDX does not suffer from resolution nor excitation problems; however, the disadvantages lies with the lengthy period of preparation (shown earlier) and possible loss of elements (e.g., soluble elements). One of the major advantages of SEM/TEM-EDXA is its ability to provide elemental information with high spatial resolution not possible with other approaches. For example, while gross chemical analyses can give an estimate of the total quantity of element in tissues, it cannot give any information on their spatial distribution within samples at the cellular level. In addition, the X-ray information derived can be qualitative and quantitative if elemental standards are used. Two restrictions of EDX include specimen damage that usually occurs during beam irradiation, which can affect X-ray emission and the sensitivity of EDX systems requiring at least $10^{-16}-10^{-18}$ g of the element of interest (concentrated at a single point/not dispersed) and minimum concentration of 0.005–0.010% although this is very dependent on the nature of the element(s) being detected.

SEM- and TEM-EDX have been used as an important technique for visualizing and quantifying the spatial microdistribution of lignins (S- and G-lignins) across wood cell walls after covalent bonding of Br ($Z = 35$) or Hg ($Z = 80$) to the aromatic rings of the lignin (Saka and Thomas, 1982). The approach was also used to confirm the differential removal of lignin during simultaneous and preferential white rot decay of wood cell walls (Otjen et al., 1988). SEM/TEM-EDX has also been used to confirm the presence of MnO_2 and other inorganic elements (e.g., calcium oxalate) during fungal (white and brown rot) attack of wood and thereby provide indirect evidence for involvement in the enzymatic and nonenzymatic mechanisms of wood decay (Blanchette, 1984; Daniel, 1994; Daniel and Bergman, 1997; Daniel et al., 1997).

SEM/TEM-EDX can also been used for studies on the spatial microdistribution of metals (e.g., CCA, copper-chromate-arsenic) and their binding in wood tissues (e.g., to lignin) and individual wood cell walls, often in relation to fungal and bacteria decay (Daniel and Nilsson, 1985, 1987, 1998; Daniel et al., 1987). Here EDX has given important information on the penetration pathways and retention of metals in wood tissues and cells as well as providing insights into fungal and bacteria metal detoxification mechanisms. More recently, the approach has been used for determining the pathways and binding of modern wood treatments such as micronized copper in wood tissues and cell walls (Matsunaga et al., 2004, 2009). Since the EDX technique requires the presence of metals, the approach is not suitable for studying microdistributions of organic treatments (e.g., chemical modifications) in wood unless a metal ion can be attached to the chemical formulation being used. For further details, readers should consult one of the many general texts on SEM/TEM-EDX available (Goldstein et al., 2003; Echlin, 2009).

ADDITIONAL TECHNIQUES FOR STUDYING WOOD CELL WALL STRUCTURE AND BIODEGRADATION

Confocal Microscopy

Compared with light microscopy, confocal laser microscopy (CLSM) has the advantage of being able to add depth of field to the viewing of sample sections after staining with a fluorochrome (e.g., 0.0025% acriflavine/acridine orange) with the approach having some similarity to that of electron tomography. CLSM, when combined with immunofluorescence and labeling of antigens, has been shown as a powerful technique for studying sections of wood samples and aspects of wood decay. It is particularly useful for studying lignin in wood samples either through its natural autofluorescence or after staining. For more details, readers are referred to other chapters in this book.

Spectroscopy: Fourier Transform Infrared Spectroscopy, X-Ray Photoelectron Spectroscopy and Time of Flight Secondary Ion Mass Spectrometry

Imaging and analysis of the chemical composition of wood surfaces and changes through wood decay can also be conducted using FTIR, Raman microscopy, and X-ray photoelectron spectroscopy (XPS). FTIR and Raman provide infrared spectra of absorption and emission and scattering of samples. FTIR has been used considerably for characterizing the chemistry of wood (Faix, 1992) and changes occurring during weathering, decay by fungi, and after chemical modification – an approach that complements EDX analyses. In particular, FTIR provides a rapid method to visualize and distinguish changes in hard and softwoods degraded by different fungal groups (e.g., white rot, brown rot, soft rot) (Pandey and Pitman, 2003). Variations include attenuated total reflection (ATR) spectroscopy, which can be applied to smaller quantities of wood and is faster than FTIR (Mohebby, 2005). FTIR imaging is also a good way for detecting stages of incipient wood decayed by brown rot fungi and allowed for fingerprinting of characteristic peaks in IR spectra (Fackler et al., 2010; Fackler and Schwanninger, 2012). The fact that FTIR is a surface technique and can be applied directly to sections without any considerable preparation has contributed to its greater use. The spatial resolution of FTIR is very dependent on the operating parameters and equipment used. Spectroscopy techniques have not so far been combined with resin embedding of samples to achieve improved resolution and quality of spectra because of the organic nature of resins.

X-ray photoelectron spectroscopy also known as electron spectroscopy for chemical analysis (ESCA) has been widely used for studies on the surface chemistry of lignocellulose materials particularly for the characterization of wood pulps but also for solid wood. XPS is a surface-sensitive quantitative spectroscopic technique detecting elements from lithium ($Z = 3$) upward and elemental compositions down to the 1000 ppm range. Spectra are derived by irradiating the wood material with a beam of X-rays and simultaneously measuring the kinetic energy and numbers of electrons escaping from the top 0–10 nm surface of the material being analyzed. The area of analysis is dependent on the instrument but lies in the region of 10–200 μm (largest area = 1–5 mm). XPS has been widely used for measuring lignin and extractives in wood and pulp materials (Duchesne et al., 2003; Fardim et al., 2005; Johansson et al., 2005; Popescu et al., 2009; Inari et al., 2011). Like ToF-SIMs, XPS is a very sensitive surface technique and therefore very susceptible to numerous contamination problems during analysis and sample preparation (Šernek et al., 2004; Johansson et al., 2005). Thus, great care is needed with application of XPS on wood/fiber materials where the migration of extractives can occur (e.g., kraft pulp fibers, softwoods containing extractives). Because it is a surface analysis technique, the preparation of wood samples/pulp fibers is extremely important not only from the point of view of avoiding deposition of volatile organic compounds but also the surface microtopography should not be too extreme. Additional problems can also occur during analysis such as charging (reminiscent of SEM) and degradation of chemical components.

TOF-SIMS is a valuable technique that can be used to simultaneously determine the spatial distribution of organic and inorganic constituents in wood. The technique is quite versatile and allows for analysis of samples down to a depth of 1–2 nm with a lateral resolution of ~1 μm (i.e., allows analysis of wood tissues down to the cellular level). TOF-SIMS is often referred to as chemical microscopy by its ability of providing total ion mass spectra of positively and negatively charged ions in samples that reflect its chemistry. For wood samples it has been used for surface detection of lignin (S- and G-lignins), carbohydrates, pectins, extractives, and metal ions (Fukushima et al., 2001). For further information including aspects on the preparation of wood samples for TOF-SIMS readers should consult Tokareva et al. (2007).

Spectroscopic methods (FTIR, XPS, TOF-SIMS) have also been used for studying aspects of wood decay (Popescu et al., 2009; Fackler et al., 2010; Mahajan et al., 2012; Xu et al., 2013).

The main aim has been to demonstrate changes in wood chemistry (i.e., in lignin, hemicelluloses, cellulose) over time during progressive decay by white and brown rot fungi. The traditional form of wood samples for spectroscopic analyses was originally in powdered form (i.e., milled wood), but microtome sections (10–100 μm) can also be used, which allows the progress of decay to be related to structural aspects of the wood material. For more information, readers are referred to a recent review on spectroscopy methods for fungal decay (Fackler and Schwanninger, 2012).

REFERENCES

Anagnost, S.E., 1998. Light microscopic diagnosis of wood decay. IAWA J. 19, 141–167.

Bardage, S.L., Daniel, G., 1997. The ability of fungi to penetrate micropores: implications for wood surface coatings. Mater. Org. 31, 233–245.

Bergander, A., Brändström, J., Daniel, G., Salmén, L., 2002. Fibril angle variability in earlywood of Norway spruce using soft rot cavities and polarization confocal microscopy. J. Wood Sci. 48, 255–263.

Blanchette, R.A., 1984. Manganese accumulation in wood decayed by white rot fungi. Phytopathology 74, 725–730.

Blanchette, R.A., Nilsson, T., Daniel, G., Abad, A., Rowell, R., Barbour, R., 1990. Biological degradation of wood. In: Rowell, R.M., Barbour, R.J. (Eds.), Archaeological Wood: Properties, Chemistry, and Preservation. American Chemical Society, Washington DC, pp. 141–174.

Bland, D.E., Foster, R.C., Logan, A.F., 1971. The mechanism of permanganate and osmium tetroxide fixation and the distribution of lignin in the cell wall of *Pinus radiata*. Holzforschung 25, 137–143.

Bozzola, J.J., Russell, L.D., 1999. Electron microscopy: Principles and Techniques for Biologists, second ed. Jones and Bartlett Publishers, Sudbury, MA.

Brändström, J., Daniel, G., Nilsson, T., 2002. Use of soft rot cavities to determine microfibril angles in wood; advantages, disadvantages and possibilities. Holzforschung 56, 468–472.

Brändström, J., Bardage, S.L., Daniel, G., Nilsson, T., 2003. The structural organisation of the S_1 cell wall layer of Norway spruce tracheids. IAWA J. 24, 27–40.

Butterfield, B.G., Meylan, B.A., 1980. Three-Dimensional Structure of Wood: An Ultrastructural Approach. Chapman and Hall, London.

Cavalier, A., Spehner, D., Humbel, B.M., 2009. Handbook of Cryo-Preparation Methods for Electron Microscopy. CRC Press, Boca Raton, FL.

Choat, B., Cobb, A.R., Jansen, S., 2008. Structure and function of bordered pits: new discoveries and impacts on whole-plant hydraulic function. New Phytol. 177, 608–626.

Daniel, G., 1994. Use of electron microscopy for aiding our understanding of wood biodegradation. FEMS Microbiol. Rev. 13, 199–233.

Daniel, G., 2003. Microview of wood under degradation by bacteria and fungi. In: Goodell, B., Nicholas, D.D., Schultz, T.P. (Eds.), Wood Deterioration and Preservation: Advances in Our Changing World. American Chemical Society, Washington DC, pp. 34–72.

Daniel, G., Bergman, Ö., 1997. White rot and manganese deposition in TnBTO-AAC preservative treated pine stakes from field tests. Holz Roh-Werkst. 55, 197–201.

Daniel, G., Duchesne, I., 1998. Revealing the surface ultrastructure of spruce fibre using Field-emission-SEM. Proceedings of the Seventh International Conference on Biotechnology in the Pulp and Paper Industry. Vancouver, Canada, pp. 81–85.

Daniel, G., Nilsson, T., 1985. Ultrastructural and TEM-EDAX studies on the degradation of CCA treated radiata pine by tunnelling bacteria. The International Research Group on Wood Preservation (IRG/WP No. 1260). Stockholm, Sweden.

Daniel, G., Nilsson, T., 1987. Comparative studies on the distribution of lignin and CCA elements in birch using electron microscopic X-ray microanalysis. The International Research Group on Wood Preservation (IRG/WP No. 1328). Stockholm, Sweden.

Daniel, G.N., Nilsson, T., 1989. Interactions between soft rot fungi and CCA preservatives in *Betula verrucosa*. J. Institute Wood Sci. 11, 162–171.

Daniel, G., Nilsson, T., 1998. Developments in the study of soft rot and bacterial decay. In: Bruce, A., Palfreyman, J.W. (Eds.), Forest Products Biotechnology. Taylor & Francis, London, pp. 37–62.

Daniel, G., Nilsson, T., Singh, A.P., 1987. Degradation of lignocellulosics by unique tunnel-forming bacteria. Can. J. Microbiol. 33, 943–948.

Daniel, G., Nilsson, T., Pettersson, B., 1991. Poorly and non-lignified regions in the middle lamella cell corners of birch (*Betula verrucosa*) and other wood species. IAWA Bull. n.s. 12, 70–83.

Daniel, G., Singh, A.P., Nilsson, T., 1996. Ultrastructural and immunocytochemical studies of the window and bordered pit membranes of *Pinus sylvestris* L. In: Donaldson, L.A., Singh, A.P., Butterfield, B.G., Whitehouse, L.J. (Eds.), Recent Advances in Wood Anatomy. Forest Research Institute Ltd, Rotorua, pp. 373–383.

Daniel, G., Nilsson, T., Volc, J., 1997. Electron microscopical observations and chemical analyses supporting Mn uptake in white rot degraded *Alstonia* and pine wood stakes exposed in acid coniferous soil. Can. J. Microbiol. 43, 663–671.

Daniel, G., Volc, J., Niku-Paavola, M.-L., 2004. Cryo-FE-SEM & TEM immuno-techniques reveal new details for understanding white-rot decay of lignocellulose. C. R. Biol. 327, 861–871.

Daniel, G., Tokoh, I., Bardage, S., 2004. The surface and intracellular nanostructure of wood fibres: electron microscope methods and applications. In: Schmitt, U., Ander, P., Barnett, J.R., Emons, A.M.C., Jeronimidis, G., Saranpää, P., Tschegg, S. (Eds.), Wood Fibre Cell Walls: Methods to Study Their Formation, Structure and Properties. Swedish University of Agricultural Sciences, Uppsala, pp. 87–104.

Daniel, G., Filonova, L., Kallas ÅM, Teeri, T.T., 2006. Morphological and chemical characterisation of the G-layer in tension wood fibres of *Populus tremula* and *Betula verrucosa*: labelling with cellulose-binding module $CBM1_{HjCel7A}$ and fluorescence and FE-SEM microscopy. Holzforschung 60, 618–624.

Donaldson, L., 2008. Microfibril angle: measurement, variation and relationships – a review. IAWA J. 29, 345–386.

Donaldson, L.A., Knox, J.P., 2012. Localization of cell wall polysaccharides in normal and compression wood of radiata pine: relationships with lignification and microfibril orientation. Plant Physiol. 158, 642–653.

Donaldson, L., Xu, P., 2005. Microfibril orientation across the secondary cell wall of Radiata pine tracheids. Trees 19, 644–653.

Donaldson, L., Radotić, K., Kalauzi, A., Djikanović, D., Jeremić, M., 2010. Quantification of compression wood severity in tracheids of *Pinus radiata* D. Don using confocal fluorescence imaging and spectral deconvolution. J. Struct. Biol. 169, 106–115.

Duchesne, I., Daniel, G., 1999. The ultrastructure of wood fibre surfaces as shown by a variety of microscopical methods – A review. Nord. Pulp Pap. Res. J. 14, 129–139.

Duchesne, I., Daniel, G., 2000. Changes in surface ultrastructure of Norway spruce fibres during kraft pulping – Visualisation by field emission-SEM. Nord. Pulp Pap. Res. J. 15, 54–61.

Duchesne, I., Hult, E.-L., Molin, U., Daniel, G., Iversen, T., Lennholm, H., 2001. The influence of hemicellulose on fibril aggregation of kraft pulp fibres as revealed by FE-SEM and CP/MAS ^{13}C-NMR. Cellulose 8, 103–111.

Duchesne, I., Daniel, G., Van Leerdam, G.C., Basta, J., 2003. Surface chemical composition and morphology of ITC kraft fibres as determined by XPS and FE-SEM. J. Pulp Pap. Sci. 29, 71–76.

Echlin, P., 2009. Handbook of Sample Preparation for Scanning Electron Microscopy and X-ray Microanalysis. Springer Science, Business Media, New York.

Fackler, K., Schwanninger, M., 2012. How spectroscopy and microspectroscopy of degraded wood contribute to understand fungal wood decay. Appl. Microbiol. Biotechnol. 96, 587–599.

Fackler, K., Stevanic, J.S., Ters, T., Hinterstoisser, B., Schwanninger, M., Salmén, L., 2010. Localisation and characterisation of incipient brown-rot decay within spruce wood cell walls using FT-IR imaging microscopy. Enzyme Microb. Technol. 47, 257–267.

Faix, O., 1992. Fourier transform infrared spectroscopy. In: Lin, S.Y., Dence, C.W. (Eds.), Methods in Lignin Chemistry. Springer-Verlag, Berlin-Heidelberg, pp. 83–109.

Fardim, P., Gustafsson, J., von Schoultz, S., Peltonen, J., Holmbom, B., 2005. Extractives on fiber surfaces investigated by XPS, ToF-SIMS and AFM. Colloids Surf. A 255, 91–103.

Filonova, L., Kallas Å.M., Greffe, L., Johansson, G., Teeri, T.T., Daniel, G., 2007. Analysis of the surfaces of wood tissues and pulp fibers using carbohydrate-binding modules specific for crystalline cellulose and mannan. Biomacromolecules 8, 91–97.

Fukushima, K., Yamauchi, K., Saito, K., Yasuda, S., Takahashi, M., Hoshi, T., 2001. Analysis of lignin structures by ToF-SIMS. Proceedings of the Eleventh International Symposium on Wood and Pulping Chemistry. Nice, France, pp. 327–329.

Goldstein, J.I., Newbury, D.E., Joy, D.C., Lyman, C.E., Lifshin, E., Sawyer, L., Michael, J.R., 2003. Scanning Electron Microscopy and X-ray Microanalysis, third ed. Springer Science, Business Media, New York.

Hildén, L., Daniel, G., Johansson, G., 2003. Use of a fluorescence labelled, carbohydrate-binding module from *Phanerochaete chrysosporium* Cel7D for studying wood cell wall ultrastructure. Biotechnol. Lett. 25, 553–558.

Hurbain, I., Sachse, M., 2011. The future is cold: cryo-preparation methods for transmission electron microscopy of cells. Biol. Cell 103, 405–420.

Inari, G.N., Pétrissans, M., Dumarcay, S., Lambert, J., Ehrhardt, J.J., Šernek, M., Gérardin, P., 2011. Limitation of XPS for analysis of wood species containing high amounts of lipophilic extractives. Wood Sci. Technol. 45, 369–382.

Inoué, T., Osatake, H., 1988. A new drying method of biological specimens for scanning electron microscopy: the *t*-butyl alcohol freeze-drying method. Arch. Histol. Cytol. 51, 53–59.

Johansson, L.-S., Campbell, J.M., Fardim, P., Hultén, A.H., Boisvert, J.-P., Ernstsson, M., 2005. An XPS round robin investigation on analysis of wood pulp fibres and filter paper. Surf. Sci. 584, 126–132.

Khalili, S., Nilsson, T., Daniel, G., 2001. The use of soft rot fungi for determining the microfibrillar orientation in the S2 layer of pine tracheids. Holz Roh-Werkst. 58, 439–447.

Kim, J.S., Daniel, G., 2012. Distribution of glucomannans and xylans in poplar xylem and their changes under tension stress. Planta 236, 35–50.

Kim, J.S., Sandquist, D., Sundberg, B., Daniel, G., 2012. Spatial and temporal variability of xylan distribution in differentiating secondary xylem of hybrid aspen. Planta 235, 1315–1330.

Mahajan, S., Jeremic, D., Goacher, R.E., Master, E.R., 2012. Mode of coniferous wood decay by the white rot fungus *Phanerochaete carnosa* as elucidated by FTIR and ToF-SIMS. Appl. Microbiol. Biotechnol. 94, 1303–1311.

Matsunaga, H., Matsumura, J., Oda, K., 2004. X-ray microanalysis using thin sections of preservative-treated wood. Relationship of wood anatomical features to the distribution of copper. IAWA J. 25, 79–90.

Matsunaga, H., Kiguchi, M., Evans, P., 2009. Microdistribution of copper-carbonate and iron oxide nanoparticles in treated wood. J. Nanopart. Res. 11, 1087–1098.

Mohebby, B., 2005. Attenuated total reflection infrared spectroscopy of white-rot decayed beech wood. Int. Biodeterior. Biodegradation 55, 247–251.

Mott, L., Shaler, S.M., Groom, L.H., 1996. A technique to measure strain distributions in single wood pulp fibers. Wood Fiber Sci. 28, 429–437.

Nakagawa, K., Yoshinaga, A., Takabe, K., 2012. Anatomy and lignin distribution in reaction phloem fibres of several Japanese hardwoods. Ann. Bot. 110, 897–904.

Nation, J.L., 1983. A new method using hexamethyldisilazane for preparation of soft insect tissues for scanning electron microscopy. Stain Technol. 58, 347–351.

Otjen, L., Blanchette, R.A., Leatham, G.F., 1988. Lignin distribution in wood delignified by white-rot fungi: X-ray microanalysis of decayed wood treated with bromine. Holzforschung 42, 281–288.

Pandey, K.K., Pitman, A.J., 2003. FTIR studies of the changes in wood chemistry following decay by brown-rot and white-rot fungi. Int. Biodeterior. Biodegradation 52, 151–160.

Popescu, C.-M., Tibirna, C.-M., Vasile, C., 2009. XPS characterization of naturally aged wood. Appl. Surf. Sci. 256, 1355–1360.

Reis, D., Vian, B., 2004. Helicoidal pattern in secondary cell walls and possible role of xylans in their construction. C. R. Biol. 327, 785–790.

Reis, D., Vian, B., Chanzy, H., Roland, J.C., 1991. Liquid crystal-type assembly of native cellulose-glucuronoxylans extracted from plant cell wall. Biol. Cell 73, 173–178.

Richter, H.G., Grosser, D., Heinz, I., Gasson, P.E., 2004. IAWA list of microscopic features for softwood identification. IAWA J. 25, 1–70.

Ruel, K., Nishiyama, Y., Joseleau, J.-P., 2012. Crystalline and amorphous cellulose in the secondary walls of *Arabidopsis*. Plant Sci. 193, 48–61.

Ruzin, S.E., 1999. Plant Microtechnique and Microscopy. Oxford University Press, New York.

Saka, S., Thomas, R.J., 1982. Evaluation of the quantitative assay of lignin distribution by SEM-EDXA-technique. Wood Sci. Technol. 16, 1–18.

Schwarze, F.W.M.R., 2007. Wood decay under the microscope. Fungal Biol. Rev. 21, 133–170.

Šernek, M., Kamke, F.A., Glasser, W.G., 2004. Comparative analysis of inactivated wood surface. Holzforschung 58, 22–31.

Spurr, A.R., 1969. A low-viscosity epoxy resin embedding medium for electron microscopy. J. Ultrastruct. Res. 26, 31–43.

Srebotnik, E., Messner, K., 1994. A simple method that uses differential staining and light microscopy to assess the selectivity of wood delignification by white rot fungi. Appl. Environ. Microbiol. 60, 1383–1386.

Tokareva, E.N., Pranovich, A.V., Fardim, P., Daniel, G., Holmbom, B., 2007. Analysis of wood tissues by time-of-flight secondary ion mass spectrometry. Holzforschung 61, 647–655.

Tokareva, E.N., Fardim, P., Pranovich, A.V., Fagerholm, H.-P., Daniel, G., Holmbom, B., 2007. Imaging of wood tissue by ToF-SIMS: critical evaluation and development of sample preparation techniques. Appl. Surf. Sci. 253, 7569–7577.

Turkulin, H., Holzer, L., Richter, K., Sell, J., 2005a. Application of the ESEM technique in wood research. Part I. Optimization of imaging parameters and working conditions. Wood Fiber Sci. 37, 552–564.

Turkulin, H., Holzer, L., Richter, K., Sell, J., 2005b. Application of the ESEM technique in wood research. Part II. Comparison of operational modes. Wood Fiber Sci. 37, 565–573.

Vazquez-Cooz, I., Meyer, R., 2002. A differential staining method to identify lignified and unlignified tissues. Biotech. Histochem. 77, 277–282.

Wheeler, E.A., Baas, P., Gasson, P.E., 1989. IAWA list of microscopic features for hardwood identification. IAWA Bull. n.s. 10, 219–332.

Xu, G., Wang, L., Liu, J., Wu, J., 2013. FTIR and XPS analysis of the changes in bamboo chemical structure decayed by white-rot and brown-rot fungi. Appl. Surf. Sci. 280, 799–805.

Yoshinaga, A., Kusumoto, H., Laurans, F., Pilate, G., Takabe, K., 2012. Lignification in poplar tension wood lignified cell wall layers. Tree Physiol. 32, 1129–1136.

Chapter 16

Rapid Freezing and Immunocytochemistry Provide New Information on Cell Wall Formation in Woody Plants

Keiji Takabe*, Jong Sik Kim**

*Division of Forest and Biomaterials Sciences, Graduate School of Agriculture, Kyoto University, Kyoto, Japan; **Department of Forest Products, Swedish University of Agricultural Sciences, Uppsala, Sweden

Chapter Outline

INTRODUCTION

Wood is the product of cell division of cambial cells followed by cell expansion and secondary wall formation. The secondary wall is composed of three layers, namely, the S_1, S_2, and S_3 layers. The main components of the secondary wall are cellulose, hemicelluloses, and lignins. They are highly organized within the cell walls by the cell and the automatic assembly of cell wall components according to their chemical nature. It is still ambiguous how each cell wall component is assembled to develop cell walls. Chemical fixatives, such as glutaraldehyde, are widely used to fix the specimen. They, however, sometimes cause artifacts such as distortion of the cell and extraction of unfixed soluble

Secondary Xylem Biology. http://dx.doi.org/10.1016/B978-0-12-802185-9.00016-4
345

materials. The alternative method to fix the specimen is rapid freezing by which the living tissues are frozen within a few milliseconds. When the specimen is frozen without ice crystal formation, it gives us more improved images of living cells and developing cell walls. In this chapter, the assembly of each cell wall component is discussed based on the images obtained after rapid freezing.

Cell wall components originate from photosynthetic products, such as sucrose, transported through the phloem. Some of the photosynthetic products are converted into various kinds of substances by successive enzymatic reactions, and transported to the cell organelle to store starch and phenolic compounds, and cell walls to give the cell rigidity. Although many enzymes are involved in the biosynthesis of cell wall components, their expression in the differentiating xylem, their localization in the cells and the cell walls, and their relation had been obscure for a long time. Recent progress in immunocytochemical techniques enables us to visualize the *in situ* localization of enzymes in the living cells and the localization of each cell wall component at electron microscopic level. In this chapter, recent advances in the immunocytochemical analysis of enzyme localization and cell wall components are summarized.

RAPID FREEZING PROVIDES NEW INFORMATION ON CELL WALL FORMATION IN WOODY PLANTS

Until now, glutaraldehyde had been widely used as a fixative to observe the specimen under a transmission electron microscope. Almost 30 years ago, the rapid freeze and freeze substitution (RFS) method was introduced to observe the specimen under a transmission electron microscope. A cooling rate of more than 10,000°C/s makes the biological specimen freeze without ice crystal formation in the cells. Inomata et al. (1992) and Samuels et al. (2002) clearly showed that the differentiating xylem cell of *Cryptomeria japonica* and *Pinus contorta*, respectively, fixed by RFS is quite different from that by conventional chemical fixation (CF) (Fig. 16.1). The plasma membrane is observed to have a very smooth profile after RFS, though it shows a wavy structure after CF. Cell organelles, such as the rough endoplasmic reticulum (rER), show very smooth profiles after RFS. The mitochondria and amyloplast are round or elliptical shapes and filled with osmiophilic substances. Golgi vesicles are fully filled with some substances after RFS, although they contain a small amount of substances after CF. Developing secondary walls after RFS are quite different from those after CF. Homogenous textures of cell walls and very smooth inner surfaces covered with amorphous substances are observable after RFS, although lamellated structures of developing cell walls are seen after CF.

The PATAg (periodic acid-thiocarbohydrazide-Ag proteinate) test developed by Thiery (1967) provides very useful information on polysaccharide distribution under a transmission electron microscope (Figs 16.2 and 16.3). Nonlignified primary and secondary walls are heavily stained by PATAg test. Particularly, the inner surface of the developing cell walls is covered with homogeneous sub-

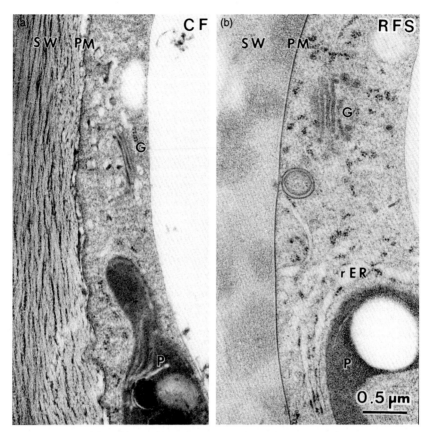

FIGURE 16.1   Differentiating compression wood tracheids of *C. japonica* fixed by (a) conventional chemical fixation (CF) and (b) rapid freeze and freeze substitution (RFS). Both tracheids are almost at the same differentiating stage. G, Golgi apparatus; P, plastid; PM, plasma membrane; rER, rough endoplasmic reticulum. Plasma membrane and cell organelle show very smooth profiles after RFS. Cell wall structure is almost homogeneous after RFS, although lamellated structure appears after CF.

stances stained very heavily by PATAg test. This staining pattern is a common feature of developing cell walls in softwood and hardwood (Fig. 16.3). Staining intensity becomes weak when the cell wall lignification proceeds, indicating that the cell wall polysaccharides are encrusted with lignin (Figs 16.2 and 16.3). The Golgi apparatus is also stained by PATAg test. The staining intensity gradually increases toward the *trans*-face of the Golgi cisternae. Golgi vesicles are also heavily stained (Fig. 16.2). These results suggest that polymerization of hemicelluloses gradually proceeds toward the *trans*-face of the Golgi cisternae. Polymerized hemicelluloses are packed into the Golgi vesicles, and transported to the inner surface of the developing cell wall by exocytosis of the vesicles, resulting in the formation of a very thin layer composed of hemicelluloses.

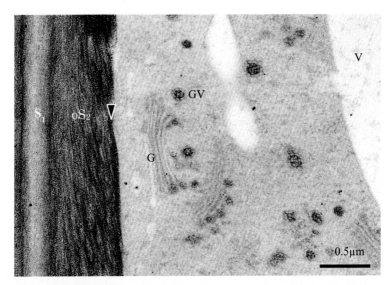

FIGURE 16.2 **Differentiating compression wood tracheid of _C. japonica_ fixed by RFS followed by staining PATAg.** G, Golgi apparatus; GV, Golgi vesicle; S_1, outer layer of secondary wall; oS_2, outer portion of middle layer of secondary wall; V, vacuole. The staining intensity of Golgi cisternae gradually increases toward the _trans_-face. The Golgi vesicles are strongly stained. The inner surface of the developing oS_2 is covered with a very thin layer (arrow head) stained very strongly. oS_2 is also stained strongly, although S_1 becomes weak. _(Courtesy of F. Sano.)_

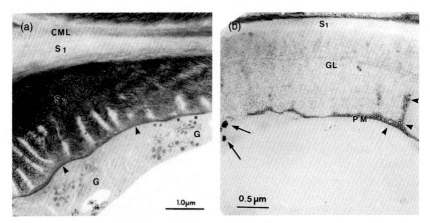

FIGURE 16.3 **Differentiating compression wood tracheid of _C. japonica_ (a) and tension wood fiber of _Eucalyptus globulus_ (b) fixed by RFS followed by staining PATAg.** CML, compound middle lamella; G, Golgi apparatus; GL, gelatinous layer; PM, plasma membrane; S_1, outer layer of secondary wall; S_2, middle layer of secondary wall. Inner surface of developing cell walls are covered with a very thin layer (arrow heads) stained very strongly in both cells. Heavily stained vesicles (arrows) appear in tension wood fiber. _(Courtesy of F. Sano and K. Kawamura.)_

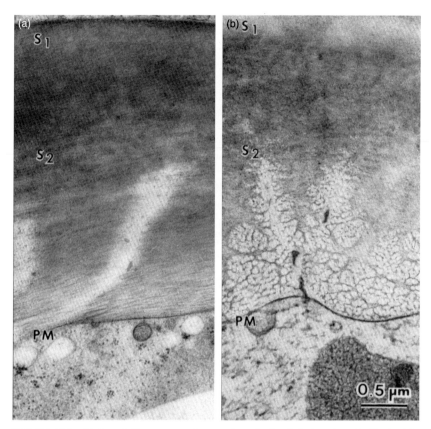

FIGURE 16.4   Differentiating compression wood tracheids of *C. japonica.* (a) Obtained by rapid freeze and freeze substitution and (b) obtained by slow freeze and freeze substitution. PM, plasma membrane; S_1, outer layer of secondary wall; S_2, middle layer of secondary wall. Ultrastructure of inner portion of the developing secondary wall is destroyed by ice crystal caused by slow freezing. *(Courtesy of F. Sano.)*

Although the RFS method has many advantages in preserving the living cells close to their native state, it has difficulty in making the specimen freeze in good condition. If the specimen freezes slowly, ice crystals grow in cells and developing cell walls, leading to the destruction of the ultrastructure of cells and developing cell walls. However, ice crystal formation gives us interesting information on water distribution within the cell wall at the ultrastructure level. Differentiating tracheids frozen slowly showed ice crystal formation at the inner part of the developing secondary walls and amorphous thin layers, indicating that newly deposited cell walls and amorphous thin layers contain a significant amount of water. In contrast, ice crystal formation could not be found at the outer portion of the secondary wall, suggesting that lignification excludes water from nonlignified and lignifying cell walls (Fig. 16.4).

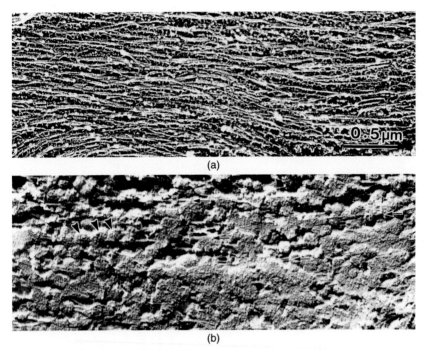

(a)

(b)

FIGURE 16.5   Freeze etched images of developing secondary wall thickening in Z. *elegans*.
(a) Nonlignified secondary wall thickening appears to have many microfibrils with cross-linkings.
(b) When the lignification proceeds, many spherical substances appear along and between the microfibrils. *(Courtesy of J. Nakashima.)*

Nakashima et al. (1997) showed the detailed three-dimensional ultrastructure of developing cell walls by rapid freezing followed by freeze etching. Cellulose microfibrils with many cross-linkings were clearly observable in nonlignified secondary wall thickening of *Zinnia elegans* (Fig. 16.5a). There were
many spaces between cellulose microfibrils, which might be filled with water.
When lignification of the secondary wall thickening proceeded, many small
spherical substances appeared along and between microfibrils (Fig. 16.5b).
Finally, microfibrils and cross-linkings were completely encrusted with amorphous substances. Dehydrogenative polymerization of monolignols might occur along and between micofibrils to form spherical structures, and finally to fill
the spaces completely (Fig. 16.6).

This information provides a new idea of cell wall formation in woody plants.
Hemicelluloses are synthesized in the Golgi apparatus and are transported to the
inner surface of the developing cell wall to form the thin layer. This layer may
be composed of hemicelluloses and a significant amount of water. Cellulose
is synthesized by terminal enzyme complexes (TCs) localized on the plasma
membrane, and then extruded into the thin layer. Crystallization of cellulose
microfibril may be affected by hemicelluloses localized in the thin layer. Newly

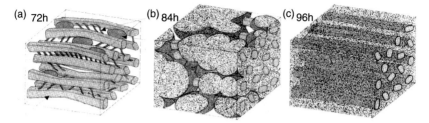

FIGURE 16.6   Model of formation of secondary wall thickening in Z. *elegans*. Times in figures are incubation times after homogenate of mesophyll cells. (a) 72 h (b) 84 h (c) 96 h.

deposited cellulose microfibrils are cross-linked with hemicelluloses to form some spaces in the cell wall. The spaces are filled with water. Lignification of the secondary wall starts at the outermost part of the S_1 layer and proceeds toward the lumen lagging behind the cell wall thickening. The hydrophilic cell wall becomes hydrophobic as lignification proceeds. Water in the hydrophilic cell wall may be extruded by lignification.

LOCALIZATION OF ENZYMES INVOLVED IN CELL WALL FORMATION REVEALED BY IMMUNOCYTOCHEMISTRY

The freeze fracture technique after rapid freezing reveals the existence of TCs localized on the plasma membrane in bacterium, algae, and higher plants. A linear TC and three rows of linear TC were observed at the end of synthesizing cellulose microfibrils in *Gluconacetobacter xylinus* (*Acetobactor xylinum*) and *Valonia ventricosa*, respectively. A rosette-type TC was found in higher plants such as *Z. elegans*. These results were, however, indirect evidences of enzyme complexes involved in cellulose synthesis. Kimura et al. (1999) apparently showed that the polyclonal antibody raised against the catalytic subunit domain of cotton cellulose synthase recognized specifically the rosette-type TCs in *Vigna angularis* by immunolabeling coupled with freeze fracture technique.

Staehelin et al. (1991) reported that immunolabeling of polygalacturonic acid/rhamnogalacturonan-I (PGA/RG-I) appeared nearly over *cis*- and medial cisternae of the Golgi apparatus, and the immunolabeling of xyloglucan (XG) over *trans*-cisternae and *trans*-Golgi network. They discussed the possibility that PGA/RG-I is synthesized in *cis*- and the medial cisternae of the Golgi apparatus, and XG in *trans*-cisternae followed by packing of XG in the *trans*-Golgi network.

Lignification proceeds in three steps. The first step is the synthesis of monolignols in the cell; the second step is the transportation of monolignols toward developing cell walls; and the last step is the dehydrogenative polymerization of monolignols within the cell wall. Many enzymes, such as phenylalanine ammonia-lyase (PAL), cinnamate 4-hydroxylase (C4H), and cinnamyl alcohol dehydrogenase (CAD), are involved in the synthesis of monolignols. Poly-clonal antibodies were raised against purified enzymes or peptides synthesized

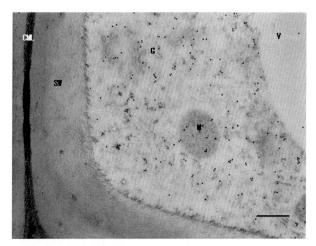

FIGURE 16.7 **Immunocytochemical localization of phenylalanine ammonia-lyase (PAL) in differentiating xylem of *P. sieboldii* × *P. grandidentata*.** CML, compound middle lamella; G, Golgi apparatus; M, mitochondrion; V, vacuole. Immunolabeling of PAL is mainly distributed in cytosol. *(Courtesy of T. Sato.)*

artificially according to the amino acid sequences of enzymes involved in monolignols synthesis. Takabe et al. (2001) and Sato et al. (2004) found that the labelings of PAL were mainly distributed in cytosol during secondary wall formation in poplar (Fig. 16.5). Ye (1997) and Takeuchi et al. (2001) investigated the immunolabeling of caffeoyl-CoA O-methyltransferase (CCoAOMT) and caffeate/5-hydroxyferulate O-methyltransferase (CAOMT), respectively, and showed their distributions in cytosol. Takabe et al. (2001) showed that CAD was also distributed in cytosol during secondary wall formation. On the contrary, Sato et al. (2004) found that the labelings of C4H were localized near the membrane of rER during secondary wall formation, indicating that hydroxylation of monolignol intermediate occurs on the membrane of rER. These results indicate that hydroxylation of monolignol intermediate occurs on the membrane of rER, and other reactions in cytosol (Figs 16.7–16.8).

Transportation of monolignols toward developing cell walls had been wrapped in mystery for a long time. Miao and Liu (2010) applied the biochemical experiment using microsomal fractions originated from leaves of *Arabidopsis thaliana* and showed that an ATP-binding cassette (ABC)-like transporter is a possible candidate for transportation of monolignols toward the cell wall. Tsuyama et al. (2013) also applied biochemical experiment using microsomal fractions originated from the differentiating xylem of poplar (*Populus sieboldii* × *Populus grandidentata*) and Japanese cypress (*Chamaecyparis obtusa*), and showed the proton (H^+) gradient-dependent transportation of coniferin across the tonoplast and endomembrane compartments (Fig. 16.7). Morikawa et al. (2010) revealed that coniferin was distributed in the lumen of differentiating tracheids in Japanese cypress (*C. obtusa*) by Raman microscopy.

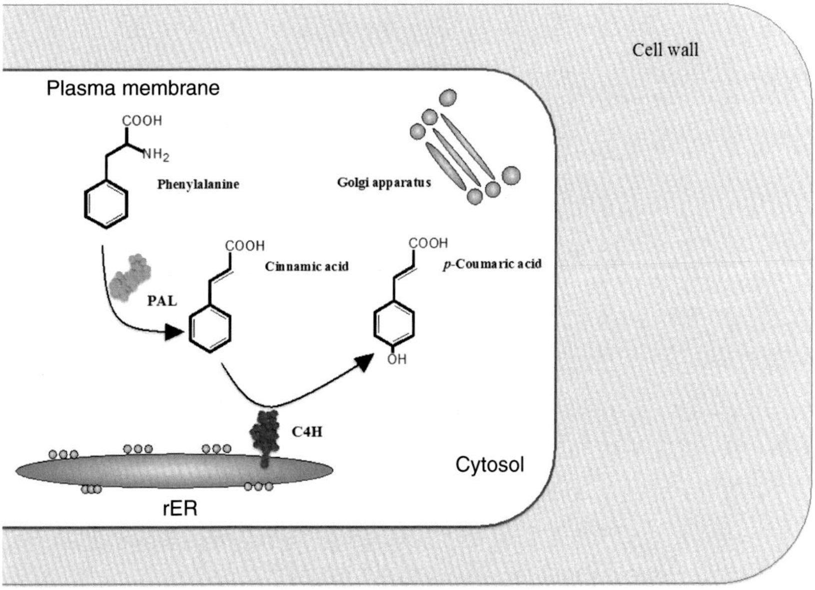

FIGURE 16.8   Model of monolignol intermediates reaction in the cell. C4H, cinnamate 4-hydroxylase; PAL, phenylalanine ammonia-lyase; rER, rough endoplasmic reticulum. Conversion of phenylalanine to cinnamic acid occurs in cytosol and the following conversion to *p*-coumaric acid occurs on the membrane of the rER.

Peroxidases (POXs) and laccases are thought to be involved in dehydrogenative polymerization of monolignols. Sato et al. (2006) found that immunolabeling of cationic POX was localized on developing secondary wall thickenings in *Z. elegans*. They suggested that cationic POX is involved in the polymerization of monolignols. Takeuchi et al. (2005) also showed that the labeling of anionic POX is localized on the Golgi vesicles and plasma membrane during secondary wall formation in poplar. Anionic POX may be synthesized in rER and transported to the plasma membrane by fusing the Golgi vesicles to the plasma membrane.

Immunocytochemistry Reveals the Deposition Process of Hemicelluloses in Normal Wood Cell Walls

Although the PATAg test coupled with RFS method provides new information on the synthesis, deposition, and distribution of polysaccharides in the differentiating xylem, it has a limitation in distinguishing each polysaccharide. Northcote et al. (1989) raised polyclonal antibody against xylan and showed that immunolabeling appeared on the secondary thickening of tracheary elements and on the Golgi apparatus. Awano et al. (1998, 2000) raised polyclonal antibody against xylan extracted from Japanese beech, and showed that

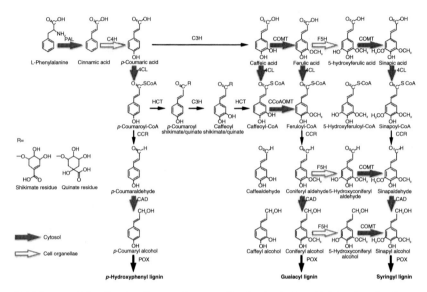

FIGURE 16.9 **Monolignol pathway and enzyme distribution in the cell.** *(Monolignol pathway: courtesy of Y. Tsutsumi.)*

most of the xylan deposits to the secondary cell wall. Interestingly, some xylan deposit intussusceptionally to the cell wall. Maeda et al. (2000) also raised the polyclonal antibody against glucomannan extracted from Japanese cypress (*C. obtusa*), and showed that glucomannan deposits appositionally to secondary cell walls.

Many monoclonal antibodies have been introduced and are commercially available. Kim et al. (2010b, 2010c) investigated the immunolocalization of galactoglucomannans (GGMs) and arabino-4-O-methylglucuronoxylans (AGXs) in differentiating the xylem of *Cryptomeria japonica* by using monoclonal antibodies against glucomannan and xylan. They found that the immunolabeling of GGMs appeared at the cell corner regions of S_1 when the secondary wall formation started. Then, the labeling appeared all over S_1. At the early stage of S_2 formation, limited GGMs labeling was observed on S_2. After that, the labeling gradually increased. Finally, GGMs labeling showed inhomogeneous distribution in the secondary walls in mature tracheid, namely, stronger labeling was observed at the boundary between S_1 and S_2. Interestingly, more abundant GGMs labeling was observed in the tracheids formed in the previous year (Fig. 16.9). Deacetylation of GGMs treated with mild alkaline led to the tremendous increase of the GGMs labeling all over the secondary walls, suggesting that some epitope of GGMs are masked with acetyl group. Increase of GGM labeling in the previous year may be caused by deacetylation of GGMs after cell wall formation is finished. The density of

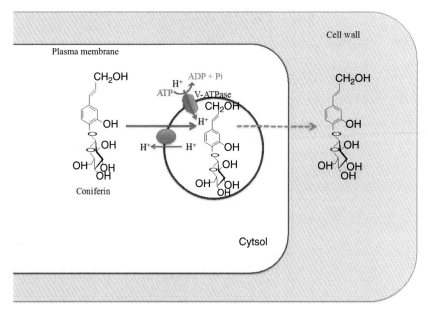

FIGURE 16.10 Proton (H⁺) gradient-dependent transportation of coniferin across the tonoplast and endomembrane compartments.

GGMs labeling in S_1 is more than that in S_2 and S_3, indicating more GGMs are deposited in S_1 (Figs 16.10 and 16.11).

AGXs labeling was observed at the cell corner region of S_1 when the S_1 formation started. Then the labeling appeared all over the developing secondary walls except the boundary between S_1 and S_2. The labeling was, however, uniformly distributed all over the secondary walls in mature tracheid. The labeling was observed on the cell corner middle lamellae after secondary wall formation started, suggesting that some AGXs penetrate through the developing secondary walls and deposit to the cell corner middle lamellae. It was also observed on the warts at the final stage of cell wall formation. The density of AGXs labeling in S_2 and S_3 is more than that in S_1, indicating more AGXs is distributed in S_2 and S_3. The deposition process of GGMs and AGXs is summarized in Fig. 16.12.

Deposition of Hemicelluloses in Pit Membrane

Kim et al. (2011a, 2011b) showed that GGMs labeling on the pit membrane is different from AGXs labeling. The GGMs labeling appeared on the torus and margo of the pit membrane when S_1 formation started. Strong GGMs labeling was observed on the surface of torus in the early stage of S_2 formation. Then the labeling decreased and finally disappeared in mature tracheid. On the contrary, AGXs labeling could not be observed on the torus and margo

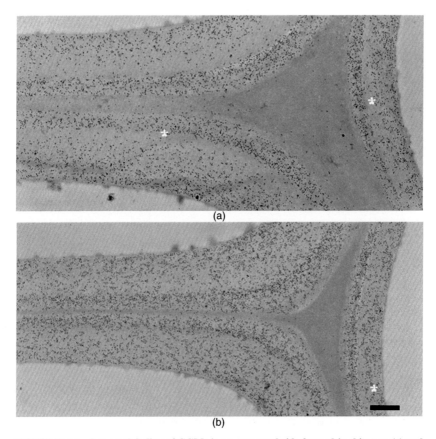

FIGURE 16.11   Immunolabeling of GGMs in mature tracheids formed in this year (a) and previous year (b). Stronger labeling appears at the boundary of S_1 and S_2 in mature tracheid formed in this year. Labeling at the outer portion of S_2 (*) is limited. Labeling intensity increases in mature tracheid formed in the previous year.

in pit membrane during cell wall formation, even though the heavy AGXs labeling appeared on the developing secondary wall. Thomas (1967), Imamura and Harada (1973), and Imamura et al. (1974a, 1974b) suggested that the pit membranes were encrusted with amorphous substances during secondary wall formation, but the encrusting substance disappeared after the cell wall formation was completed. Degradation of GGMs may occur in the final stage of cell wall formation.

No AGXs labeling on the pit membrane during secondary wall formation suggests that deposition of AGXs is well controlled by the cell. It is generally accepted that hemicelluloses, such as AGXs, are synthesized in the Golgi apparatus and transported to the developing cell wall by exocytosis of the Golgi vesicles. GGMs are deposited to the developing secondary walls and pit

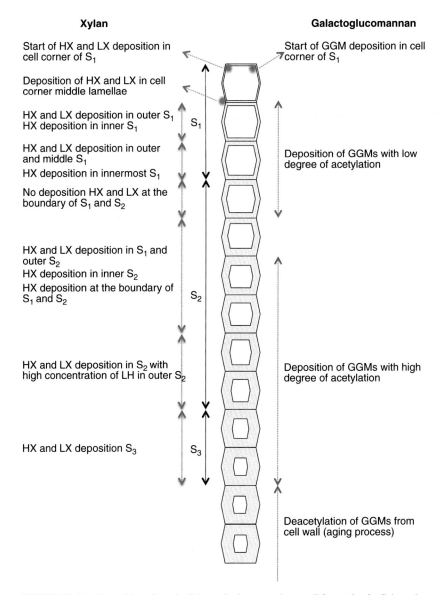

FIGURE 16.12   Deposition of hemicelluloses during secondary wall formation in *C. japonica*.

membranes. On the contrary, AGXs are not deposited to the pit membranes but to the developing secondary walls. Trafficking of vesicles containing AGXs may be well controlled by the cell (Fig. 16.9).

Deposition of Polysaccharides in Compression Wood Cell Walls Revealed by Immunocytochemistry

When coniferous trees are inclined, they form compression wood at the lower side of the stem to return the stem to the original position. Compression wood is quite different from normal wood. Compression wood tracheid is characterized by the rounded transverse surface, thicker secondary walls composed of S_1 and S_2, and lack of S_3. It contains more lignin and abundant galactan, but less cellulose. Ultraviolet (UV) microscopy revealed the heavy UV absorbance at the outer portion of S_2, indicating more lignin in the outer portion of S_2. Kim et al. (2010a) revealed that the immunolabeling of GGMs, AGXs, and galactan are heterogeneously distributed within the secondary walls. GGMs labeling could not be observed on the primary cell wall in cambial and expansion cells. They were first detected at the cell corner region of S_1 when the S_1 formation started. Then they distributed all over S_1 during the S_1 formation stage, and then little was observed on the outer portion of S_2 during the early stage of S_2 formation. After that, the labelings were again distributed on the inner portion of S_2 during the later stage of secondary wall formation. AGXs labelings were similar to GGMs labeling except the cell corner middle lamellae. The labelings were observed at the cell corner middle lamellae when the secondary wall formation proceeded. On the contrary, galactan labeling was found on S_1 when S_1 formation started, and then the abundant labeling appeared on the outer portion of S_2 during the early stage of S_2 formation. The labeling was not found on the inner portion of S_2 during the later stage of S_2 formation. These results indicate the inhomogeneous distribution of polysaccharides in compression wood cell walls. GGMs and AGXs are mainly distributed in S_1 and the inner portion of S_2. In contrast to GGMs and AGXs, galactan is abundantly distributed in the outer portion of S_2(Fig. 16.13).

CONCLUSIONS

The RFS method provides very improved images of living cells and developing cell walls under a transmission electron microscope. Rapid freezing followed by fixation with osmium tetroxide at $-80°C$ prevents not only the distortion of the cell organelle, but also the extraction of water-soluble substances from cells and cell walls during fixation. The inner surface of developing cell walls are covered with a very thin layer composed of newly deposited hemicelluloses and subsequent amount of water. Lignification excludes water from nonlignified and lignifying cell walls.

Immunocytochemistry enables us to visualize the enzyme localization within the cells and cell walls. C4H is localized on rER, and several enzymes involved in monolignols synthesis are localized in cytosol. POXs are localized on the plasma membrane and lignifying cell walls. Immunocytochemistry allows us to visualize the enzymes involved in cell wall formation.

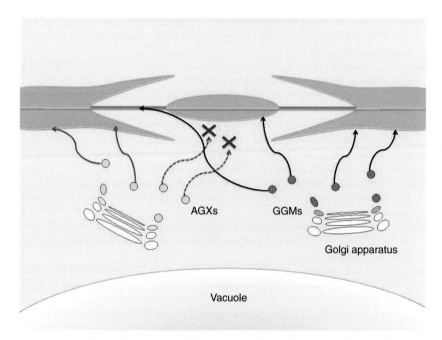

FIGURE 16.13   Deposition of GGMs and AGXs in bordered pit. GGMs are deposited in the developing secondary wall and pit membrane. GGMs deposited in the pit membrane are decomposed in the final stage of cell wall formation. Although AGXs are deposited in the developing secondary wall, they do not deposit in the pit membrane during secondary wall formation.

Monoclonal antibodies raised against each cell wall component are powerful tools to visualize the subcellular localization of cell wall components within developing and developed cell walls. GGMs and AGXs are deposited during secondary wall formation. Galactan is deposited specifically to the outer portion of S_2 in compression wood. GGMs are deposited on the pit membrane in the early stage of secondary wall formation, but decomposed in the final stage of cell wall formation. AGXs are not deposited to pit membrane during cell wall formation. Because various kinds of monoclonal antibodies will be available, more precise information on cell wall formation will be obtained in the near future.

ACKNOWLEDGMENTS

The authors are most grateful to the many colleagues who have contributed to the works summarized in this chapter. Particularly, Fumie Sano provided us beautiful electron micrographs obtained by RFS method.

REFERENCES

Awano, T., Takabe, K., Fujita, M., 1998. Localization of glucuronoxylan in Japanese beech visualized by immunogold labeling. Protoplasma 202, 213–222.

Awano, T., Takabe, K., Fujita, M., Daniel, G., 2000. Deposition of glucuronoxylans on the secondary cell wall of Japanese Beech as observed by immuno-scanning electron microscopy. Protoplasma 212, 72–79.

Imamura, Y., Harada, H., 1973. Electron microscopic study on the development of the bordered pit in coniferous tracheids. Wood Sci. Technol. 7, 189–205.

Imamura, Y., Harada, H., Saiki, H., 1974a. Further study on the development of the bordered pit in coniferous tracheids. Mokuzai Gakkaishi 20, 157–165.

Imamura, Y., Harada, H., Saiki, H., 1974b. Embedding substances of pit membranes in softwood tracheids and their degradation by enzymes. Wood Sci. Technol. 8, 243–254.

Inomata, F., Takabe, K., Saiki, H., 1992. Cell wall formation of conifer tracheid as revealed by rapid-freeze and substitution method. J. Elec. Microsc. 41, 369–374.

Kim, J.S., Awano, T., Yoshinaga, A., Takabe, K., 2010a. Immunolocalization of β-1-4-galactan and its relationship with lignin distribution in developing compression wood of *Cryptomeria japonica*. Planta 232, 109–119.

Kim, J.S., Awano, T., Yoshinaga, A., Takabe, K., 2010b. Temporal and spatial immunolocalization of glucomannans in differentiating earlywood tracheid cell walls of *Cryptomeria japonica*. Planta 232, 545–554.

Kim, J.S., Awano, T., Yoshinaga, A., Takabe, K., 2010c. Immunolocalization and structural variations of xylans in differentiating earlywood tracheid cell walls of *Cryptomeria japonica*. Planta 232, 817–824.

Kim, J.S., Awano, T., Yoshinaga, A., Takabe, K., 2011a. Temporal and spatial diversities of the immunolabeling of mannan and xylan polysaccharides in differentiating earlywood ray cells and pits of *Cryptomeria japonica*. Planta 233, 109–122.

Kim, J.S., Awano, T., Yoshinaga, A., Takabe, K., 2011b. Distribution of hemicelluloses in warts and the warty layer in normal and compression wood tracheids of *Cryptomeria japonica*. J. Korean Wood Sci. Technol. 39, 420–428.

Kimura, S., Laosinchai, W., Itoh, T., Cui, X., Linder, C.R., Brown, Jr., R.M., 1999. Immunogold labeling of rosette terminal cellulose-synthesizing complexes in the vascular plant *Vignaangularis*. Plant Cell 11, 2075–2085.

Maeda, Y., Awano, T., TakabeK, Fujita, M., 2000. Immunolocalization of glucomannan in the cell wall of differentiating tracheids in *Chamaecyparis obtusa*. Protoplasma 213, 148–156.

Miao, Y.C., Liu, C.J., 2010. ATP-binding cassette-like transporters are involved in the transport of lignin precursors across plasma and vacuolar membranes. Proc. Natl. Acad. Sci. USA 107, 22728–22733.

Morikawa, Y., Yoshinaga, A., Kamitakahara, H., Wada, M., Takabe, K., 2010. Cellular distribution of coniferin in differentiating xylem of *Chamaecyparis obtusa* as revealed by Raman microscopy. Holzforschung 64, 61–67.

Nakashima, J., Mizuno, T., Takabe, K., Fujita, M., Saiki, H., 1997. Direct visualization of lignifying secondary wall thickenings in *Zinnia elegans* cells in culture. Plant Cell Physiol. 38, 818–827.

Northcote, D.H., Davey, R., Ray, J., 1989. Use of antisera to localize callose, xylan and arabinogalactan in the cell-plate, primary and secondary walls of plant cells. Planta 178, 353–366.

Samuels, A.L., Rensing, K.H., Douglas, C.J., Mansfield, S.D., Dharmawardhara, D.P., Ellis, B.E., 2002. Cellular machinery of wood production: differentiation of secondary xylem in *Pinus contorta* var. *latifolia*. Planta 216, 72–82.

Sato, T., Takabe, K., Fujita, M., 2004. Immunolocalization of phenylalanine ammonia-lyase and cinnamate-4-hydroxylase in differentiating xylem of poplar. C. R. Biol. 327, 827–836.

Sato, Y., Demura, T., Yamawaki, K., Inoue, Y., Sato, S., Sugiyama, M., Fukuda, H., 2006. Isolation and characterization of a novel peroxidase gene ZPO-C whose expression and function are closely associated with lignification during tracheary element differentiation. Plant Cell Physiol. 47, 493–503.

Staehelin, L.A., Giddings, T.H., Levy, S., Lynch, M.A., Moore, P.J., Swords, K.M.M., 1991. Organisation of the secretory pathway of cell wall glycoproteins and complex polysaccharides in plant cells. In: Hawes, C.R., Coleman, J.O.D., Evans, D.E. (Eds.), Endocytosis, Exocytosis and Vesicle Traffic in Plants. Cambridge University Press, pp. 183–198.

Takabe, K., Takeuchi, M., Sato, T., Ito, M., Fujita, M., 2001. Immunocytochemical localization of enzymes involved in lignification of the cell wall. J. Plant Res. 114, 509–515.

Takeuchi, M., TakabeK, Fujita, M., 2001. Immunolocalization of *O*-methyltransferase and peroxidase in differentiating xylem of poplar. Holzforschung 55, 146–150.

Takeuchi, M., Takabe, K., Fujita, M., 2005. Immunolocalization of an anionic peroxidase in differentiating poplar xylem. J. Wood Sci. 51, 317–322.

Thiery, J.P., 1967. Miseenévidence des polysaccharides sur coupes fines enmicroscopieélectronique. J. Microsc. 6, 987–1018.

Thomas, R.J., 1967. The ultrastructure of differentiating and mature bordered pit membranes from cypress (*Taxodium distichum* L. Rich). Wood Fiber 4, 87–94.

Tsuyama, T., Kawai, R., Shitan, N., Matoh, T., Sugiyama, J., Yoshinaga, A., Takabe, K., Fujita, M., Yazaki, K., 2013. Proton-dependent coniferin transport, a common major transport event in differentiating xylem tissue of woody plants. Plant Physiol. 162, 918–926.

Ye, Z.-H., 1997. Association of caffeoyl coenzyme A 3-O-methyltransferase expression with lignifying tissues in several dicot plants. Plant Physiol. 108, 1341–1350.

Chapter 17

Distribution of Cell Wall Components by TOF-SIMS

Dan Aoki*, Kaori Saito**, Yasuyuki Matsushita*, Kazuhiko Fukushima*
*Department of Biosphere Resources Science, Graduate School of Bioagricultural Sciences,
Nagoya University, Furo-cho, Chikusa-ku, Nagoya, Japan; **Division of Diagnostics
and Control of the Humanosphere, Research Institute for Sustainable Humanosphere,
Kyoto University, Uji, Kyoto, Japan

Chapter Outline

CURRENT SITUATION OF MICROSCOPIC ANALYSES

Most of the components in plants are heterogeneously distributed depending on their purposes. Such distribution conditions of organics and inorganics are essential information for discussing biosynthesis, transportation, metabolism, and the mechanical properties of plants. Today, we can use, for example, X-ray, IR, Raman, and UV–Vis spectroscopies to obtain microscopic information about periodic structures, elements, interatomic bonds, and molecules. Furthermore, imaging mass spectrometry provides us microscopic mass information.

Imaging mass spectrometry makes a distribution map of chemicals by means of mass spectrometric data. Each pixel of the image contains a mass spectrum. In the case of making a distribution map of a chemical A, the ion counts of A in each spectrum in each pixel are used as a function of the brightness as shown in Fig. 17.1.

Today, matrix-assisted-laser-desorption/ionization mass spectrometry (MALDI-MS) and secondary ion mass spectrometry (SIMS) might be the mainstream techniques of the imaging mass spectrometry, and time-of-flight (TOF)

Secondary Xylem Biology. http://dx.doi.org/10.1016/B978-0-12-802185-9.00017-6

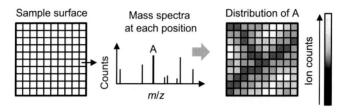

FIGURE 17.1 A schematic diagram of the imaging mass spectrometry.

type mass detector is widely used for both. They differ mainly in the sample pretreatment, mass range, and mass resolution of the spectrum, and a spatial resolution of the resultant image. In this chapter, a view of the recent researches related to biochemical components in plant by using (TOF-) SIMS is described.

GENERAL ASPECTS OF TOF-SIMS

TOF-SIMS is one of the most sensitive surface analyses giving us positional information about many inorganic and organic substances. The essential features of the measurement are the spatial resolution, wide-range elemental and molecular information by TOF detector, high sensitivity to trace-amount chemicals at the very surface, and there is no need to do any pretreatment to the sample surface for the measurements. Figure 17.2 displays a schematic illustration of TOF-SIMS measurements. The words "secondary ions" mean the ionized and sputtered particles by the primary ion bombardment. They occur only from a few nanometers of the sample surface, and the primary ion beam diameter is smaller than 1 μm.

What needs to be emphasized here is that the secondary ion yield for each chemical can vary by several orders of magnitude. Furthermore, the yield of a particle strongly depends on the chemical state of the surrounding area; we call the phenomena as a matrix effect. Consequently, great attention should be given

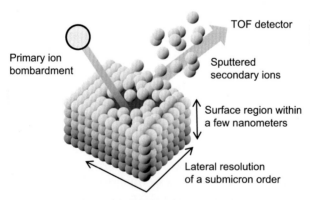

FIGURE 17.2 A schematic illustration of TOF-SIMS measurements.

to discuss the quantity of the target components with TOF-SIMS data. The more detailed fundamentals of TOF-SIMS are described elsewhere (Vickerman and Briggs, 2001; Thellier et al., 2001).

PLANT ANALYSES BY TOF-SIMS

In TOF-SIMS measurements, it is nearly impossible to detect large macromolecules as they are and we can only detect their fragment ions. Therefore, when we want to visualize the position or evaluate the molecular weight of huge biomacromolecules, we should consider the parallel use of other methods such as MALDI-MS, fluorescent microscopy, immunoelectron microscopy, and so on. Secondary xylem, the major part of woody plants, consists of polymers, low-molecular-weight compounds, and inorganics. As has been mentioned, the polymer components are detected as fragment ions in TOF-SIMS measurements and the difference of their unit structures and relative quantity are discussed. The fragmentation behavior of biomacromolecules is important, and the major components of the cell wall, cellulose, hemicellulose, and lignin, are introduced.

TOF-SIMS studies on inorganics and extractives in plants are also summarized in this chapter. Inorganic ions have good mass-reliability and their detection sensitivity is higher than that of organics. Inorganics showing variable distribution conditions in various plant tissues are an interesting target from the view point of plant physiology. Low-molecular-weight compounds, such as extractives, are detectable as a molecular ion and their assignments are easier than that of macromolecules if the purified sample is available. However, the mass-overlapping problem between different compounds should be borne in mind continually to analyze biological samples. Finally, a cryo-TOF-SIMS approach is briefly mentioned as a progressive topic.

Polysaccharides

Polysaccharides are the major part of the plant cell wall. Positive TOF-SIMS spectra of amorphous cellulose regenerated from *N,N*-dimethylacetamide/LiCl solution and a tentative fragmentation scheme of cellulose are shown in Fig. 17.3. The polymerization unit of cellulose is anhydroglucose ($C_6H_{10}O_5$) and the unit is usually detected as a protonated ion $[C_6H_{10}O_5]$ $[H^+]$ (*m/z* 163.06), a dehydrated and protonated ion $[C_6H_8O_4][H^+]$ (*m/z* 145.05), another dehydrated ion $[C_6H_6O_3]$ $[H^+]$ (*m/z* 127.04), and many small fragments (Fardim and Durán, 2003). Alkali metal adduct ions are also detectable. In Fig. 17.3, ions of *m/z* 185.04, 167.03, and 149.02 are Na^+ adduct ions of the mentioned fragment ions of *m/z* 163.06, 145.05, and 127.04, respectively. The mass difference between the protonated and the sodium adduct ions is ≈ 22 due to the difference of $^1H^+$ and $^{23}Na^+$. Usually, only cellulose ions smaller than the glucose unit are detectable from wood samples in SIMS measurements probably because of the high crystallinity of the cellulose in the cell wall. In case of regenerated amorphous cellulose, however,

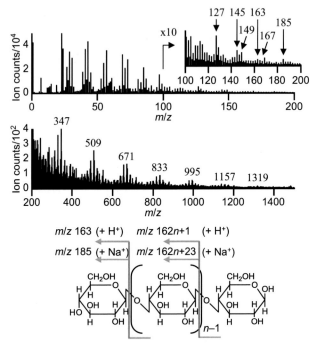

FIGURE 17.3 A positive TOF-SIMS spectrum of regenerated amorphous cellulose and a tentative fragmentation scheme leading to formation of characteristic ions at m/z $162n + 1$ $(+H^+)$ and $162n + 23$ $(+Na^+)$ from cellulose ($n = 1, 2, 3...$).

oligomer ions, like $[C_6H_{10}O_5]_n[Na^+]$ (m/z $162.05n + 22.99$), are also detectable. In Fig. 17.3, oligomer ions up to m/z 1319.41 ($n = 8$) are distinguishable. These fragment oligomers of $[C_6H_{10}O_5]_n[Na^+]$ are also detected by MALDI-TOF-MS (Lunsford et al., 2011).

Hemicellulose has several polymerization units and the mass spectrum should contain structural information of the macromolecular hemicellulose; nevertheless, few attempts have been made at a structural heterogeneity evaluation of hemicellulose by using TOF-SIMS. The following situation might be one of the difficulties of the TOF-SIMS study of hemicellulose.

Xylan is one of the most abundant hemicelluloses. Concerning a purified xylan, a series of fragment ions is similar to that of cellulose. Therefore, $[C_5H_8O_4][H^+]$ (m/z 133.05) and monodehydrated ion $[C_5H_6O_3][H^+]$ (m/z 115.04) can be obtained as characteristic ions derived from the anhydroxylose ($C_5H_8O_4$) unit (Fardim and Durán, 2003; Tokareva et al., 2011). However, hemicellulose in woody plants consists of several types of monosaccharides and the composition ratio depends on the sample. Regarding this, Goacher et al. (2011) reported a possibility that m/z 115.04 and 133.05 ions did not represent xylan in whole wood samples and the peaks were probably contaminated with other organic compounds. This kind of "ion overlapping" problem should be remembered when discussing complex materials.

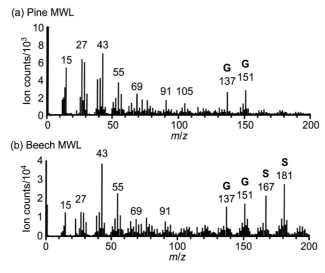

FIGURE 17.4 Positive TOF-SIMS spectra of (a) pine MWL and (b) beech MWL. The assignments G and S indicate the original building block of the fragment ion derived from.

Lignin: Fragmentation Behavior

The TOF-SIMS spectrum of lignin is more complicated than that of polysaccharides because of the fact that lignin is composed of the plural building blocks (G, guaiacyl; S, syringyl; H, p-hydroxyphenyl) and has various bonding patterns within the blocks. The representative fragment ions were reported earlier (Kleen, 2000; Fukushima et al., 2001). Figure 17.4 displays the typical positive TOF-SIMS spectra of milled wood lignin (MWL) (Björkman, 1956, 1957) obtained from softwood pine (Fig. 17.4a) and hardwood beech (Fig. 17.4b). In Fig. 17.4, ions of m/z 15 (CH_3), 27 (C_2H_3), 43 (C_3H_7 and/or C_2H_3O), 55 (C_4H_7 and/or C_3H_3O), and 69 (C_5H_9 and/or C_4H_5O) are typical organic fragment ions. Characteristic ions derived from the G-unit (m/z 137 and 151) and S-unit (m/z 167 and 181) are designated as G and S, respectively.

Exact chemical formulas of these lignin-fragment ions were investigated by a series of isotope-labeling experiments (Saito et al., 2005a). Figure 17.5 shows three major building blocks of lignin (Fig. 17.5a–c) and the determined characteristic fragment ions derived from G-unit (Fig. 17.5d–f). A series of similar fragment ions are also detectable from the S-unit and H-unit. Their m/z values are 167 and 181 for S-unit, and 107 and 121 for H-unit; the mass difference is simply due to the number of their methoxy group (Saito et al., 2005b, 2006).

To elucidate the fragmentation behavior of lignin, several dimer and polymer model compounds have been synthesized and investigated (Saito et al., 2005a, 2005b, 2006; Matsushita et al., 2013). As polymer model compounds, dehydrogenation polymers (DHPs) connected only by β-O-4′ type linkage (Kishimoto et al., 2005, 2006) and DHPs randomly polymerized by

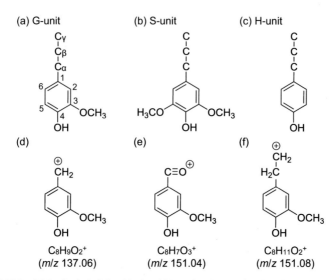

(a) G-unit   (b) S-unit   (c) H-unit

(d) $C_8H_9O_2^+$ (m/z 137.06)   (e) $C_8H_7O_3^+$ (m/z 151.04)   (f) $C_8H_{11}O_2^+$ (m/z 151.08)

FIGURE 17.5   Major lignin building blocks of (a) guaiacyl, (b) syringyl, and (c) p-hydroxyphenyl; and (d–f) chemical formulas of characteristic fragment ions derived from G-unit of lignin.

horseradish peroxidase and H_2O_2 (Wayman and Obiaga, 1974) were used. Figure 17.6 illustrates molecular structures of dimer model compounds and DHPs (β-O-4' type). To make the atom position clear, Arabic numerals and Greek characters are assigned to the benzene ring and its side-chain, respectively.

As a result of TOF-SIMS measurements using the model compounds, a tentative fragmentation behavior of lignin might be described as follows. The β-O-4' type dimers and polymers (Fig. 17.6c, i, k, and l) gave all of the three typical fragment ions (Fig. 17.5d–f). However, if the C_α has a double-bonded oxygen (Fig. 17.6a and b), the fragment ion m/z 151.08 (Fig. 17.5f) is not detected (Saito et al., 2005a). This phenomenon is simply interpreted as a result of a strong C=O bond. In the case of the C_β having CH_2OH as a C_γ-unit, the fragment ion m/z 151.08 was detected from the dimer (Fig. 17.6c), but it was not detected from the polymer (Fig. 17.6j) (Saito et al., 2005b, 2006). The difference of the fragmentation mechanism between dimers and polymers is still not clear.

The 5–5' type dimer does not produce any of the typical three fragment ions probably because of the strong 5–5' linkage. The spectrum mainly consists of small fragment ions and the dimer ion m/z 330 (Saito et al., 2005b). The linkages between benzene ring and neighboring carbon seem to be more stable than carbon–carbon linkages at the side-chain. In the case of the β-5' type dimer (Fig. 17.6g), it was experimentally proved that the fragment ion m/z 137.06 (Fig. 17.5d) was derived only from the A ring (Matsushita et al., 2013). The fragment ions derived from the B ring (Fig. 17.6g) always had C_β (from A ring) at C5' position. In this instance, the A ring does not produce a fragment ion m/z 151.08 (Fig. 17.5f).

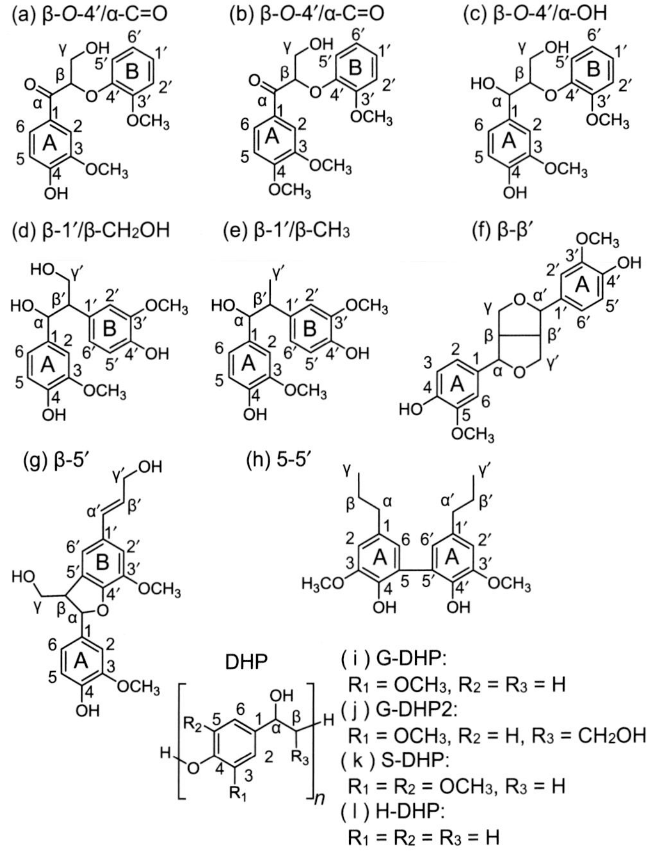

FIGURE 17.6 **Molecular structures of (a–h) dimer and (i–l) polymer (β-O-4′ type DHP) model compounds.** Dimer model compounds are shown with the carbon number, the bonding type, and the structure such as (a) a dimer having the β-O-4′ type bonding and the α-C=O structure. DHP model compounds are shown as a repeating unit and the functional group pattern such as (i) G-DHP consists of guaiacyl type unit.

Similarly, the ion m/z 151.08 should be derived only from the B ring of β-1′ type dimer (Fig. 17.6d and e) (Saito et al., 2005b). Concerning the β-β′ type dimer (Fig. 17.6f), the β-β′ linkage might be more stable than the α-β linkage because the fragment ion m/z 151.08 was scarcely detected in the spectrum (Saito et al., 2005b). From these reports, the fragment ion m/z 151.08 (Fig. 17.5f) is more difficult to occur than the fragment ion m/z 151.04 (Fig. 17.5e). In other words, the presence of the fragment ion m/z 151.08 means that the specific structure exists in the lignin at the position. Further investigation of the fragmentation behavior of lignin may provide more detailed information about the microscopic heterogeneity of the lignin structure.

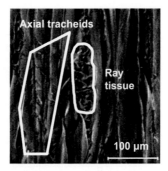

FIGURE 17.7   A positive TOF-SIMS total ion image of a tangential section of *P. densiflora*. ROIs for axial tracheid and ray tissue are illustrated by white solid lines.

Lignin: Microscopic and Semiquantitative Evaluation

As mentioned earlier, the secondary ion yield for each chemical is not proportional to their actual abundance. For example, the ion count ratio of cellulose and lignin is not the same with their abundance ratio. However, the relative ion intensity might be useful to track and surmise the transition of their abundance.

Figure 17.7 shows a positive TOF-SIMS image of the tangential section of *Pinus densiflora* and ROIs (regions of interest) for axial tracheids and a ray tissue to obtain specific mass spectra for each selected area (Zheng et al., 2014a). Relative intensities of lignin were estimated by Eq. (17.1) using ROIs in sapwood, heartwood, and the transition zone between sapwood and heartwood.

$$\begin{aligned} &\text{Relative intensity of G} - \text{lignin} \, (\% \, \text{of cellulose ions}) \\ &= ([m/z137] + [m/z151]) / ([m/z127] + [m/z145]) \times 100 \, (\%) \end{aligned} \quad (17.1)$$

where the square brackets mean the ion counts at the m/z value.

Here, Zheng et al. (2014b) used cellulose-derived fragment ions of m/z 127 and 145 as cell-wall standard ions. The result is illustrated in Fig. 17.8. The G-lignin ion intensities in the axial-tracheid ROI showed similar values among sapwood, transition zone, and heartwood; however, those in the ray-tissue ROI increased from sapwood to heartwood. This lignification transition of ray tissues was further confirmed by microchemical analysis using laser microdissection and thioacidolysis method.

One of the important structural parameters of lignin in hardwood, S to G ratio (S/G value) can be evaluated by TOF-SIMS also (Saito et al., 2011; Zhou et al., 2011). S/G values are evaluated by Eq. (17.2).

$$S/G \, \text{value} = ([m/z167] + [m/z181]) / ([m/z137] + [m/z151]) \quad (17.2)$$

where the square brackets mean the ion counts at the m/z value.

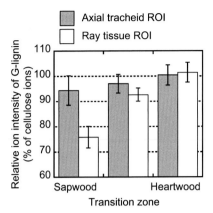

FIGURE 17.8   **TOF-SIMS relative ion intensity of lignin G-unit (m/z 137 and 151) to cellulose (m/z 127 and 145) in regions of axial tracheids and ray tissues for sapwood, transition zone, and heartwood.** $N = 3$ for each ROIs.

Saito et al. (2011) confirmed the relationship between the secondary ion counts and the actual molar ratio of S and G unit using monomer (coniferyl alcohol and sinapyl alcohol), dimer (Fig. 17.6c and its analog having two more methoxy groups at the 5 and 5′), and polymer (Fig. 17.6j and its analog having one more methoxy group at the R_2) model compounds. As a result of TOF-SIMS measurements, the calibration curve showed a good linearity between S/G value (TOF-SIMS) and the molar ratio of S and G unit in the sample. Especially in the case of polymer model compounds, the S to G molar ratio and S to G ratio in TOF-SIMS ion counts were very similar. This result indicates that TOF-SIMS can be used for microscopic relative-evaluation of S/G value in plant tissues.

Figure 17.9 displays positive TOF-SIMS images of the cross-section of maple (Saito et al., 2012). In Fig. 17.9a, ROIs 1 and 2 indicated by arrows and ROIs 3–6 enclosed by white solid lines were used to evaluate the S/G value.

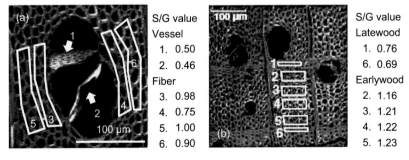

FIGURE 17.9   **Positive TOF-SIMS total ion images of the cross-section of maple.** Regions indicated by white solid lines and arrows are fiber and vessel regions, respectively. S/G value comparison between ROIs of (a) vessel and fiber and (b) latewood and earlywood. The values shown on the right side of the images were calculated by Eq. (17.2).

In Fig. 17.9b, ROIs 1 and 6 correspond to latewood region, and ROIs 2–5 correspond to earlywood region. As a result of ROI evaluation, the S/G values in the vessel were lower than those of fiber, and the S/G values in latewood region were lower than those of earlywood region.

Inorganics

At the end of twentieth century, the applicability of (TOF-)SIMS for wood chemistry had already been recognized. Inorganic metals were investigated earlier than the mentioned polymer components because of their mass-reliability and high ion yield (Chryssoulis et al., 1995; Grignon et al., 1997; Kuhn et al., 1995; Lazof et al., 1994; Mangabeira et al., 1999; Ripoll et al., 1992).

As a pioneering work regarding the cambial region and adjacent phloem and xylem regions, Kuhn et al. (1997) studied the distribution and transport of Mg, K, and Ca in Norway spruce (*Picea abies* [L.] Karst.) by using stable isotope tracers. Follet-Gueye et al. (1998) reported a change of Ca concentration in beech (*Fagus sylvatica* L.) correlating with a cambium preactivation.

In a recent study, inorganic metals were investigated simultaneously with organic compounds such as lignin and carbohydrates. Tokareva et al. (2007) studied the distribution of lignin, polysaccharides, and inorganic metals and showed a clear difference between them in regions of the bordered pit tori and an inner layer of ray parenchyma cells. Saito et al. investigated the chemical differences between sapwood, transition zone, and heartwood of *Chamaecyparis obtusa* using several inorganic metals and lignin ions (Saito et al., 2008a). The characteristic Al localization in maple tree was also reported in comparison with other inorganic metals (Saito et al., 2014).

Heavy metals in plants were also investigated. Kaldorf et al. (1999) investigated selective deposition of heavy metals in maize (*Zea mays* L. var. Honeycomb) colonized by fungus. Brabander et al. (1999) revealed intraring variability of toxic heavy metals in red oak (*Quercus rubra*) to estimate long-term records of ambient levels of metals. Mangabeira et al. (2004, 2006, 2011) focused on Cr and studied the distribution and its alteration effect.

Extractives

Plants contain various extractives. Conventional extraction processes are accompanied with loss of their positional information. TOF-SIMS measurements should be a good approach to analyze extractives having a known molecular formula. In most cases, a secondary ion derived from an extractive molecule has a specific m/z value, and the value is larger than fragment ions of the structural unit of polymer components in plant cell wall mentioned earlier. Consequently, the extractive ions are easy to recognize in the mass spectrum. Some early studies were conducted in a research area of pulp and paper chemistry (Fardim and Durán, 2000, 2002; Kleen, 2000; Kleen et al., 2003). In this section, researches regarding the microscopic positional information of extractives in wood tissues are introduced.

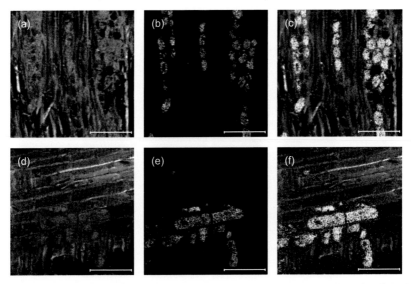

FIGURE 17.10   Positive TOF-SIMS images of *Diospyros kaki*. (a and d) Total ion, (b and e) *m/z* 218, and (c and f) overlay image of *m/z* 218 on total ion. An image set of (a–c) shows the tangential section at the edge of the blackened heartwood region, and an image set of (d–f) illustrates the radial section at the pith region. Scale bars are 100 μm.

The distribution of naphthalene derivatives of *Diospyros kaki* was investigated (Matsushita et al., 2012). As a result of gas chromatography–mass spectrometry analysis using macroscopically cut blocks of the wood, the compound was detected at the edges of the blackened heartwood and at the pith region. TOF-SIMS total ion images and the ion from the compound (*m/z* 218) are displayed in Fig. 17.10. The localization of the compound in parenchyma cells and their neighboring axial elements is obviously visualized.

Positional information of heartwood substances can be a help for felling date determination of the wood. A discolored ancient wood was measured by TOF-SIMS to discriminate the visually indistinguishable sapwood–heartwood boundary by means of chemical information (Saito et al., 2008b). The wood of Hinoki cypress (*C. obtusa*) felled approximately 1300 years ago had still retained a few kinds of heartwood substances such as hinokinin and hinokiresinol. These specific ions were detected from heartwood and the boundary region between sapwood and heartwood. The identification of the sapwood region allows us to surmise the felling date of the tree by using dendrochronological analysis. Consequently, the date of the outermost annual ring of the region measured by TOF-SIMS was 705 A.D., and the felling date was estimated to be around 740 A.D. based on the average number of sapwood rings in the Hinoki wood in Japan.

The distribution of water and ferruginol, a heartwood substance of Sugi (*Cryptomeria japonica*) was studied (Imai et al., 2005; Kuroda et al., 2008, 2014). Figure 17.11 shows the distribution of ferruginol molecular ion (*m/z* 285). The accumulation of ferruginol started at the middle of the intermediate wood, in the

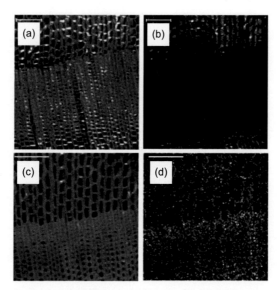

FIGURE 17.11 Positive TOF-SIMS images of cross-section of *C. japonica*. (a and b) Intermediate wood and (c and d) heartwood. Images are (a and c) total ion and (b and d) *m/z* 285. Scale bars are 100 μm.

earlywood near the annual ring boundary (Fig. 17.11a and b). The accumulation continued for several years, and finally ferruginol distributes over the whole region in the heartwood (Fig. 17.11c and d). The accumulation timing of ferruginol was discussed in consideration of the distribution of ray parenchyma cells and water in the heartwood-forming xylem by using cryo-scanning electron microscopy (SEM) observations.

Cryo-TOF-SIMS to Approach the Living-State Visualization

The phrase "cryo-TOF-SIMS" referred to here means the TOF-SIMS measurements of frozen-hydrated samples. As has been mentioned, SIMS measures only a few nanometers of the sample surface and SIMS requires a very high degree of vacuum. Due to these SIMS characteristics, the measurements of frozen-hydrated biological samples still have been an uphill and active topic. The method should have a possibility of chemical mappings of frozen-hydrated samples with a submicron lateral resolution, and it might be a useful data to discuss the biological mechanisms with living-state mass visualization.

There are several reports of cryo-(TOF-)SIMS experiments using higher plants and showing the distribution of inorganics. Dérue et al. (1999a, 1999b, 2006) analyzed water, Na^+, Mg^+, K^+, and Ca^+ ions in the flax (*Linum usitatissimum* L.). Dickinson et al. (2006) investigated *Pteris vittata*, human fibroblast cells, and a polar ice, and obtained distribution images of positive (Cr^{3+}, Ca^{2+}) and negative (AsO_2^-, CN^-, PO_2^-) ions. Metzner et al. (2008, 2010a, 2010b) reported a

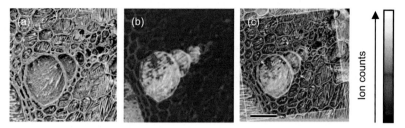

FIGURE 17.12   (a) A cryo-SEM image, (b) a cryo-TOF-SIMS image of K^+ ion ratio ($[^{41}K^+]/$ ($[^{39}K^+] + [^{41}K^+]$)), and (c) their overlaid image. Scale bar is 100 μm.

series of cryo-TOF-SIMS studies using climbing bean (*Phaseolus vulgaris* L.). Metzner et al. used stable isotope tracers (e.g., $H_2{}^{18}O$, ^{26}Mg, ^{41}K, and ^{44}Ca) to display fluxes of the nutrients effectively, and the distribution conditions were illustrated by using an overlay method using cryo-SEM images (Fig. 17.12). This SIMS-SEM-overlay method is widely used in TOF-SIMS studies to compensate for the lateral resolution of TOF-SIMS. Figure 17.12 shows a clear penetrating flux of the ^{41}K isotope tracer in water with detailed histological information provided by a cryo-SEM image. As another report using an isotope tracer, Iijima et al. (2011) visualized lateral water transport pathways of soybean (*Glycine max* (L.) Merr.) using deuterium-labeled water (D_2O). Time-dependent deuterium ion intensities suggested the water transport pathways.

Kuroda et al. (2013) studied a heartwood substance ferruginol in a frozen-hydrated *C. japonica* sample. Figure 17.13 displays three measurement surfaces at nearly the same lateral position of the sample block. At first, the sample

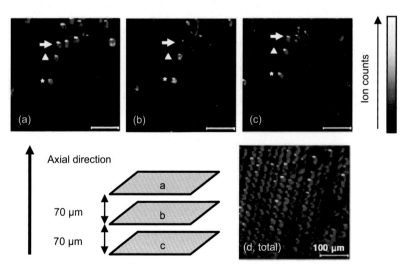

FIGURE 17.13   Three-dimensional cryo-TOF-SIMS images of (a–c) ferruginol and (d) total ion. Face b and c were 70 μm below face a and b, respectively. Arrows, triangles, and stars mean the same cell. Scale bars are 100 μm.

was measured by cryo-TOF-SIMS at the surface (a), and then the sample was cut by a sliding microtome to make the next surface (b) and measured by cryo-TOF-SIMS again. Furthermore, the same procedure was repeated and an image of surface (c) was obtained. In comparison with Fig. 17.11 measured at dry condition, single cell-level localization of ferruginol in a frozen-hydrated sample was revealed. Moreover, they found that the ferruginol distributed unevenly in a depth dimension of the same tracheid cell.

CONCLUSIONS AND PROSPECTS

This chapter summarized (TOF-)SIMS studies concerning plants. Of course, the mass is not perfect information and the lateral resolution of the SIMS imaging does not extend to that of electron microscopy. We should use other microscopic, spectroscopic, and chromatographic measurements in tandem with SIMS measurements; nevertheless, TOF-SIMS giving mass spectrometric information in a submicron lateral resolution should be a valuable tool for investigating the heterogeneous distribution of many chemical components in various samples.

REFERENCES

Björkman, A., 1956. Studies on finely divided wood. Part I. Svensk Papperstidning 57, 477–485.

Björkman, A., 1957. Lignin and lignin-carbohydrate complexes. Indust. Eng. Chem. 49, 1395–1398.

Brabander, D.J., Keon, N., Stanley, R.H.R., Hemond, H.F., 1999. Intra-ring variability of Cr, As, Cd, and Pb in red oak revealed by secondary ion mass spectrometry: Implications for environmental biomonitoring. PNAS USA 96, 14635–14640.

Chryssoulis, S.L., Stowe, K.G., Niehuis, E., Cramer, H.G., Bendel, C., Kim, J.Y., 1995. Detection of collectors on concentrator mineral grains by time of flight secondary-ion mass spectrometry (TOF-SIMS). Trans. Inst. Min. Metall., Sect. C 104, C141–C150.

Dérue, C., Gibouin, D., Lefebvre, F., Rasser, B., Robin, A., Le Sceller, L., Verdus, M.C., Demarty, M., Thellier, M., Ripoll, C., 1999a. A new cold stage for SIMS analysis and imaging of frozen-hydrated biological samples. J. Trace Microprobe Tech. 17, 451–460.

Dérue, C., Gibouin, D., Lefebvre, F., Demarty, M., Verdus, M.C., Ripoll, C., 1999b. SIMS imaging of frozen-hydrated biological samples with a Cameca IMS 4f. Biol. Cell 91, 281.

Dérue, C., Gibouin, D., Demarty, M., Verdus, M.-C., Lefebvre, F., Thellier, M., Ripoll, C., 2006. Dynamic-SIMS imaging and quantification of inorganic ions in frozen-hydrated plant samples. Microscop. Res. Tech. 69, 53–63.

Dickinson, M., Heard, P.J., Barker, J.H.A., Lewis, A.C., Mallard, D., Allen, G.C., 2006. Dynamic SIMS analysis of cryo-prepared biological and geological specimens. Appl. Surf. Sci. 252, 6793–6796.

Fardim, P., Durán, N., 20003. Surface characterisation of unbleached *Eucalyptus* kraft pulp using XPS and TOF-SIMS. In: Proceedings of the Sixth European Workshop on Lignocellulosics and Pulp, 07–311.

Fardim, P., Durán, N., 2002. Surface chemistry of eucalyptus wood pulp fibres: effects of chemical pulping. Holzforschung 56, 615–622.

Fardim, P., Durán, N., 2003. Modification of fibre surfaces during pulping and refining as analysed by SEM, XPS and ToF-SIMS. Colloids Surf. A 223, 263–276.

Follet-Gueye, M.-L., Verdus, M.-C., Demarty, M., Thellier, M., Ripoll, C., 1998. Cambium pre-activation in beech correlates with a strong temporary increase of calcium in cambium and phloem but not in xylem cells. Cell Calcium 24, 205–211.

Fukushima, K., Yamauchi, K., Saito, K., Yasuda, S., Takahashi, M., Hoshi, T., 2001. Analysis of lignin structures by TOF-SIMS. Proceedings of the Eleventh International Symposium on Wood and Pulping Chemistry, 327–329.

Goacher, R.E., Jeremic, D., Master, E.R., 2011. Expanding the library of secondary ions that distinguish lignin and polysaccharides in time-of-flight secondary ion mass spectrometry analysis of wood. Anal. Chem. 83, 804–812.

Grignon, N., Halpern, S., Jeusset, J., Briançon, C., Fragu, P., 1997. Localization of chemical elements and isotopes in the leaf of soybean (*Glycine max*) by secondary ion mass spectrometry microscopy: critical choice of sample preparation procedure. J. Microsc. 186, 51–66.

Iijima, M., Yoshida, T., Kato, T., Kawasaki, M., Watanabe, T., Somasundaram, S., 2011. Visualization of lateral water transport pathways in soybean by a time of flight-secondary ion mass spectrometry cryo-system. J. Exp. Bot. 62, 2179–2188.

Imai, T., Tanabe, K., Kato, Y., Fukushima, K., 2005. Localization of ferruginol, a diterpene phenol, in *Cryptomeria japonica* heartwood by time-of-flight secondary ion mass spectrometry. Planta 221, 549–556.

Kaldorf, M., Kuhn, A.J., Schroöder, W.H., Hildebrandt, U., Bothe, H., 1999. Selective element deposits in maize colonized by a heavy metal tolerance conferring arbuscular mycorrhizal fungus. J. Plant Physiol. 154, 718–728.

Kishimoto, T., Uraki, Y., Ubukata, M., 2005. Easy synthesis of -*O*-4 type lignin related polymers. Org. Biomol. Chem. 3, 1067–1073.

Kishimoto, T., Uraki, Y., Ubukata, M., 2006. Chemical synthesis of -*O*-4 type artificial lignin. Org. Biomol. Chem. 4, 1343–1347.

Kleen, M., 2000. Surface chemistry of kraft pulp fibres during TCF bleaching studied by TOF-SIMS. Proceedings of the Sixth European Workshop on Lignocellulosic and Pulp, 41–44.

Kleen, M., Kangas, H., Laine, C., 2003. Chemical characterization of mechanical pulp fines and fiber surface layers. Nord. Pulp Pap. Res. J. 18, 361–368.

Kuhn, A.J., Bauch, J., Schröder, W.H., 1995. Monitoring uptake and contents of Mg, Ca and K in Norway spruce as influenced by pH and Al, using microprobe analysis and stable isotope labelling. Plant Soil, 168–169, 135–150.

Kuhn, A.J., Schröder, W.H., Bauch, J., 1997. On the distribution and transport of mineral elements in xylem, cambium and phloem of Spruce (*Picea abies* [L.] Karst.). Holzforschung 51, 487–496.

Kuroda, K., Imai, T., Saito, K., Kato, T., Fukushima, K., 2008. Application of ToF-SIMS to the study on heartwood formation in *Cryptomeria japonica* trees. Appl. Surf. Sci. 255, 1143–1147.

Kuroda, K., Fujiwara, T., Imai, T., Takama, R., Saito, K., Matsushita, Y., Fukushima, K., 2013. The cryo-TOF-SIMS/SEM system for the analysis of the chemical distribution in freeze-fixed *Cryptomeria japonica* wood. Surf. Interface Anal. 45, 215–219.

Kuroda, K., Fujiwara, T., Hashida, K., Imai, T., Kushi, M., Saito, K., Fukushima, K., 2014. The accumulation pattern of ferruginol in the heartwood-forming *Cryptomeria japonica* xylem as determined by time-of-flight secondary ion mass spectrometry and quantity analysis. Ann. Bot. 113, 1029–1036.

Lazof, D.B., Goldsmith, J.G., Rufty, T.W., Linton, R.W., 1994. Rapid uptake of aluminum into cells of intact soybean root tips. Plant Physiol. 106, 1107–1114.

Lunsford, K.A., Peter, G.F., Yost, R.A., 2011. Direct matrix-assisted laser desorption/ionization mass spectrometric imaging of cellulose and hemicellulose in *Populus* tissue. Anal. Chem. 83, 6722–6730.

Mangabeira, P., Mushrifah, I., Escaig, F., Laffray, D., França, M.G.C., Galle, P., 1999. Use of SIMS microscopy and electron probe X-ray microanalysis to study the subcellular localization of aluminium in *Vicia faba* roots cells. Cell. Mol. Biol. 45, 413–422.

Mangabeira, P.A.O., Labejor, L., Lamperti, A., de Almeida A-AF, Oliveira, A.H., Escaig, F., Severo, M.I.G., da, C., Silva, D., Saloes, N., Mielke, M.S., Lucena, E.R., Martins, M.C., Santana, K.B., Gavrilov, K.L., Galle, P., Levi-Setti, R., 2004. Accumulation of chromium in root tissues of *Eichhornia crassipes* (Mart.) Solms. in Cachoeira River—Brazil. Appl. Surf. Sci., 231–232, 467–501.

Mangabeira, P.A., Gavrilov, K.L., de Almeida A-AF, Oliveira, A.H., Severo, M.I., Rosa, T.S., da, C., Silva, D., Labejof, L., Escaig, F., Levi-Setti, R., Mielke, M.S., Loustalot, F.G., Galle, P., 2006. Chromium localization in plant tissues of *Lycopersicum esculentum* mill using ICP-MS and ion microscopy (SIMS). Appl. Surf. Sci. 252, 3488–3501.

Mangabeira, P.A., Ferreira, A.S., de Almeida A-AF, Fernandes, V.F., Lucena, E., Souza, V.L., dos Santos, Jr., A.J., Oliveira, A.H., Granier-Loustalot, M.F., Barbier, F., Silva, D.C., 2011. Compartmentalization and ultrastructural alterations induced by chromium in aquatic macrophytes. Biometals 24, 1017–1026.

Matsushita, Y., Jang, I.-C., Imai, T., Takama, R., Saito, K., Masumi, T., Lee, S.-C., Fukushima, K., 2012. Distribution of extracts including 4,8-dihydroxy-5-methoxy-2-naphthaldehyde in *Diospyros kaki* analyzed by gas chromatography-mass spectrometry and time-of-flight secondary ion mass spectrometry. Holzforschung 66, 705–709.

Matsushita, Y., Ioka, K., Saito, K., Takama, R., Aoki, D., Fukushima, K., 2013. Fragmentation mechanism of the phenylcoumaran-type lignin model compound by ToF-SIMS. Holzforschung 67, 365–370.

Metzner, R., Schneider, H.U., Breuer, U., Schroeder, W.H., 2008. Imaging nutrient distributions in plant tissue using time-of-flight secondary ion mass spectrometry and scanning electron microscopy. Plant Physiol. 147, 1774–1787.

Metzner, R., Schneider, H.U., Breuer, U., Thorpe, M.R., Schurr, U., Schroeder, W.H., 2010a. Tracing cationic nutrients from xylem into stem tissue of French bean by stable isotope tracers and cryo-secondary ion mass spectrometry. Plant Physiol. 152, 1030–1043.

Metzner, R., Thorpe, M.R., Breuer, U., Blümler, P., Schurr, U., Schneider, H.U., Schroeder, W.H., 2010b. Contrasting dynamics of water and mineral nutrients in stems shown by stable isotope tracers and cryo-SIMS. Plant, Cell Environ. 33, 1393–1407.

Ripoll, C., Jauneau, A., Lefébvre, F., Demarty, M., Thellier, M., 1992. SIMS determination of the distribution of the main mineral cations in the depth of the cuticle and the pecto-cellulosic wall of epidermal cells of flax stems: problems encountered with SIMS depth profiling. Biol. Cell 74, 135–142.

Saito, K., Kato, T., Tsuji, Y., Fukushima, K., 2005a. Identifying the characteristic secondary ions of lignin polymer using ToF-SIMS. Biomacromolecules 6, 678–683.

Saito, K., Kato, T., Takamori, H., Kishimoto, T., Fukushima, K., 2005b. A new analysis of the depolymerized fragments of lignin polymer using ToF-SIMS. Biomacromolecules 6, 2688–2696.

Saito, K., Kato, T., Takamori, H., Kishimoto, T., Yamamoto, A., Fukushima, K., 2006. A new analysis of the depolymerized fragments of lignin polymer in the plant cell walls using ToF-SIMS. Appl. Surf. Sci. 252, 6734–6737.

Saito, K., Mitsutani, T., Imai, T., Matsushita, Y., Yamamoto, A., Fukushima, K., 2008a. Chemical differences between sapwood and heartwood of *Chamaecyparis obtusa* detected by ToF-SIMS. Appl. Surf. Sci. 255, 1088–1091.

Saito, K., Mitsutani, T., Imai, T., Matsushita, Y., Fukushima, K., 2008b. Discriminating the indistinguishable sapwood from heartwood in discolored ancient wood by direct molecular mapping of specific extractives using time-of-flight secondary ion mass spectrometry. Anal. Chem. 80, 1552–1557.

Saito, K., Kishimoto, T., Matsushita, Y., Imai, T., Fukushima, K., 2011. Application of TOF-SIMS to the direct determination of syringyl to guaiacyl (S/G) ratio of lignin. Surf. Interface Anal. 43, 281–284.

Saito, K., Watanabe, Y., Shirakawa, M., Matsushita, Y., Imai, T., Koike, T., Sano, Y., Funada, R., Fukazawa, K., Fukushima, K., 2012. Direct mapping of morphological distribution of syringyl and guaiacyl lignin in the xylem of maple by time-of-flight secondary ion mass spectrometry. Plant J. 69, 542–552.

Saito, K., Watanabe, Y., Matsushita, Y., Imai, T., Koike, T., Sano, Y., Funada, R., Fukazawa, K., Fukushima, K., 2014. Aluminum localization in the cell walls of the mature xylem of maple tree detected by elemental imaging using time-of-flight secondary ion mass spectrometry (TOF-SIMS). Holzforschung 68, 85–92.

Thellier, M., Dérue, C., Tafforeau, M., Le Sceller, L., Verdus, M.-C., Massiot, P., Ripoll, C., 2001. Physical methods for in vitro analytical imaging in the microscopic range in biology, using radioactive or stable isotopes (review article). J. Trace Microprobe Tech. 19, 143–162.

Tokareva, E.N., Pranovich, A.V., Fardim, P., Daniel, G., Holmbom, B., 2007. Analysis of wood tissues by time-of-flight secondary ion mass spectrometry. Holzforschung 61, 647–655.

Tokareva, E., Pranovich, A.V., Holmbom, B.R., 2011. Characteristic fragment ions from lignin and polysaccharides in ToF-SIMS. Wood Sci. Technol. 45, 767–785.

Vickerman, J.C., Briggs, D., 2001. ToF-SIMS: surface analysis by mass spectrometry. IM Pub. and SurfaceSpectra Ltd, Chichester.

Wayman, M., Obiaga, T.I., 1974. The modular structure of lignin. Can. J. Chem. 52, 2102–2110.

Zheng, P., Aoki, D., Yoshida, M., Matsushita, Y., Imai, T., Fukushima, K., 2014a. Lignification of ray parenchyma cells in the xylem of *Pinus densiflora*. Part I: Microscopic investigation by POM, UV microscopy, and TOF-SIMS. Holzforschung 68, 897–905.

Zheng, P., Aoki, D., Matsushita, Y., Yagami, S., Fukushima, K., 2014b. Lignification of ray parenchyma cells in the xylem of *Pinus densiflora*. Part II: Microchemical analysis by laser microdissection and thioacidolysis. Holzforschung 68, 907–913.

Zhou, C., Li, Q., Chiang, V.L., Lucia, L.A., Griffis, D.P., 2011. Chemical and spatial differentiation of syringyl and guaiacyl lignins in poplar wood via time-of-flight secondary ion mass spectrometry. Anal. Chem. 83, 7020–7026.

Index

Printed in the United States
By Bookmasters